The Nature of Plant Communities

Most people can readily identify a forest, or a grassland or a wetland – these are the simple labels we give different plant communities. The aim of this book is to move beyond such simple descriptions to investigate the 'hidden' structure of vegetation, asking questions such as how do species in a community persist over time? What prevents the strongest species from taking over? And, are there rules that confer stability and produce repeatable patterns? Answers to these questions are fundamental to community ecology and to the successful management of the world's varied ecosystems, many of which are currently under threat. In addition to reviewing and synthesising our current knowledge of species' interactions and community assembly, this book also seeks to offer a different viewpoint – to challenge the reader, and to stimulate ecologists to think differently about plant communities and the processes that shape them.

J. Bastow Wilson was a professor of botany at the University of Otago, New Zealand. He taught ecology from 1971 until his retirement in 2013, when he was awarded the title of Emeritus Professor. He was elected a fellow of the Royal Society of New Zealand in 1997, in recognition of his prominence and global leadership in plant ecology and vegetation science. In 1999 he joined the board of chief editors of the *Journal of Vegetation Science* (including its sister-journal, *Applied Vegetation Science*), and in 2000 he became the chair of the chief editors until his retirement. In honour of his services to plant ecology, Wilson was made an honorary life member of the International Association for Vegetation Science in 2013. Over the course of his career Wilson made sustained, insightful and significant contributions to our understanding of how plant communities function, with his research published in over 230 scientific papers. He passed away in April 2015 after a short illness.

Andrew D. Q. Agnew, now retired, taught students plant ecology and taxonomy in Dundee, Baghdad, Nairobi and Aberystwyth. He has a deep interest in the flora of Kenya, and has published widely on the flora and vegetation of that country. Agnew was a long-term colleague of Wilson, and together they published many scientific papers on vegetation dynamics and plant community ecology.

Stephen H. Roxburgh is an ecologist with the Commonwealth Scientific and Industrial Research Organisation in Canberra, Australia. He has published more than 100 scientific papers and reports on a range of ecological topics including plant community structure and the maintenance of biological diversity, vegetation patterns and dynamics, and greenhouse gas and carbon accounting. He was a former PhD student of Wilson and worked closely with him over the last weeks of his life to help bring to completion *The Nature of Plant Communities.*

The Nature of Plant Communities

J. BASTOW WILSON
University of Otago

ANDREW D. Q. AGNEW
Aberystwyth University

STEPHEN H. ROXBURGH
CSIRO Land and Water

CAMBRIDGE
UNIVERSITY PRESS

CAMBRIDGE
UNIVERSITY PRESS

University Printing House, Cambridge CB2 8BS, United Kingdom

One Liberty Plaza, 20th Floor, New York, NY 10006, USA

477 Williamstown Road, Port Melbourne, VIC 3207, Australia

314–321, 3rd Floor, Plot 3, Splendor Forum, Jasola District Centre, New Delhi – 110025, India

79 Anson Road, #06–04/06, Singapore 079906

Cambridge University Press is part of the University of Cambridge.

It furthers the University's mission by disseminating knowledge in the pursuit of
education, learning, and research at the highest international levels of excellence.

www.cambridge.org
Information on this title: www.cambridge.org/9781108482219
DOI: 10.1017/9781108612265

First published 2019

Printed in the United Kingdom by TJ International Ltd. Padstow Cornwall

A catalogue record for this publication is available from the British Library.

Library of Congress Cataloging-in-Publication Data
Names: Wilson, J. Bastow, author. | Agnew, A. D. Q. (Andrew D. Q.), author. |
    Roxburgh, Stephen H., 1966– author.
Title: The nature of plant communities / J. Bastow Wilson, Andrew D.Q. Agnew, Stephen H. Roxburgh.
Description: New York, NY : Cambridge University Press, 2019. | Includes bibliographical
    references and index.
Identifiers: LCCN 2018060511 | ISBN 9781108482219 (hardback : alk. paper)
Subjects: LCSH: Plant communities.
Classification: LCC QK911 .W54 2019 | DDC 581.7/82–dc23
LC record available at https://lccn.loc.gov/2018060511

ISBN 978-1-108-48221-9 Hardback

To
Frederick E. Clements

Contents

Preface

The landforms of the earth are mostly clothed with vegetation. We, the authors, have been able to study such vegetation during our full working lives, and it has been enormously rewarding. Like the architectural heritage of the built environment, landscape has the power to be uplifting. This reaction is personal, but nevertheless real. We like being in plant communities. We also like trying to find out how they work: using science to seek the processes that mould vegetation, searching for general patterns and attempting the formulation of community-level theories. Such study is a homage to nature.

Books exist describing the plant communities of parts of the world, or the whole of it. Other books assume that there are rules governing the assembly of communities. We hope to examine plant communities in general terms, but without preconceptions about them. The term 'reductionist' means that we start with the basic attributes of plants, and do not believe the more complex concepts of community ecology without good evidence. We have deliberately concentrated on areas where we feel we have a particular contribution to make to the literature. In a book with such a wide scope, it has been possible to mention only a small fraction of the literature for any particular topic. We have therefore included the work that strikes us as particularly useful or novel, even if occasionally there are some problems with it. We are not generally seeking to summarise the field as a textbook would. Rather, we are often putting forward another view, another emphasis, hoping to stimulate ecologists and their postgraduate students to think of plants, plant communities and the processes that shape them in a different way.

The overall conclusions have been the most difficult. Our argument from first principles has not led us to an overarching theory, but then the closest anyone has come to this is F. E. Clements, to whom we dedicate this book. He saw much and understood much, but his conclusions were mainly descriptive rather than predictive. There is only one recent realistic and comprehensive theory, C-S-R, and the real world turns out to be too complex for it to be more than a guide. At present, community ecologists can only see through a glass darkly.

We have provided a glossary limited to terms that will be less familiar, or have been used in a variety of ways in the literature.

We are very grateful to those who have commented on parts or all of our drafts: Nicholas Adams, Mike Austin, Ryan Bailey, Forbes Boyle, Gretchen Brownstein, Amadou S. Camara, David R. Causton, Arthur O. Chater, Peter Chesson, Jennifer Costanza,

Robert M. M. Crawford, Alastair H. Fitter, Tony D. Fox, Jason D. Fridley, Kelly Frogley, Gareth W. Griffiths, Ruth F. Griffiths, Paul L. Guy, Richard Hartnup, Christine V. Hawkes, Michael J. Heads, Lee Anne Jacobs, Todd Jobe, Dane Kuppinger, William G. Lee, Jess Long, Elizabeth Matthews, Jeffrey Ott, David M. Paterson, Robert K. Peet, Aimee Pritchard, Janice Martin, Norman Mason, Charles E. Mitchell, Jill Rapson, Nitin Sekar, Amanda Senft, Matthew Simon, John Baron Steel, Brooke Wheeler, Jacqueline White, Peter S. White and Lizzie Wilberforce. Monica A. Peters kindly drew Figure 2.6. We thank Janie Mason and Raewyn Stedman for support.

Glossary

Abundance: Any measure of the amount of a species present, e.g. calorific value (ideal), biomass (the practical optimum), relative cover frequency, cover, local frequency.

Allee effect: Population growth rate is low in sparse populations, probably due to difficulty in contacting a mate but possibly due to problems in attracting a disperser.

Allogenic: Due to causes outside the community.

Alpha (niche/guild): Using particular resources, i.e. different resources than those used by other species in the local community.

Alterative stable states (ASS): Two or more vegetation/environment states in the same underlying habitat, each state locally stable but able to be shifted to another state by a large perturbation.

Altruistic facilitation: A *reaction* by one species that increases the survival/growth/ reproduction of another but is disadvantageous to the plant causing the reaction.

Annuation: Variation in species composition from year to year.

Apomixis: The production of a seed by a plant without meiosis or gamete fusion, therefore potentially identical in genotype to the mother plant. Babies without sex.

Arbuscular: Arbuscular mycorrhizae (= AM = VAM = endotrophic mycorrhizae) are a type of fungus-root association in which the fungal hyphae are extensive within the root, and indeed form arbuscules within the root cell.

ASS: *Alternative stable states*.

Assembly rules: Restrictions on the observed patterns of species presence or absence that are based on the presence or abundance of one or other groups of species (not simply the response of individual species to the environment).

Autogenic: Caused by the plant community. In 'autogenic disturbance', the plants disturb each other. 'Autogenic environmental heterogeneity' as opposed to underlying (i.e. abiotic) heterogeneity.

Autosuccession: Succession where climax species regenerate directly, without specialist pioneer species.

Barro Colorado Island: A 1,500-ha island formed when a valley in Panama was dammed to form part of the Panama Canal. It has been used as an example of an area of mainland converted into island status. Numerous ecological studies have been conducted there. The establishment of permanent plots for tree demography by Stephen Hubbell in 1982 is particularly important.

Beta (niche/guild):   Tolerant of particular environmental (non-resource) conditions within particular spatial range(s), i.e. differing in tolerances and therefore spatial distribution from other species.

Beta-niche filtering: The exclusion of some species because they cannot tolerate the physical environment.

Bibury: The site of roadside vegetation in southern England that has been monitored yearly since 1958, originally by Arthur J. Willis and E. W. Yemm, and since by other ecologists from Sheffield University. The original aim was to investigate possible use of weed killer to control woody invasion, but the control plots have proved to have the most lasting interest.

BioDepth: A collaborative, multi-site research programme funded by the European Community into the effects of species richness on yield, invasion resistance, etc.

C: Carbon.

C: In C-S-R theory, C habitats are ones where productivity is high and disturbance low, so that competition is intense, and C species are strong competitors able to persist in the vegetation of such sites. These concepts depend on the theory's suggestion that competition is more intense in more productive sites.

Cedar Creek: Cedar Creek Natural History Area, an experimental field area comprising oldfields, belonging to the University of Minnesota, and the site of many experiments by David Tilman and others.

Challenge: A test by an organism of a new habitat.

Chequerboarding: A situation where pairs of species are mutually exclusive across islands/plots.

Circular interference network: Interference abilities of species are A > B, B > C and C > A (or a more complex pattern with circularity).

Climax: A stable endpoint of succession.

Community: The set of one or more species existing within a particular area at a particular time.

Community matrix: A summary of all possible pairwise species interactions in an equilibrium community, expressed as the effect that a small change in the equilibrium abundance of one species has on the equilibrium density of another, whilst holding the abundances of all the other species present at their respective equilibrium values.

Competition: An interaction between individuals, brought about by a shared requirement for a resource in limited supply, leading to a reduction in the survivorship, growth and/or reproduction of at least some of the competing individuals concerned.

C-S-R: The theory of J. P. Grime that habitats and species can be arranged in a triangle, comprising a tradeoff between competition (C), stress (S) and disturbance/ruderality (R).

CWD: Coarse woody debris, i.e. litter of branches and whole trees.

Cyclic succession: Two or more vegetation states replace each other in a cycle that is repeated over time.

Disseminule: A sexually or asexually produced dispersal unit capable of developing into a new individual or ramet.

Disturbance: A marked change in the environment for a limited period, often with the removal of plant material.

Ecesis: The process by which a plant or animal becomes established in a new habitat.

Ecotone: A sharp change in community composition, i.e. more rapid than either side of the ecotone.

Embryophyte: Any plant from liverworts 'upwards', i.e. bryophyte, pteridophyte, gymnosperm or angiosperm.

Epigenetics: Changes caused by modification of gene expression, rather than alteration of the genetic code itself.

Equal chance: Each of the species present in the local habitat pool has an equal interference ability, and therefore an equal chance of establishing at any point.

Facilitation: A *reaction* by one species that increases the survival/growth/reproduction of another.

Florula/florule: The flora of a local environment.

Genet: A genetic individual, i.e. the derivative of one zygote. It may be one recognizable 'individual', or it may comprise many plants produced by vegetative reproduction or by apomixis.

Guild: A group of species that are similar in some way that is ecologically relevant, or might be.

Humped-back curve: In this context a relation where species richness is maximal at intermediate levels of productivity (or standing crop plus litter).

Inertia: Tendency not to change.

Interference: This covers *competition*, *allelopathy*, parasitism, pest transmission and other interactions in which the primary effect of one plant on another is negative.

Leaf: 'Although no satisfactory definition of a leaf is possible, I shall assume that we all know what we are talking about': F. G. Gregory, cited by L. Croizat in his 'Principia Botanica'.

Lyapunov (= Liapunov) stability: The ability of a community, after an infinitely small pulse perturbation, to return to its original state in infinite time (assuming external factors and the species pool remain constant).

Macrophyte: Macroscopic plant, e.g. tree.

Mesocosm: A medium-sized experimental community.

Microcosm: A small experimental community.

Mutualism: An interaction between two plants that increases the survival/growth/reproduction of both.

Mycorrhiza (pl. mycorrhizae):   A close association between a fungus and a root. The two major types, found in hosts from many families, are ectotrophic and arbuscular. Other types are associated with the Ericales (heaths, etc.) and Orchidaceae (orchids).

N: Nitrogen, generally as nitrate or ammonium salts.

Nesting: A situation where an island/plot contains all the species found in the next most species-poor island/plot, plus additional species.

Niche: A region as 'n-dimensional hyperspace' where the dimensions are all the environmental, resource or behavioural (e.g. phenology, foraging) parameters that permit an organism to live.

Niche complementarity: The tendency for coexisting species that occupy a similar position along one niche dimension to differ along another.

Ombrotrophic (= ombrogenous): Dependent on precipitation for its water and mineral nutrients (a type of mire).

Oskar: A suppressed tree seedling in the understorey, small but old.

Outbreak: A sudden increase in the abundance of a species over a few years followed by a decrease also over a few years.

P: Phosphorus, generally as phosphate.

Paludification: The process of peat bog formation.

Park Grass: The world's longest-running ecological experiment and therefore most important. Established at Rothamsted Agricultural Station in the English Midlands in 1856. There were originally 20 plots with different fertiliser treatments, though most have been subdivided since. They have been monitored, with varying degrees of detail, ever since.

Patch: Small area of vegetation within a larger matrix.

Plant: A photosynthetic organism with chlorophyll a, or a close saprophytic or parasitic relative.

Podzolization: A process of soil formation, particularly in cool, humid regions, in which the upper layers are leached of minerals, which are then concentrated in lower layers.

Polycarpic (of a ramet): Reproducing sexually more than once in a life cycle, generally in more than one year.

Press perturbation: A change in the physical or biotic environment of a community that continues to be applied.

Primary succession: Vegetation change starting in conditions not influenced by any previous vegetation on the site, e.g. bare sand, unvegetated water.

Pseudogamy: Pollen is needed for seed development, and fertilises the endosperm, but the embryo itself is produced by apomixis.

PSU: Photosynthetic unit: the leaflet, simple leaf, cladode, unit of green stem, etc.

Pulse perturbation: A change in the physical or biotic environment of a community that is applied and immediately removed (as immediately as it can be).

Quadrat: A vegetation sample of specified shape and area or volume.

R: In C-S-R theory, R habitats are ones where there is frequent disturbance, i.e. removal of plant material, and R species are the ruderals typical of such habitats. They are similar to 'r' species of r-K theory.

Ramet: A vegetatively produced plant unit, such as a strawberry 'plant'.

Raunkier life form: A classification of species according to the position of their resting buds during the most unfavourable season.

Reaction: The change in local environment caused directly by a plant. 'Ecological engineering' is similar, but perhaps has implications that the change benefits the species/community, comprising a *switch*. 'Niche construction' seems to be a synonym, with 'positive niche construction' causing a *switch* and 'negative niche construction' comprising *facilitation*.

Redox potential: Reduction–oxidation potential, the tendency to donate or accept electrons.

Relay floristics: A caricature of F. E. Clements' concepts of succession, in which vegetation must pass through a relay of seral stages, each stage facilitating the next, ending in the climax.

Reliability: Lack of temporal variation in a community.

Resilience: The degree and speed of recovery of a community after a *pulse perturbation*.

Resistance: Lack of change in a community when a *pulse perturbation* is applied.

RGR: Relative growth rate = growth per unit time per unit biomass. It is equivalent to the r of zoological population models. We do not make a distinction between vegetative and sexual reproduction, nor between an increase in numbers and in 'individual' size, so we use RGR to cover all population growth. An RGR of 0.0 means no change, and RGR <0.0 means that the population is shrinking and one >0.0 means that it is expanding.

RGR$_{max}$:   RGR, at the phase of growth when RGR is highest for the species (young) in environmental conditions that produce the highest RGR in that species (which is hard to achieve in practice).

Rheotrophic (= minerotrophic):   Receiving water, with mineral nutrients, that has flowed through/over mineral soil (a type of mire).

RYM (Relative Yield of Mixtures):   In plant competition experiments, the yield of the mixture compared to the mean of the monocultures.

RYT (Relative Yield Total):   In plant competition experiments, the sum of each species' yield in mixture compared to its yield in monoculture.

S: In C-S-R theory, S habitats are ones where productivity is low, and S species are ones typical of such habitats.

Secondary succession: Vegetation change starting after a disturbance where vegetation has previously existed and where the effects of that vegetation remain, typically in soil development.

Seral stage: A vegetation state in succession-to-climax before the climax.

SLW: Specific leaf weight = leaf weight per unit area, e.g. cm^2/g. It is the reciprocal of SLA = specific leaf area, and references to SLA have been converted to SLW for uniformity.

Somatic mutation: The occurrence of a mutation in the somatic (i.e. non-reproductive) cells of an organism, resulting in a genetically mosaic individual.

Spatial mass effect: Some species are present in an area where they cannot reproduce, or cannot reproduce fast enough for a self-sustaining population, but the population can remain because of an influx of disseminules from a nearby habitat where the species can maintain itself.

Species diversity: *Species richness* and the evenness of abundance among species in an area (e.g. quadrat) of specified size and shape. With the right combination of indices, diversity = richness × evenness.

Species richness: The number of (plant) species present in an area (e.g. quadrat) of specified size and shape.

Stability: see *Liapunov stability*.

Stratum (pl. strata):   A vertical layer, usually in an above-ground plant canopy but also below-ground.

Subvention: A positive interaction between plants, including *benefaction*, *facilitation* and *mutualism*.

Succession: Sequential change in vegetation.

Succession to climax: From any starting point there is one pathway of vegetational change to a 'climax', with no reversals.

Switch: A positive feedback between a species or community and its environment, in which the species/community changes the environment by reaction in a way that gives it relative benefit over alternative species/communities.

Synusia (pl. synusiae):   A *stratum* or other vertical rôle in a forest such as an epiphyte, aerial partial parasite or liana.

University of Otago Botany Lawn: Established c. 1965 with the sowing of an *Agrostis capillaris*/*Festuca rubra* mix. Thirty-six other species have arrived since then through natural dispersal. Since 1965, the lawn has been mown to a height of c. 2.7 cm, fortnightly in the growing season and monthly in winter.

Vascular plant: Pteridophyte, gymnosperm or angiosperm.

1 Plants Are Strange and Wondrous Beings

1.1 Introduction

Our aim in this book is to present a fresh perspective on the structure and functioning of plant communities, and especially the forces that limit the coexistence of some species and promote the coexistence of others. We hope this perspective will stimulate ecologists to think afresh about plants and plant communities and, even if they sometimes disagree with us, help them to form their own synthesis.

None of the existing theories of ecological communities has proved truly satisfactory; evidence for MacArthur's deductive theory of limiting similarity is sparse (Wilson and Stubbs 2012; see Section 5.6) and Grime's comprehensive, inductive C-S-R theory is far from being precisely predictive (Wilson and Lee 2000; see Section 6.5). Moreover, theories of the structure of communities are uninformative if no attempt is made to understand the mechanisms involved. Our approach is therefore reductionist, building from basic processes to generalisations about communities, always requiring solid evidence from the real world.

We have concentrated on areas where we feel we have a special insight to offer, and naturally some of these are areas where we have made contributions to the literature. These include species' environmental reaction (the effects of a plant or a community on its habitat, including via litter), interference (negative effects of species on each other by any means: competition, allelopathy, autogenic disturbance, interaction via hetero trophs, etc.), facilitation, switches (*sensu* Howard Odum) and assembly rules (*sensu* Wilson 1999a). There are topics we have not dwelt upon because, whilst interesting in their own right, they do not seem to advance our discussion of the structure and functioning of communities. Examples are species–area curves, the productivity–richness humpback curve, and ordination techniques and their use. Perhaps we could best characterise this book as a monograph on the core principles of plant community structure, approached bottom-up, i.e. from the individual plant to the whole community.

We shall emphasise terrestrial vascular plants because more is known about them, and the processes involved are generally found in them. However, most of the principles must apply also to lower plants, down to macroalgae and plankton (Tilman 1981; Wilson et al. 1995b; Steel et al. 2004), and we shall take examples from any group of organisms when we fancy. We shall not often discuss animals, though; this book is about plants.

1

1.1.1   Organisation of This Book

In this chapter we give the background to our topic, the nature of plants. Chapter 2 discusses the manifold ways in which plants can interact with each other. Many of these interactions tend to make species exclude each other, yet vegetation almost always comprises mixtures of species – the 'Paradox of the Plankton' – so in Chapter 3 we examine how this happens. Chapter 4 is an account of the collective ecological behaviour of species' mixes, i.e. of whole-community processes, which is in fact the overall topic of our book. Whilst those four chapters pull together the nuts and bolts of our enquiry, in Chapter 5 we attempt to synthesise by looking for precepts to species' coexistence in mixtures, i.e. assembly rules – the restrictions imposed on species' coexistence. In Chapter 6 we examine existing models of plant community structure. Finally, in Chapter 7 we put forward the main processes that structure plant communities as we see them, which we believe is the closest it is possible to come to an overarching theory, at least at this stage, and may always be.

1.1.2   The Plant in the Ecosystem

The five essentials of any ecosystem are: (1) input of energy, mainly from the sun via photosynthesis; (2) the capital of energy in the biomass of living organisms; (3) transfer of energy between trophic levels, e.g. from plants to herbivores; (4) cycling, especially of elements and (5) allogenic rate regulation, i.e. the control of the rates of these and other processes by environmental factors such as temperature (Reichle et al. 1975). The plant cover, i.e. the vegetation, is the major contributor to all of these essentials, in some cases the sole contributor.

Very rarely does a single plant species persist on its own, even when a gardener or farmer tries to make it so. Our subject is therefore multispecies communities. Every language uses terms to divide the plant cover of landscapes into communities; terms such as grassland (tussock, pasture, meadow), forest (conifer, deciduous, evergreen) and scrub (evergreen, summer-deciduous, krummholz). The ability to classify, subjectively or using objective methods (e.g. the British National Vegetation Classification; Rodwell 1991–2000), shows that species' mixes occur as repeated patterns. This is no surprise, since each habitat within the landscape supports only those species that are physiologically able to tolerate the particular environmental conditions (see Section 1.5.2). Even when a species can tolerate the environment, it might be excluded by the strength of interference from other species (see Section 1.5.3). Further, there might be adaptation to, or even intolerance of, the presence of other particular species or species' groups; these are the assembly rules (see Section 1.5.4).

Phytosociologists have invoked repeated patterns to rationalise the concepts of *fidelity* and *constancy* in plant associations. The issue is whether these 'associations'/'communities' have definable properties, whether there are distinct associations that can be classified and named. We are not interested in where one community ends and another begins – clearly, there are sometimes discontinuities between species' mixes, sometimes not – except in so far as the question increases our understanding of the processes that structure communities. Our definition of 'community' is therefore empirical:

The set of one or more species existing within a particular area at a particular time.

Our journey begins with a discussion of the nature of higher plants and an overview of the core concepts that form the basis of plant community ecology.

1.2 The Nature of Land Plants

Land plants, attached by roots or rhizoids and generally autotrophic, have quite different ecological properties from animals, and ideas developed for other trophic levels can seldom be applied to plants. We find two basic consequences of plant morphology:

1. *Physical movement and stasis.* Once established, land plants are sedentary.[1] However, plant organs have a limited length of life, so the plant must continually produce new modules, and thus inevitably explore new space.
2. *The problem of the individual in plants*: Applying the concept of an 'individual' to vegetatively reproducing plants is problematic. Even with seed reproduction, size is plastic and there may be somatic mutation. The great uncertainty as to whether plants can recognise themselves, and recognise kin, raises more questions. The concept of 'individual' is not generally useful for plants.

1.2.1 Physical Movement and Stasis

Animals of most species move around but, having grown, maintain approximately the same adult body, replacing organs cell-by-cell or molecule-by-molecule until death, even though in some fish, reptiles and molluscs the body continues to enlarge. Plants are always indeterminate in size, that size depending on the environment.

Plant architecture is modular, and to a considerably greater extent than in animals. Indeed, the plant has been seen as a population of modules (Harper 1977). Sometimes the repeated modules are discrete, such as leaves, flowers and vascular bundles in monocotyledonous trees (e.g. *Cordyline* spp., cabbage trees), though they can be hierarchically arranged, e.g. inflorescences comprising flowers, themselves comprising carpels, stamens, petals and sepals. Sometimes the modules are continuous, such as a year's growth of *Welwitschia mirabilis* leaves. In other cases the modules are adjoined to neighbours as in syncarpous ovaries, or to earlier modules, such as a year's addition of new phloem and xylem in a temperate dicotyledonous tree. Again, a hierarchy is possible, for example modules could be distinguished radially between parenchyma rays. Perhaps the exception is in *Wolffia* spp. (watermeal), where the plant comprises only one module, though even this has a stamen as another module. Modular growth is universal in plants.

---

[1] 'Sessile' in zoological terminology.

The living cells of these modules have a limited functional lifespan.[2] Because plants are essentially sedentary, defence from herbivores can be only by structure and chemistry, not by running away. This puts a selective premium on cell walls that are low in food value to herbivores but also strong enough to support cell turgor, resulting in cell walls basically of cellulose and lignin. These compounds cannot be recycled within the plant, so aged and dysfunctional modules are generally discarded and replaced by new ones, sometimes several times during the life of a plant.

Over a broad spatial extent plants are sedentary. Yet, replacement modules commonly are formed distally on the stem, or on side branches, so a plant must continually move, explore and expand into adjoining space,[3] and in the process may interfere with its neighbours. Vegetative reproduction, which often serves for resource foraging, is just an extreme form of this general phenomenon of plant expansion. Roots must also grow to explore for immobile phosphorus and, as a result, need to replace the root cap module. In this sense, plants move, whereas animals stay within their adult body. The plant's litter – flower parts, fruit and associated structures, leaves, shed branches and eventually the whole plant – is part of its movement and its effect on neighbours, i.e. part of its extended phenotype. This mandatory movement, often not recognised when plants are described as sedentary, results in autogenic disturbance, and is a major topic in Chapter 3. Some colonial, sedentary animals are similar to plants in that they must occupy new space to stay alive – some Tunicata (tunicates), corals and Porifera (sponges) – though the modules causing mandatory growth in corals are not discarded as leaves are, but sequestered like the xylem in the heartwood of trees. These few animals have similar movements to plants.

1.2.2   The Problem of the Individual Plant

It is difficult to recognise plant individuals (Firn 2004). We see three main problems and one unsolved issue, that of self/kin recognition.

1.2.2.1   The Plant as a Genetic Mosaic

It is well known that due to their basic growth pattern plants can comprise a genetic mosaic, a chimera, one example being sectorisation in trees, another being variegation (Gill et al. 1995). Animals often isolate their gametangial cells early in development, which reduces the risk of genetic transmission errors. In plants, the reproductive cells differentiate late, and the many cell divisions between the original zygote and the production of gametes allow the accumulation of somatic mutations: both base-pair changes and chromosomal rearrangements. Scofield (2006) estimated c. 41,000 cell divisions per metre of height growth in *Quercus rubra* (northern red oak), which allows 1,230,000 cell divisions by the time it has reached 30 m height. Occasionally, the results

---

[2] The photosynthetic rate of a leaf declines from quite early in its life, often even before it is fully expanded.
[3] Even cactuses may increase in size during their life (de Kroon and van Groenendael 1997).

of somatic mutation are clearly visible and adaptive, as when a branch of a *Eucalyptus melliodora* (yellow box) tree has resistance to herbivory that the rest of the tree does not (Padovan et al. 2013).

The rate of somatic mutation is more difficult to estimate. Most mutations will be deleterious, but some will survive, at least if recessive. For technical reasons the rate has been calculated mainly for sublethal, deleterious mutations, which is not directly relevant here. Bobiwash et al. (2013) estimated for *Vaccinium angustifolium* (low-bush blueberry) in Canada approximately three sublethal, partially dominant somatic mutations within each shrub. Klekowski and Godfrey (1989) estimated from a field survey of *Rhizophora mangle* (red mangrove) trees in the Caribbean region a mutation rate for albinism (selectively neutral in a heterozygote) of $6-7 \times 10^{-3}$ mutations per haploid genome per generation. Gross et al. (2012) provided evidence for considerable somatically generated genetic variation within a clone of the vegetatively reproducing shrub *Grevillea rhizomatosa* (Gibraltar grevillea) in New South Wales, Australia.

Somatic mutations certainly exist, and some survive beyond the cell in which they appear. Epigenetic changes, notably via DNA methylation, could also occur locally within the plant. What our science needs to know, and does not, is the extent to which somatic mutations arise that might be adaptive in some niches, changing the genotype of branches, sectors or the whole plant; mutations that are potentially passed on to future generations. The limited evidence available suggests that evolutionary change and divergence can occur as fast in populations of apomicts as in sexual ones, potentially leading to ecotypic adaptation that might be due to selection acting on somatic mutation (Pellino et al. 2013). Somatic mutation has considerable, unrecognised implications in plant ecology.

1.2.2.2   Production of New Modules

In vegetatively reproducing plants, a genet comprises ramets that are at first dependent, later semi-independent but supporting each other when in need, and eventually independent, perhaps with the group then splitting into patches (Harberd 1962; Marshall 1996). These patches can then become genetically differentiated through somatic mutation, so they do not technically comprise a clone. Similar issues arise with apomictic seeds: the lineage is initially identical in genotype but may later form divergent genetic lines by somatic mutation. Root grafting and occasional branch grafting between different trees is the converse situation: distinct genets that are physiologically interdependent (Dallimore 1917; Fraser et al. 2006).

All plants reproduce asexually in one way or another. There is no basic distinction between the apomictic seeds of *Taraxacum* spp. (dandelion), plantlets from the leaf margin of *Kalanchoe daigremontiana* (mother of thousands), separating rhizome fragments of *Elytrigia repens* (couch grass), root suckers of *Populus tremuloides* (aspen), and a bud that produces new leaf modules. These all replicate an original genotype, but after many mitotic divisions can accumulate mutations. Plant ecologists must cease aping animal ecologists in dealing with individuals, and deal with modules within genets, though even the latter is not fixed in its genes (see Section 1.2.2.1).

1.2.2.3   Size Plasticity

Another problem with applying the traditional animal 'individual' concept to plants is that whilst most animals are relatively predictable in size at a particular age, individuals of one plant genotype can differ in biomass by several orders of magnitude (Harper 1977). The modules within one plant can differ plastically too, if they are in different microenvironments.

1.2.2.4   Self-Recognition and Defining the Individual

There is some, be it controversial, evidence that the root system of an individual genet or even ramet can distinguish between:

(a) self/others, i.e. its own roots/shoots ('self') versus those of other individuals of the same species ('others') or

(b) kin/stranger, i.e. its siblings or offspring ('kin') versus non-kin of the same species, i.e. no more related to it than the population in general ('strangers').

Some self/others studies have found when plants are grown in contact with roots or whole plants of an identical genotype (i.e. self), they differ, often in root growth or shoot:root ratio, from those grown with other genotypes. Thus, Falik et al. (2003) found in a split-root-system experiment that when a plant was growing with its own genotype it produced significantly less root than when growing (competing?) with roots of another genotype. Similar effects have been seen in kin/stranger comparisons. For example, Murphy and Dudley (2009) found that plants of *Impatiens* cf. *pallida* (jewel-weed) growing with the roots of a plant from the same selfed family had lower leaf:root ratios than those grown with strangers.

Some self/other work suggests that self-recognition can wane. Thus, Gruntman and Novoplansky (2004) found a root mass response to self/non-self tillers of *Buchloe dactyloides* (buffalo grass) just after the tillers had been separated, but the effect decreased with time, until after 60 days, self tillers were no different in root mass from stranger tillers. This implies a mechanism somehow based on the physiological state of the plant, not its genotype. However, a physiological basis does not explain the self versus kin results of Murphy and Dudley (2009) and others. One possible explanation for these effects is that if plants are genetically different, there can be niche differentiation, leading to overyield. This is difficult to test because we do not know the factor in which they might differ. Another explanation is that if a plant of one genotype/family grows larger than that of another, it is able to take more than 50 per cent of the pot's resources, giving a greater total biomass. Bhatt et al. (2011) were inclined to discount the latter explanation for their results with *Cakile edentula* (sea rocket) on the grounds that two plants growing together were not significantly more different from each other in biomass when they were from a different selfed family than when they were from the same one. However, non-significance is always poor evidence of no effect, especially since F was here 2.53.

If self/kin/stranger recognition is real, how could it work? One possibility is by the chemical composition of root exudates moving through the soil solution or via common mycorrhizal networks. Biedrzycki and Bais (2010) found that plants of *Arabidopsis thaliana* growing in water previously inhabited by the same plant differed in root morphology from those grown in water previously inhabited by self/kin or by strangers, and the latter two differed from each other. The self versus stranger effects were suppressed by the presence of exudation-inhibitor sodium orthovanadate, but self vs kin effects were not. Atmospheric volatiles could conceivably carry a signal that is decodable only by a plant of the same genotype, or by its kin (e.g. Karban et al. 2013). However, a plant could hardly identify itself uniquely via chemicals. A system similar to the *S* alleles involved in pollination self-incompatibility would be possible. One locus would not enable identification to an individual genotype, and kin would not necessarily carry the same allele, but trends may still exist. Crepy and Casal (2015) suggested that kin were recognised by light-spectrum signals (R:FR and blue). It seems unlikely that the spectrum could identify genotypes, but they suggested that leaf positioning affected how the signal was received. An explanation in terms of the microflora associated with a plant/genotype is also possible.

Other studies have found no self or kin effects, or found them in some species but not others (e.g. Lepik et al. 2012). This is reasonable: self-recognition might operate in some species but not others.

Such self- or kin-recognition would have huge impact on plant community assembly, as indeed it does for social animals. The effects seem sporadic, and some ecologists explain them away in terms of root exploration of soil volumes, or of nutrient (NPK) competition (Nord et al. 2011). Scientists are always cautious about accepting results when it is difficult to imagine what mechanism has caused them. The ecological significance with respect to communities of these self- and kin-recognition experiments is still unknown.

1.2.2.5   Conclusion

Because of modular structure, plastic variation in size, vegetative reproduction, apomixis and somatic mutation, the concept of the 'individual' is not always valid for plants, demographically, ecologically or genetically, and indeed the concept of 'self' is contentious. Plants can be viewed as colonies of modules.

1.2.3   Species and the Plant Community

For an ecologist, a meaningful species must have a unique phenotype, with consequently unique environmental tolerances and reactions. The name of a taxonomic species allows us to predict much of a plant's morphology, physiology and growth. 'Morphospecies' can predict only the attributes already known.

1.3   Reaction

1.3.1   The Concept

The plant is affected by its environment, but it also changes its environment. Clements (1904, 1916) coined the term 'reaction' for 'the effect which a plant or a community exerts upon its habitat. . . . Direct reactions of importance are confined almost wholly to physical factors', and listed 20 such factors. Gleason (1927) agreed with Clements, as he almost always did (Section 6.2). The acidification caused by *Sphagnum* species in bogs is a well-known example, but for example, shading is a reaction too, on the light environment. Eviner and Chapin (2003) gave a comprehensive list. With more knowledge, biotic reaction would now be included. Reaction is the very basis of community assembly.

Any organism must cause reaction. The effect varies from slight to major, but plants especially cause reaction because of their bulk, their surface area, their absorption of resources, their aerial and below-ground secretion or leakage of materials, and their production of litter. These reactions modify local light availability, micro- and macro-climate, environmental chemistry and geomorphological processes, and thus the whole ecosystem. A species' response to its own reaction can be negative, i.e. altruistic facilitation (Clements 1916; this volume Section 4.3), or positive (a switch: Wilson and Agnew 1992; this volume Section 4.5). The near-synonyms 'ecosystem engineering' (Jones et al. 1994) and 'niche construction' (Odling-Smee 1988) were coined more recently, but we use 'reaction' because:

(a)   'Reaction' has priority.
(b)   'Niche construction' was coined, and is often used, more narrowly to include both ecological change (reaction) and consequent evolutionary change (Post and Palkovacs 2009), even with genotype/environment feedback that makes it a switch in evolutionary time (Scott-Phillips et al. 2014), whereas we are addressing here *environmental* change. Elsewhere, 'niche construction' is used more broadly, e.g. to include breeding system and herbivory response (Shuker 2014). Moreover, niche construction can be negative, a quite different process that leads to succession-towards-climax or to cyclic succession (Sections 4.3 and 4.4). However, 'niche construction' is useful in some contexts, and we shall sometimes refer to it.
(c)   Both 'ecosystem engineer' and 'niche construction' imply that only a few species have such effects, whereas we (with Clements 1904 and Gleason 1927) emphasise that all species have them.

Each species, with its unique phenotype, necessarily differs from others in its resource requirements, acquisition efficiencies, by-products, and phenologies of production and litter deposition, and thus in its reactions. These reactions are the basis of the great majority of types of plant interference and facilitation (Chapter 2), and indeed are behind the great majority of ecological processes. Since species are almost always spatially aggregated, the result will be autogenic heterogeneity in environment and resources, adding to the intrinsic heterogeneity (Section 7.3).

1.3.2   Reaction on Physical Factors

Reaction can readily be seen in the light regime beneath and beside different species, though much more is known of species' differences in total light transmittance than of changes in spectral composition (Section 2.2.4). Temperature, relative humidity and O_2/CO_2 concentrations can all change along with the light environment, and do.

Soil reactions are caused by foliar leaching, the decay of above- and below-ground litter, nutrient and water uptake, root exudation and occasionally changes in the soil atmosphere. They generally occur slowly, but for example it is clear that a few species, such as *Picea abies*, *Calluna vulgaris* (heather) and *Sphagnum* spp., differ strongly from other species in their community in the magnitude of their reaction on pH. The clearest evidence for autogenic heterogeneity in reactivity comes from forests, where the sheer size of trees (relative to scientists) makes their patches large and easy to sample. For example, Pelletier et al. (1999) found that in a mixed-species forest in Quebec, Canada, soil of the forest floor was different beneath different species. For example, extractable soil Ca was lower below *Fagus grandifolia* (American beech). In most such observational studies there is a chicken-and-egg problem: perhaps the soil differences are determining which species grows at a point, not the reverse. Fujinuma et al. (2005), finding higher ammonium-acetate extractable Ca and Mg beneath *Tilia americana* (basswood) than beneath *Acer saccharum* (sugar maple) in a 1–ha patch within Michigan forest, discussed this issue, but considered that preexisting soil hetero-geneity at that scale was unlikely. Pelletier et al. (1999) went two steps further: (a) they used spatial statistics to remove spatial correlations, attempting to examine the effects of individual trees, and (b) they offered evidence that *F. grandifolia* produces litter which, from its Ca, lignin, polyphenol and tannin contents, was likely to reduce soil Ca. The study of Ehrenfeld et al. (2001) produced evidence in another way. They found higher pH below two exotic species in a deciduous forest in New Jersey, USA, than beneath the native *Vaccinium* spp., but they also grew the species in the greenhouse on field soil and found similar pH differences to that observed in the forest.

The ideal evidence is from long-term randomised experiments, and Binkley and Valentine (1991) reported that in a 50-year-long replicated experiment in Connecticut, USA, soils under *Picea abies* (Norway spruce) were lower in pH, with less than half the exchangeable Ca, Mg and K, and higher in Al, than under *Fraxinus pensylvanica* (green ash). Changes in pH can be accompanied by, or effect, changes in nutrient availability, e.g. decreased pH can increase P availability. The experiments in Augusto et al.'s (2002) review show the effect of *P. abies* lowering pH to be quite general; indeed there are hints that it may be more generally true of gymnosperms. Challinor (1968) in a 30-year experiment with four tree species in North America found under *P. abies* the soils had greater pore space, higher total soil N and exchangeable K, and higher exchangeable Ca at the surface. Sartori et al. (2007) found higher extractable K after seven years beneath *Larix decidua* (larch) than beneath five *Populus* (poplar) species/hybrids/cultivars. However, the experiment of Alriksson and Eriksson (1998) with five tree species growing for 23 years showed, apart from a difference in pH in the litter +organic layer, which can be seen as a difference in the litter itself, only a difference in exchangeable Mg in the uppermost of five mineral soil layers.

Most of these changes will come via differences in the species' litter and possibly root exudates, but the process is difficult to observe. Even genotypes within a species can differ in their litter decomposition (Madritch et al. 2006), and thus possibly in their reaction on the soil. Experiments with soil litter bags normally last 2–5 years, whilst it might take 50 years to see the effects, and such experiments are usually established to examine the litter, not the underlying soil.

1.3.3   Biotic Reaction

Reaction can also be indirect, via the soil biota, potentially giving heterotroph-generated autogenic heterogeneity. Bezemer et al. (2010) found differences in bacteria, fungi and enchytraeid worms beneath two grassland forbs. Viketoft et al. (2005) demonstrated that the nematode communities differed markedly below field monocultures of 12 grassland herbs, both in total numbers and in species' composition; and in follow-up work Viketoft (2008) reported similar effects for six of these species from a greenhouse experiment, where the soil could be defaunated by alternate freezing and heating, then reinoculated with nematodes and microflora, to give a uniform starting point. Nematode community abundance and species' composition differed between plant species. Whether these differences affect the plant species differentially is not known, but there are fascinating hints in the work of van Ruijven et al. (2003) who demonstrated negative impacts of *Leucanthemum vulgare* (oxeye daisy) on invasion success in experimental mixtures, with results implicating nematode populations associated with the *Leucanthemum* as a potential explanation.

1.3.4   Conclusion

The unique characteristics of species result in unique reactions on their environment, and those reactions are the forces behind almost every process determining community organisation, which is the central topic of this book. Yet many aspects of physical reaction are hardly known, and in spite of recent work even less so for biotic reaction.

1.4   Niche and Guild

Grinnell (1904) and Elton (1927) introduced 'niche', both defining it as a zone within habitat space, outlined by physical and trophic parameters. Hutchinson (1957) formalised this as 'a region in n-dimensional hyperspace' where the dimensions are all the environmental, resource or behavioural (e.g. phenology, foraging) parameters that permit an organism to live.[4]

Niche and guild are central concepts in community organisation and assembly. They are closely related: a guild is a group of species with similar niches.

---

[4] Hutchinson uses 'environmental variables' to describe the axes of the hyperspace, which implies only beta niche (see below), but his subsequent discussion clearly includes resources, making the hyperspace one in alpha+beta niche.

1.4.1   'The Species Is the Niche' and Empty Niches

Some have defined niches by the species occupying them, so that 'the idea of an ecological niche without an organism filling it loses all meaning' (Levins and Lewontin 1985) and 'there is no such thing as an empty niche' (Chase and Leibold 2003).[5] It is even claimed that empty niches cannot exist under Hutchinson's definition, though Hutchinson (1959) used the term. If the species is the niche, whether that is the fundamental niche or the realised niche is not clear.

Others accept the idea of an empty niche (e.g. Tilman and Lehman 2001; Kueffer et al. 2009). In fact it is a necessary concept in theory, even if a difficult one. There must be regions of hyperspace that higher plants could never fill: floating in the air, growing on the ice at the South Pole or living at 100°C in hydrothermal steam vents. We presume these are not empty niches. However, there must be niches that could be occupied by a species, but are not occupied by any of the current species' pool due to limitations of evolution or dispersal, perhaps combined with the Allee effect (Colwell and Rangel 2009; McInerny and Etienne 2012). We can illustrate the absurdity of 'the species is the niche' by observing innovative invaders. Did no niche for a cactus exist in central Australia until *Opuntia stricta* (prickly pear cactus) was introduced (Dodd 1940)? Was there no niche for a cactus-eating insect before the moth *Cactoblastis cactorum* was introduced for biological control of *O. stricta*? These were empty niches that were later filled. Indeed, the empty niche is an important concept in considering species' invasions. Land itself was presumably an empty niche once.

Tilman (1997) claimed to find evidence of empty niches. He sowed seeds of up to 54 species into native grassland at Cedar Creek, Minnesota, USA. Many became established. However, this did not cause extinctions among the species originally present since the proportion of original species lost was not correlated with the number of species added ($r = +0.16$, $R^2 = 2.6$ per cent, not significant). Even more interestingly, the total visually estimated 'cover' of those species originally present did not decrease as more species were added ($r = 0.04$, $R^2 = 0.16$ per cent, not significant). These R^2 values are impressively low. Tilman concluded that the added species occupied empty niches. However, this study illustrates some of the problems in plant ecological work. Firstly, the species' composition probably covaried with the species' richness. Secondly, cover was measured subjectively, 'estimated by eye'. Objective measurements of abundance – presence/absence, local frequency, point quadrats or sorted biomass[6] – are available, are used and are essential in experimental work (unfortunately this issue will repeatedly mar results that we report in this book, though we use such work only when there are no alternatives). Thirdly, the concept of adding the cover of different species to give 'total cover' is illogical since it counts leaves of different species that lie under each

---

[5] Odling-Smee et al. (2003) believed that for Hutchinson 'a niche cannot exist without an occupant', as did Colwell and Rangel's (2009), but we see no reason to understand Hutchinson's words thus.
[6] Ideally calorific value.

other but not leaves of the same species that lie under each other: it does not give the total cover of the community, or any meaningful value (Wilson 2011).

However, 'the species is the niche' concept does remind us that there is really just niche space, which may be divided up into niches in various ways. For example, a niche can be split in two when an invader takes up residence, subject to rules such as limiting similarity (MacArthur and Levins 1967; this volume Section 5.6).

1.4.2   Guilds

'Guild' was coined as an ecological term by Drude (1885), 'Artengenossenschaft', for a group of species moving from one region to another, such as exotic species. It was used thus by Clements (1904, 1905) and Wilson (1989a). Perhaps independently of Drude, Schimper (1898, 1903) used the term 'Artengenossenschaft'/'guild' to mean a synusia (a structural unit within a community, such as a stratum) in a forest. Tansley (1920) used it in the same way, writing of 'guilds of the same dependent life form, such for instance as lianes'. Root (1967) ignored these established usages, and with animals in mind redefined the guild as a 'group of species using similar resources in a similar way'. This is not directly useful for plants, since almost all use the same resources (the sun's energy, water, CO_2, N, P, K and minor elements) in very similar ways. The guild is a category that is intended to be ecological rather than taxonomic, and Wilson (1999b), equally ignoring established usages, or ignorant of them, defined it as: 'a group of species that are similar in some way that is ecologic-ally relevant, or might be'. 'Or might be' is necessary here because the ecologist hardly ever knows at the beginning of an investigation whether the guilds being used are the real ones, and often not even at the end (but see the discussion of intrinsic guilds in Section 5.7.3).

1.4.3   Types of Niche and Guild

Two types of niche can be distinguished. The alpha (α) niche represents the resources used, i.e. the role of the species, e.g. the species' rooting depth, which affects the nutrients and water available to it. The beta (β) niche is the range of physical environ-mental conditions under which the fitness of a species is maintained, such as its temperature tolerance and therefore its potential geographical limits. Many methods of calculation of niche width and overlap can be used for both alpha and beta niches, and there are areas of character- and concept-overlap. However, much ecological discussion has been confused by failing to take the distinction into account. There are two corresponding basic types of guilds, alpha and beta, causing even more confusion (Wilson 1999b).

The distinction of the fundamental niche (i.e. the species' environmental tolerance) versus the realised niche (i.e. that after restriction or extension by other biota) was known to Clements (1907, p. 291) as 'Alternation due to competition'. It was mentioned by Tansley (1917) and Gleason (1917), and formalised as 'fundamental' and 'realised' by Hutchinson (1957). The realised niche of a species is usually considerably narrower than the fundamental one because of interference, herbivory, diseases, the availability of pollinators,[7] etc. The distinction between fundamental and realised is important, and too often ignored in modelling and predicting species' distributions (Veloz et al. 2012).

The plant is not a passive player in the niche: some niche characteristics, both alpha and beta, can be modified by the plant's reaction and that of its associates – the 'niche construction' of Odling-Smee et al. (2003) – expanding its realised niche in a particular locality. The ability of a plant to fit into a niche can be modified by its morphological and physiological plasticity.

Alpha versus beta niche, fundamental versus realised niche, even plant versus environment, are all important distinctions, but all are fuzzy at the edges.

1.4.3.1   Alpha Niche and Guild

The axes of the alpha niche are controlled by the morphology and physiology of the plant, its growth and its chemistry.

1. Morphology and its plasticity influence or are influenced by resource foraging and capture (light, water source, nutrients), persistence (storage organs, wood), autogenic disturbance (Section 2.5), heat budget (convective, transpirative, radiative), physical defence against herbivores (glands, hairs, thorns), pollination and dispersal biology. Examples of such niche differentiation are synusiae in forest (i.e. the strata, epiphytes, lianas, etc.), rooting depth and the parasitic habit.

2. Phenology: Species differ in their innate seasonal cycle and their response to environmental signals, and thus in timing of growth and of flowering ($\pm$ pollination)/fruiting (dispersal) in temperate vegetation and in leaf flushing (one or more times during the year) in tropical forests. Flower opening, nectar production and leaf opening or orientation can also differ through the day.

3. Physiology: The chemical functioning of a plant ultimately controls all its processes, but we may list as physiological examples: light requirement, photosynthesis type (C_3, C_4 or Crassulacean acid metabolism [CAM]), P sources via root phosphatase and organic acid exudate, N sources (N_2, NH_4, NO_3, amino acids), mycotrophy, chemical defence against herbivores and pathogens, and pollinator attraction.

4. Plants can also create niches de novo for other species, for example for parasites and epiphytes: i.e. niche construction.

Thus, species in the same niche, or in the same alpha guild, are similar in their resource use. For example, within northern European forests, species that are in the same alpha guild might be the trees *Tilia cordata* (linden), *Quercus petraea* (sessile oak) and *Fagus sylvatica* (beech). They are using similar resources: the light at the top of the canopy during the summer half-year, as well as nutrients and water from the full profile of the soil. They will therefore tend to exclude each other (Section 3.2). Species that could be

---

[7] It is not clear whether restrictions caused by other trophic levels should be included as biotic environment in the definition of the beta niche, or whether they should be regarded as part of the biotic force that restricts a species to a realised niche smaller than its fundamental niche. We use the latter definition here.

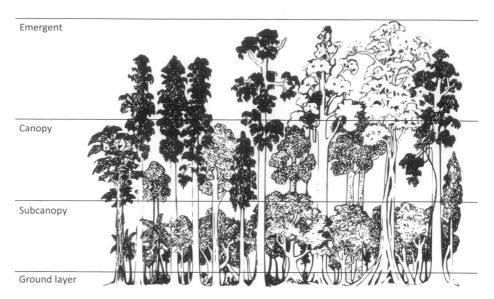

Figure 1.1 Stratification: profile of a New Zealand Podocarp-Broadleaf forest, showing four main strata.
Adapted and redrawn from Halkett (1991).

in different alpha guilds are *Tilia cordata*, the hemi-parasite *Viscum album* (mistletoe), the liana *Hedera helix* (ivy) and the ground herb *Anemone nemorosa*. They use different light/support/nutritional resources. This might enable them to co-occur; indeed, finding them together would suggest that they could be from different alpha guilds.

The most obvious alpha guilds in plant communities are stratum guilds (Schimper 1898, 1903; Figure 1.1). Almost all plant communities are structured vertically. Above-ground, the greater the vegetation cover, the more predictable is the vertical change in microclimate. Highly structured forests may have a stratum of separated, emergent trees, a more continuous upper canopy, then subcanopy trees, shrubs, tall herbs, creeping herbs and bryophytes, lianas and epiphytes (including lichens, bryophytes and vascular plants). The strata represent specialisation to vertical differences in light, temperature, water, CO_2 and nutrient resources. There is also stratification in grasslands, for example in the wet grasslands of Tierra del Fuego (Díaz Barradas et al. 2001) and even in lawns (Roxburgh et al. 1993). Naturally, stratification by primary producers can also be paralleled by stratification in consumer communities.

Similar stratification would be expected below ground because shoot litter is initially deposited on the soil surface, maintaining the structure and mineral nutrient status of the upper soil. Most water arrives at the soil surface and percolates down, acidified by organic acids and CO_2, hydrolysing the mineral fragments in the soil and making nutrients such as phosphate available. Plant roots and soil respiration can affect this, for example by releasing acids. Water deep in the soil, including artesian water, is available to deep roots and may rise up by capillarity and hydraulic lift. Temperature is buffered at depth. All this can lead to stratification of root systems. Succulents of New and Old World deserts have surface roots that can take up water and nutrients in ephemeral rainstorms, in contrast to deeper-rooted shrubs (Whitford 2002). Dodd et al. (1984) surveyed 43 woody species from savannah in South West Australia, and Timberlake and Calvert (1993) 96 shrubs and trees of Zimbabwe savannah, both finding that there were indeed species with consistently shallow roots and others with deep taproots, though most of the species had both lateral superficial roots and descending taproots. Herbaceous communities can also be stratified below ground (Weaver and Clements 1929, p. 213; Cody 1986).

1.4.3.2   Beta Niche and Guild

Beta niche axes are the environmental features of the locality and its biota, that is to say the habitat. This is related to Chesson's (2008) concept of 'environment' as a factor that does not form a feedback loop, i.e. is not affected by the organisms themselves, though in practice there is environmental reaction (Section 1.3). We distinguish six aspects of the beta niche:

1. **Climate** affects the ambient environment at all spatial scales: insolation, temperature, soil water availability, atmospheric humidity, snow, wind (which also acts as a pollen vector), aeration, CO_2, and, especially in ombrotrophic systems, also mineral nutrients. Climate affects morphology, for example as a partial determinant of the Raunkiaer life form (life form can also differ between alpha niches, for example in forest stratification).

2. **Geomorphology** describes the soil substratum, varying with altitude, slope and aspect.

3. **Soil chemistry**, notably the mineral nutrient content as modified by pH and waterlogging, affects system functioning (nutrient availability, cycling). In the short term, mineral nutrients often have dominant effects. Indeed, Grime (1988) suggests that all environmental stress is via nutrient limitation. Toxicity is caused by salt, H_2S, heavy metals and NPK in high/unbalanced concentrations, all affected by pH (Wilson 1993). There can be oxygen deficit affecting, e.g., root respiration.

4. **Disturbances** allogenic to plant communities are caused by climate, geomorphology and by the actions of heterotrophic biota. They can occur on a large scale, for example the landslips caused by earthquakes, but also on small scales, right down to worm casts. A species' occurrence in relation to a disturbance, or under different severities of disturbance, is due to its tolerance of changing environmental conditions, and is therefore part of its beta niche.

5. **Animals** that interact with plants – herbivores, pollinators and dispersers – differ geographically, affecting which plant species can persist at a site.

6. **Niche construction** can occur when plants change their environment ('Reaction': Section 1.3), gathering additional resources for themselves and changing the resource availability, both positively and negatively, for other species. An example is increased input to the ecosystem of water and mineral nutrients from precipitation, or via a pump from deeper soil layers. This often comprises a positive

feedback switch (Section 4.5). A species, once established, might create a beta niche (e.g. weathering to increase nutrient supply) that can allow it to persist.

All these aspects are variable through time: by the minute for many climatic aspects, through yearly to century-scale and beyond. The beta niche can be determined by: (a) extreme events (e.g. unusually severe frosts); (b) the mean value over some period (e.g. Pigott and Huntley 1981) and (c) duration of exposure to environmental conditions, such as the length of growing season required for amortization of the cost of leaf production, the production of disseminules, or development of floral initials for the succeeding year. Duration can be quantified as accumulation indices, such as degree-days or dry/wet exceedance values (e.g. Silvertown et al. 1999).

The species in one beta guild are similar, though never identical, in their beta niche, i.e. in their ecophysiology and therefore their tolerance of environmental conditions along gradients in space or time, such as the 'guilds of edaphic and topographical specialists' of Hubbell and Foster (1986). After a geographic species pool has passed through an environmental filter (see Section 1.5.2), the remaining species will be in a beta guild; they have overlapping beta niches. For example, all subarctic saltmarsh species would be in the same broad-scale beta guild because they occur in the same climatic and soil conditions, though there will be fine-scale beta guild differences, e.g. due to elevational zonation. An example of species occurring in different beta guilds might be the temperate, mesic tree *Tilia cordata*, the subalpine *Pinus contorta* (lodgepole pine), arid land trees/shrubs of *Prosopis* spp. (mesquite), the tropical *Cinchona officinalis* (quinine) and the mangrove *Avicennia marina*. They occur in different environmental conditions (climate and/or soil), so they are necessarily found apart in space or time. Díaz et al. (1998) recorded abundances of 'plant functional types' (PFTs) of 100 species along a climatic gradient in Argentina and found that vegetative traits differed between climatic zones, demonstrating that beta guilds are filtered out from the available species' pool.

1.4.3.3 Alpha vs Beta Niches and Guilds: Conclusions

The differences between alpha niche and beta niche are fourfold (Table 1.1):

(a) Spatial scale: this will vary with the size of the ramets, but alpha-niche differences can be between plants that are growing closely enough to interact with each other (Chapter 3), whilst beta-niche differences must be on a scale large enough for the environment to vary.

(b) Factor: alpha niches differ in resource use whilst beta niches differ in environmental conditions (it is trivial to say that one species is using nitrogen in the arctic and another in the tropics, and therefore they are using different resources).

(c) Related to (b): characters defining alpha niches are related primarily to resource acquisition and use, whereas characters defining beta niches are related primarily to stress tolerance.

(d) Effect of niche similarity: more crucial than the three differences above, the net results are opposite: species of the same beta niche will tend to co-occur because they have the same environmental tolerances; species of the same alpha niche will have no such tendency to co-occur, and if exclusion by interference is operating they will tend *not* to co-occur.

Table 1.1 Differences between the concept of the alpha niche and the beta niche

	Alpha niche	Beta niche
(a) Spatial scale	Local (near or below the size of a ramet)	Between patches or regional
	Within a community	Between communities
(b) Factor	Resource	Conditions, environmental
(c) Characters	Related to use of different resources	Related to tolerance of stress
(d) Effect of niche similarity	Species will tend to exclude each other, and therefore perhaps not co-occur	Species will occur together, because of similar tolerance of environmental conditions

However, as with most ecological concepts, there is no distinct boundary between alpha and beta niche. Normally, different environments cannot be present at one point in space at one time, but clearly there are often vertical differences in factors such as light in the canopy and pH in the soil. These have to count as alpha niche because they will allow co-occurrence. Disturbance is difficult to categorise because large-scale disturbances are clearly beta, but small-scale disturbances can be seen as within-community and hence alpha, depending on the spatial grain examined. The characters defining alpha and beta niches sometimes overlap, e.g. low growth is a feature of the ground alpha niche in a forest but also the beta niche of arctic plants.

The difference between alpha and beta guilds parallels that between alpha and beta niches: species that are in the same beta guild and therefore have similar environmental tolerances will generally co-occur; species that are in the same alpha guild and therefore use similar resources will tend to exclude each other unless other mechanisms of coexistence intervene (Chapter 3).

1.4.4   Interactions between Niche Axes

There are often interactions between niche axes, especially those of the beta niche, and with the fundamental/realised niche distinction. Pigott (1970) reported that near its eastern limit in Europe the niche of *Ilex aquifolium* (holly) contracts, becoming increasingly restricted to forest, whilst *Cirsium acaule* (stemless thistle) at its northern limit in England becomes confined to southern (warm) aspects. On the other hand, Diekmann and Lawesson (1999) found four potential examples where species had wider ecological amplitudes towards their range margin in northern Sweden, and suggested that climatic stress in that region excludes important competitors, that is to say, stressful conditions allow the realised niche to expand.

1.4.5   Functional Types

The concepts 'plant functional type' (PFT) and 'guild' can be essentially identical (Wilson 1999b; Blondel 2003). The current use of PFTs as the predicted variate in models assumes knowledge of the appropriate attributes to characterise the types. In spite of the term 'functional', which implies alpha guilds, most workers have apparently intended to model beta guilds. However, the characters they have chosen have often been alpha niche ones. For example, Kleyer (2002) formed guilds ('functional types') 'to relate unique PFTs to landscape specific habitat factors and to generalise syndrome-environment relations across landscapes'. He used characters such as annual versus biennial versus perennial, plant height, regeneration from detached shoots, having narrow leaves and longevity of seed pool, that are as likely to contribute to differences in resource use as to defining environmental tolerances (i.e. as likely to vary *within* a community as between). A distinction between 'response' and 'effect' guilds obscures the issue, because there is far more to the alpha niche of a species than its reaction on the environment. As with the niche, this situation has arisen from a failure to consider the purpose of the ecologist in designating the guilds, what type of guilds they will therefore be – alpha or beta – and what characters are therefore appropriate.

1.5   The Accession of Species into Mixtures

Communities are determined by four sets of processes, the first two comprising geographical and physical limitations and the second two biological ones (Figure 1.2).

1.5.1   Step A. The Availability of Species

In ecological time the plant community can comprise only species present vegetatively or as seed in the region, i.e. the geographical species pool (which we here define by dispersal availability, separating environmental filtering as Step **B**). This is an elusive concept, since it is not clear from what habitats and from what distance a species is allowed to arrive, and what time is allowed for it to disperse. For example, a gap in the distribution of southern beech spp. and other species such as the subshrub *Kelleria laxa* in the South Island of New Zealand (Figure 1.3) is explained by panbiogeographers via geological movements 10 million years ago (Heads 1989), by some molecular biologists via glaciations over the last 200,000 years (Leschen et al. 2008), by palaeoecologists via dispersal limitation since the last glaciation c. 18,000 years BP (Wardle 1964), and by some ecologists via current environmental filtering (Haase 1990).

The dispersal of propagules of every type is leptokurtic, most travelling surprisingly short distances but with rare long-distance events (e.g. Carey and Watkinson 1993), perhaps because the species has two dispersal mechanisms. The result is 'infiltration invasion', a combination of longer-distance 'guerrilla' invasion and short-distance 'phalanx' dispersal from those nuclei (Wilson and Lee 1989; cf. Egler 1977). This can occur over a range of kilometres (Lee et al. 1991; Figure 1.4) to centimetres, as when most of the tillers of a grass tussock are produced within the leaf sheath, but a few are pushed greater distances by animal hooves (Harberd 1962). Allee effects, metapopulation dynamics and density-dependent dispersal can cause small populations to grow slowly, or become extinct. However, in the absence of one of these three mechanisms the number of disseminules arriving will not affect whether a species can establish or not; unlike others we do not see a general role for 'propagule pressure'. Dispersal does

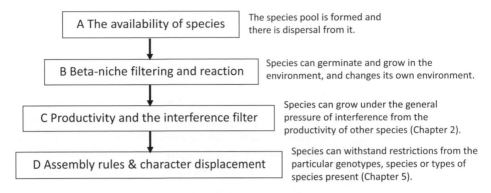

A The availability of species	The species pool is formed and there is dispersal from it.
B Beta-niche filtering and reaction	Species can germinate and grow in the environment, and changes its own environment.
C Productivity and the interference filter	Species can grow under the general pressure of interference from the productivity of other species (Chapter 2).
D Assembly rules & character displacement	Species can withstand restrictions from the particular genotypes, species or types of species present (Chapter 5).

Figure 1.2 Four steps in the accession of species into mixtures.

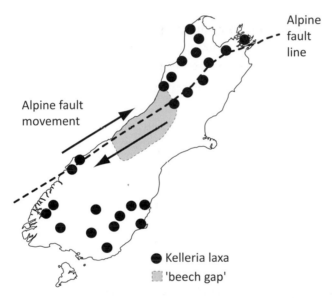

Figure 1.3 Disjunct distribution (●) of the subshrub *Kelleria laxa* in South Island, New Zealand, interpreted as an originally contiguous distribution torn apart by tectonic movement along the Alpine Fault in the last 10 million years. After Heads (1989), with the 'beech gap' from Wardle (1964).

not immediately result in community change: temporal inertia (Section 3.12) and priority effects (Section 5.4.2) delay changes. In fact, continued dispersal may be required to maintain a species, the spatial mass effect (Section 3.11).

1.5.2   Step B. Beta-Niche Filtering and Reaction

The second step in community assembly is challenge to the germination, growth and reproduction of available species by the environmental (i.e. abiotic) conditions of the site, resulting in elimination of those species that have dispersed into an area that is

Figure 1.4 Infiltration invasion by *Olearia lyallii* in the Auckland Islands.
Photograph reproduced from Lee et al. (1991), with permission of John Wiley and Sons Inc.

beyond their beta niche (see Section 1.4.5), i.e. beyond their tolerance in terms of temperature, water deficit/excess, soil chemistry, etc. (in stress habitats, some species enter dormancy and germinate only upon signals that indicate conditions favourable for their establishment, e.g. on saltmarshes [Alexander and Dunton 2002] and in arid lands [Pake and Venable 1996]). Sophisticated methods can be used to record the environmental response surface (e.g. Evangelista et al. 2008), but correlating species' composition with the environment remains an 'easy task' (Warming 1909), and we shall generally take relations with the extrinsic environment for granted in this book. However, such correlations are complicated by reaction (Section 1.3), which almost necessarily affects the establishment and survival of other plants, of the same or different species (Chapter 2). When undertaking analysis, care is therefore required to separate the contributions of beta niche environmental filtering, from filtering due to biotic interactions (Bennett and Pärtel 2017; see also Section 5.2.4).

The step from beta niche filtering (**B**) through filters **C** and **D** comprises the transition from the fundamental niche to the realised niche.

1.5.3   Step C. Productivity and the Interference Filter

Plant productivity, or more precisely net primary productivity (NPP), is total (gross) photosynthesis less any plant respiratory losses; it is thus a measure of the net rate at which carbon (and hence energy) is 'fixed' into plants. It is frequently said to be an important factor in community ecology. It is, but it is not always clearly understood. Net primary productivity is usually expressed on an annual basis (e.g. tC ha^{-1} year^{-1}), although daily, monthly and seasonal integration intervals are also common. Whilst relatively straightforward to define theoretically, NPP is very difficult to measure in practice (Roxburgh et al. 2004) and all existing methods for its measurement involve compromises. It has not helped matters that the relationship between NPP (a rate) and living biomass (a stock) is often confused. The situation can be clarified via the 'leaky bucket' analogy; the amount of water in the bucket at any given time (the biomass) is a joint function of the rate at which water is added from the top (NPP) and the rate at which it is being lost from the leak (losses via litterfall, mortality and herbivory but also minor fluxes such as root exudates and volatile organic carbon emissions). This analogy is useful as it illustrates how gains in productivity can be relatively impermanent, such as when growth is invested in short-lived modules (leaves, fine roots) rather than long-term structures (such as wood). It also highlights the fallacy of using change in biomass, on its own, as a proxy for productivity, given biomass (and thus biomass change) is a joint function of both gains from NPP, and the losses. In the context of this book the most tractable and relevant interpretation of NPP is the net rate at which carbon is fixed by plants, and made available to the next trophic level, i.e. herbivores or decomposers. One outcome of the productivity of the existing species, the rate of production of new modules, is to challenge by interference (competition, etc.) immigrants that have passed the environmental filter (**B**). This interference filter, the 'resistance controlled by productivity' of Kelly et al. (2011), depends not on the identity of the associates, but on their combined productivity. Operation of the interference filter can be seen in the ready horticultural cultivation of many species outside their natural edaphic and/or climatic range, and in their suppression by cover of weeds should the horticulturalist absent him- or herself from his or her duties.

1.5.4 Step D. Community Assembly Rules

The general ability of a species to enter the community might not be enough for its establishment. There could be further restrictions based on the identity of the associates, i.e. assembly rules: 'restrictions on the observed patterns of species presence or abundance that are based on the presence or abundance of one or other species or groups of species ...' (Wilson 1999a). The basis for them could be any of the plant-to-plant interactions we list in Chapter 2. We discuss assembly rules in Chapter 5, but we note that most are based on the niche/guild concept (see Section 1.4).

1.5.4.1 Other Trophic Levels

Plants interact with all other trophic levels. Examples are interactions with decomposers (Wardle 2002), above- and below-ground herbivores, mycorrhizal fungi (Chapter 2), endophytic fungi and bacteria, the phyllosphere and rhizosphere microflora, pollinators and dispersers, and even ants, which defend against herbivores or pathogenic fungi (Grostal and O'Dowd 1994). Top carnivores have indirect effects via the herbivore guilds. A species might be unable to maintain its population because it is beyond the environmental range of a specialised pollinator or disperser, or because of the impact of a pathogen or herbivore. For example, herbivores often prevent the establishment of woody species in grassland (Section 2.7). This would be a type of assembly rule, though since our thrust is plant communities, we generally discuss such effects only when they mediate plant–plant interactions.

1.5.4.2   Within-Species' Variation: Plastic and Genetic

We generally take species as fixed units to limit the scope of this book, but assembly rules could be based on intraspecific variants, whether plastic or genetic (Hart et al. 2016). The plasticity of plants could make intraspecific genetic adaptation difficult: when might plants need to change genetically when they can change plastically? One answer to this paradox has been 'genetic assimilation' (Pigliucci and Murren 2003) – the incorporation of plastic changes into the genotype – but the mechanism for this is unclear. Bradshaw's (1965) answer was that plants are genetically 'sewn into their winter underwear', i.e. plastic responses to an environmental shock would be too slow. Plastic responses can be caused by interference and facilitation, and this could cause or negate assembly rules. Within-species genetic variation leading to variability in the importance of intraspecific competition relative to interspecific competition has also been observed (Ehlers et al. 2016), with implications for species coexistence and community assembly. Epigenetic variation can exist too. Certainly there is considerable genetic variation that affects functional traits, both between and within populations (Carlucci et al. 2015), and, not surprisingly, this affects the growth and reproduction of neighbouring plants (Genung et al. 2012; Grady et al. 2016).

1.5.4.3   Character Displacement

A plant's genotype not only affects neighbours, but it can possibly be affected by neighbours. Turkington and Harper (1979b) suggested that microscale genetic adaptation to neighbours occurred in a North Wales pasture, perhaps in response to the spectral quality of transmitted light (Thompson and Harper 1988). This would be character displacement. However, later work suggested that the differences in this case were plastic, or caused by the strains of symbiont *Rhizobium leguminosarum* being different (Evans and Turkington 1988; Chanway et al. 1989). Character displacement at large scales as well as small is almost impossible to prove because the evidence must involve comparisons between locations, and those locations might differ in other ways (Strong 1983); it is one of the big unknowns of plant ecology. If found, character displacement would be evidence that species' interactions were a strong force in selecting between genotypes within species on a short evolutionary timescale, implying deterministic community structure and assembly rules. This would suggest that even stronger assembly rules occur based on the larger genetic differences between species. Urban (2011) considered the balance between adaptation to neighbours (character displacement) and gene flow, and suggested that character displacement 'might' occur 'often'. There are very few documented examples, at least in plants (Beans 2014).

1.6   Conclusions

We have described the basic material of plant communities. We argued that plants are colonies of modules. The animal concept of 'individual' is misleading for plants because of vegetative reproduction and apomixis, because plants are hugely plastic, and because somatic mutation means that plants are potentially genetic mosaics. Growth, vegetative reproduction, apomictic reproduction and sexual reproduction have similar implications demographically and genetically: an increase in the number of modules. Sexual reproduction differs only in how genetic variation is achieved. The disposable photosynthetic modules of plants must constantly be replaced, so a plant moves locally in space and its extended phenotype moves through litter production.

The huge majority of plant stands have more than one species, and we have outlined the basic processes **A–D** through which these communities establish and develop. The interactions between species' pool, dispersal and niche are all important in this process (Figure 1.2). Our concern in this book is to examine how species fit together to form communities. The basic concepts for this are the niche and the guild, so the distinction between alpha and beta niches and guilds is vital. Our conclusion, which we hope the reader shares, is that there is enormous complexity in the life of plants in spite of the simplicity implied in their common sedentary habit, modular structure and autotrophic nutrition. They truly are wondrous beings.

In the next chapter we examine the processes involved when one species interacts with another, initiating community development.

2 Interactions between Species

2.1 Introduction

In Chapter 1 we described the modular, plastic plant and its niche. The plant alters the environment around it, the 'reaction' of Clements (Section 1.3), in a way unique to each species, effecting a nexus of relations within the community. In this chapter we attempt to inventory the reactions and other species–species processes causing these interspecific relations, i.e. the mechanisms available for the development and organisation of communities. Plant reactions on the physical environment can be important at all scales, from geomorphology down to adjacent leaves and miniscule root hairs, and can cause interference (negative), facilitation (positive) or neutral effects. The plant also produces litter, which supports decomposer communities, affects plants directly (Section 2.4) and causes autogenic disturbance (Section 2.5). Parasitism is one of the very few types of a direct plant–plant interaction (Section 2.6). Plants are almost the sole energy basis for heterotrophs, being negatively affected by them in some ways, but relying on them for many functions; we consider only cases where one plant affects another via a heterotroph (Section 2.7). Finally, there are many possible combinations of effects; we confine ourselves to discussing herbivory and fire, because of their importance in many communities (Section 2.8).

Classifications of interactions between organisms can be confusing. 'commensalism' and 'amensalism' are clear, but found more frequently in textbooks than in real scientific work; often 'competition' is written of as if it were the only negative interaction between plants; 'mutualism' is commonly used without any evidence that the relations are mutual; litter effects are usually considered only for their effects on nutrient cycling, and autogenic disturbance is ignored as an interaction type. We hope to redress those imbalances.

2.2 Interference: Negative Effects between Plants

The literature is confused on 'interference'. Generally, plant ecologists regard competition (defined below) as one kind of interference, whilst animal ecologists regard interference as one kind of competition. Our usage recognises the precedence of Clements (1907) and the widespread influence of Grime (1979). Even when terms have been defined they have often been used carelessly. For example, 'competitive ability' and 'competitive exclusion' are used when there is no evidence that the negative effects were exclusively, or even mainly, due to competition. Of course, when we are discussing models or processes of competition, or when we are citing authors who claim to be, we shall write 'competition'.

Classifying interference is problematic, as in some cases the underlying processes can generate both negative and positive outcomes (e.g. the effects of litter; Section 2.4). The first three kinds of interference shown in Box 2.1 primarily depend on reaction, and are reviewed in this section. The remaining four classes of interference in Box 2.1 are discussed separately in following sections, either because they are more complex and can involve both positive and negative outcomes (e.g. litter effects, autogenic disturbance, heterotrophic disturbance), and/or because the interactions involve direct effects of one plant upon another (e.g. parasitism). A switch process (Section 4.5) involves interference because it is based on reaction in an environmental factor (often a non-resource one) and by definition it can have a relatively negative effect on (an)other species; succession-towards-climax (Section 4.3) and cyclic succession (Section 4.4) both involve interference. The issue of whether interference ability can be circular or whether species form a transitive interference hierarchy is considered in Section 3.5.

2.2.1   Competition

We use the definition: 'Competition is an interaction between individuals, brought about by a shared requirement for a resource in limited supply, leading to a reduction in the survivorship, growth and/or reproduction of at least some of the competing individuals concerned' (Begon et al. 1996), which is similar to: 'The tendency of neighbouring plants to utilise the same quantum of light, ion of mineral nutrient [or] molecule of water ...'[1] (Grime 2001). The central role that reaction on limiting resources plays in competition was emphasised by Clements (1907, p. 253):

> Competition arises from the reaction of one plant upon the physical factors about it, and the effect of these modified factors upon its competitors. In the exact sense, two plants, no matter how close, do not compete with each other as long as the water content and the nutrient material, the heat and light, are in excess of the needs of both. When the immediate supply of a single necessary factor falls below the combined demands of the plants, competition begins.

Niklas and Hammond (2013) quite correctly termed this 'competitive interference' – 'Competitive interference occurs when plants consume resources that would be otherwise available to their neighbours' – as opposed to 'non-competitive interference'.

Here we can consider only selected topics out of the many that have occupied plant ecologists since the masterly investigations of Clements et al. (1929). The suggestion that there is no observable competition in existing communities is covered in Section 6.4.

---

[1] Grime added competition for space, but see Section 2.2.1.1.

> **Box 2.1** Types of plant–plant interference
>
> Competition = via preemption of resources from the environment: Section 2.2.1.
> Allelopathy = via toxic substances: Section 2.2.2.
> Spectral interference: Section 2.2.3.
> Negative litter effects: Section 2.4.
> Autogenic disturbance: Section 2.5.
> Parasitism: Section 2.6.
> Via heterotrophic disturbance: Section 2.7.

2.2.1.1   Factors for Which Competition Occurs

Competition by definition occurs for resources, i.e. molecules or types of energy necessary for plant growth/maintenance that can be absorbed by either one plant or another, but not both. This clearly includes light, water and macronutrients (NPK) and more rarely/ theoretically C (generally CO_2, occasionally sugars, bicarbonate ions, etc.), micronutrients (Ca, S, Mg, etc.), and perhaps radiant heat and soil oxygen.

Light (here photosynthetically active radiation) is the resource for which competition has most often been analysed. It differs from most other resources in that it is available instantaneously, disappearing if not used, and its source is unidirectional. Competition for mineral nutrients, mainly N, P and K, is well established by experiments in pots and in the field (Clements et al. 1929; Wilson and Newman 1987). The main issue is that because N is more mobile in soils than P it is possible for two plants to come into competition for N before they compete for P (Bray 1954). The mobility of K is intermediate. Competition could in theory occur for the other essential elements. Plants might also compete with microorganisms, e.g. for iron.

The role of competition for water between trees and their understorey is well known, particularly in savannah. This can be two-way, the trees reducing the growth of the understorey and vice versa (Knoop and Walker 1985). In an experiment in the Sonoran Desert the removal of all other plants surrounding a tussock of *Hilaria rigida* increased its water potential, which led to the plant remaining green for longer into the dry season and growing more (Robberecht et al. 1983). In seasonally dry tropical forest in Brazil the removal of lianas increased the water potential of trees and also their growth rates (Perez-Salicrup and Barker 2000), though the authors did not discuss how much of this was due to the obvious increase in light due to removal of the liana canopy. Like mineral nutrients, but in contrast to light, soil water remains available for some time if not absorbed, though not indefinitely.

Competition could theoretically occur for CO_2 as a carbon source. Plants can reduce the CO_2 levels around themselves through photosynthesis, and this could affect the growth of neighbours (Oliver and Schreiber 1974). However, there is also production of CO_2 by respiration of soil and litter (Bazzaz and McConnaughay 1992) and considerable air mixing (Reicosky 1989). Depletion of CO_2 has occasionally been recorded from dense canopies: Bazzaz and McConnaughay (1992) recorded a decrease by 9 ppmv (parts per million by volume) below a stand of *Abutilon theoprasti* (velvet-leaf),

Buchmann et al. (1996) a decrease of up to 26 ppmv in the understorey of temperate forests with a dense field layer, and Reicosky (1989) a decrease of c. 80 ppmv in canopies of *Zea mays* (maize) and *Glycine max* (soybean). This decrease in CO_2 concentration would reduce the growth of most species. Clements et al. (1929) listed CO_2 as one of the resources for which competition would occur. It is probably more important than has usually been assumed, but information is needed on the depletion in natural vegetation, and of the response of the species to the CO_2 concentration change observed.

Ambient temperature is a rate-regulator, not a resource, but Clements et al. (1929) suggested that plants may compete for radiant heat. Clements et al. (1929) also suggested that plants could compete for soil O_2 in waterlogged conditions. In a sense O_2 is a resource because a molecule absorbed by one plant is not available for another, but it is also related to redox potential, which is an environmental factor.

Ecologists often write of competition for physical space. 'Space' as shorthand for all resources misleads the young and impressionable and obscures the differences between resources in their mechanisms of competition.[2] But some writers have explicitly seen space as a resource separate from light, water, NPK, etc., for example, 'competition is defined as "the tendency of neighbouring plants to utilize the same quantum of light, ion of a mineral nutrient, molecule of water, or volume of space"' (Grime 1979, 2001); 'weeds competing for light and space in the first year of growth, rather than moisture or nutrient stress' (Sage 1999). Yodzis (1986) envisaged that: 'competition for space is so different from what we normally think of as consumptive competition that it makes more sense. . . to think of it as a completely different category of competition. Certainly space is quite different from any other resource'. Yes it is, but there is never competition for it, for it is never 'in limited supply'. Chiarucci et al. (2002) measured for the first time the percentage volume occupancy of eight plant communities (four in each of Italy and New Zealand, comprising four grasslands and four shrublands). Only 0.44 to 2.89 per cent of the available volume within the canopy was occupied by plant tissue and they concluded: 'physical space is probably never limiting by itself in terrestrial higher-plant communities, so that competition for space . . . is not likely to exist'. Clements (1916, p. 72) understood this of course: 'In a few cases, such as occur when radish seeds are planted closely, it is possible to speak of mechanical competition or competition for room. . . . However [this] seems to have no counterpart in nature. There is no experimental proof of mechanical competition between root-stocks in the soil, and no evidence that their relation is due to anything other than competition for the usual soil factors – water, air and nutrients'. Actually, there was no evidence for almost 80 years that it occurs even between close radishes, and when Wilson et al. (2007a) investigated, the effect was small (7 per cent) and arguable. Even Tilman (1982), though devoting a chapter to competition for space after disturbance, notes that physical space 'may be irrelevant . . . it would seem better to study explicitly the resources supplied'. Indeed.

---

[2] The attempt by neighbouring plants to use the same space can result in contact and damage, but this is not competition, it is autogenic disturbance (Section 2.5).

The competitive abilities of the plant are determined by its characters, but not necessarily those visible in monoculture, rather those possibly changed by plasticity in mixture. Plastic responses to associates differ between responding species and between species causing the response (Bittebiere et al. 2012). This is well known above-ground, for example in the process of etiolation, but it occurs in less visible characters too, for example root spread and depth (Miller et al. 2007) and shoot:root ratio (Nagashima and Hikosaka 2011), and these responses need further exploration.

2.2.1.2   Cumulative Effects

The terms 'asymmetric interference' and the more limited 'asymmetric competition' are used in two senses. Firstly, they can mean that the interference effect of Species **A** on Species **B** is greater than that of **B** on **A** (e.g. Ives and Hughes 2002). Species differ by definition in characters and therefore very probably in their interference effect, but the degree of A/B asymmetry is an important feature of the community matrix (Section 4.8).

The second meaning of 'asymmetric interference' is that 'larger plants are able to obtain a disproportionate share of the resources (for their relative size) and suppress the growth of smaller individuals', 'size asymmetry' (Weiner et al. 1997). If this happens, the effects of interference will increase the differences in interference ability, cumulatively. Since earlier-arriving or earlier-germinating plants are usually larger, this gives 'niche preemption' (Wedin and Tilman 1993). This process will occur between species too, leading to exclusion by interference. Many authors have declared that this is a basic phenomenon, but this ignores the mechanism.

The classic study is that of Black (1958). He established swards of *Trifolium subterraneum* (subterranean clover), using large seeds (mean mass 10.0 mg) and small seeds (mean mass 4.0 mg), all of cultivar Bacchus Marsh and therefore probably of near-identical genotype. In monoculture, the two sizes of seed and a 50:50 mixture of them all produced essentially identical sward mass throughout the experiment. However, in the mixture the initial seed mass was crucial: plants from large seeds suffered no mortality and came to form 93 per cent of the mixture biomass, whereas more than half of the plants from small seeds died, and those remaining contributed only 7 per cent of the biomass. The distribution of leaf area showed the cause: plants from small seeds held their leaf area a little lower than those from large seeds. In monoculture this did not matter, but in mixture their share of the light was disproportionately small, so that by the end of the experiment they contributed 10 per cent of the leaf area but that was held so low that it captured only 2 per cent of the light. This happened because of a specific mechanism (Figure 2.1): (a) in *T. subterraneum* greater growth results in the leaf area being held higher, and (b) in competition for light the height of the leaf area is crucial. Thus, the results of competition changed the competitive abilities, giving positive feedback and hence cumulation. 'Height asymmetry' would here be more appropriate than 'size asymmetry'.

Competition for light will not always yield a cumulative growth response. In Hirose and Werger's (1995) experiment, the taller species of a wet meadow community had little advantage from the higher position of their leaves in the canopy because, though

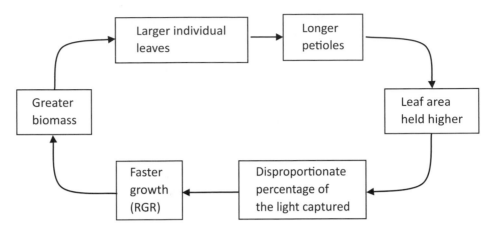

Figure 2.1 In *Trifolium subterraneum*, the results of competition affect competitive ability. RGR: relative growth rate.

holding their leaves higher, they had less of their biomass in leaf area and more in stems, and Berntson and Wayne (2000) failed to find biomass size asymmetry in competition for light between *Betula alleghaniensis* (yellow birch) seedlings, apparently because of height plasticity in small, shaded plants.

Cumulative effects would not be expected in below-ground competition, because there is no equivalent to the positive feedback of growth → height → light → growth. A plant that is larger with longer roots, able to access deep soil resources, generally has no advantage because: (a) the deep roots are hardly likely to shield roots nearer the surface from NPK rising up through the soil, especially since NPK generally does not rise and (b) nutrients are generally much more available near the surface, where small plants can easily root. Uptake of water is sometimes from aquifers, but it is hard to envisage a root shield comparable to the above-ground canopy. In terms of allometric growth, if a plant is larger by ×2 times in each dimension, it will be ×2^3 = ×8 times larger in volume and probably also in biomass. Therefore, its NPK requirements will be ×8. If most nutrients are available near the surface where all plants can reach, the nutrient resource will be approximately proportional to soil area rather than soil volume, and the larger plant's root system will cover an area only ×2^2 = ×4 times larger to cope with the ×8 times larger demand. The large plant will be the one deficient in nutrients. Thus, Rajaniemi (2003) could find only very ambivalent evidence of size asymmetry in below-ground interference, and all other investigations have found none, with in fact the larger plants in such experiments usually being at a disadvantage due to greater within-plant interference (Wilson 1988b; Weiner et al. 1997). This fits with Wang et al.'s (2010) finding that the competitive effect was closely associated with size-related traits under high-nutrient conditions when most competition will be for light, and with root-related traits under low-nutrient conditions when competition for nutrients will be more important. More definitive evidence for the critical role of competition for light in cumulative interference comes from the experiment of Hautier et al. (2009) adding NPK fertiliser to greenhouse mesocosms: fertilisation led to decreased species richness after

12 months, but this effect was removed by supplementary lighting in the understorey, both decreasing competition for light to shift it to below-ground competition, and removing the height advantage in the remaining light competition.

We could think of possible exceptions. When nutrients are available patchily, small plants can preferentially grow in a high-NPK patch, whereas a large plant might need to include in its root system the nutrient-poor matrix to its disadvantage. However, if nutrients were available intermittently in time and space, such as in animal micturition patches, perhaps a large plant could take up enough to satisfy the whole of itself from a few roots in the rich patch, whereas a small plant outside the patch could not reach in before the resource supply disappears (Figure 2.2). The results from Tamme et al. (2016), using experimental grassland assemblages growing in soil with patchily distributed nutrients, support this hypothesis. This argument applies only when the patches are transient, so it seems reasonable that Blair (2001) could find no evidence that experimentally imposed spatial heterogeneity in nutrients led to size-asymmetric below-ground interference.

Thus, asymmetric interference, and the resulting cumulation effects, will occur only when there is a specific feedback mechanism, such as that via height/competition for light feedback. Statements such as: 'Earlier arrival or faster initial development of a species leads to space occupancy both above and below ground and contributes to species success' (Körner et al. 2008) are misleading generalisations.[3] However, competition for light is widespread and almost always cumulative, so exclusion by interference and reduced opportunity for coexistence that it gives have huge implications for community structure (DeMalach et al. 2016), which we shall point out often.

2.2.1.3   Size Distribution and Self-Thinning

Competition is basically the same in plants and in animals. One owl eats a mouse, so another owl cannot and perhaps dies. One plant takes up water from the soil, so another plant cannot and perhaps dies. However, since animals are usually quite aplastic, interference normally affects density rather than individual size. Plants are plastic, so interference affects plant size first, before density. Nevertheless, plant interference often does result in some decrease in density, and this has been generalised as an approach to a straight line on a log–log plot of mean individual plant weight against density. The slope of such a line has been claimed from geometric theory and from observations to be any of −0.5, −0.75, −1.33 and −1.5 (e.g. Dcng et al. 2012). When there is competition for light and it is, as usual, cumulative, there is a particular reason to expect mortality: a dominance hierarchy will result in which the small/short plants lag further and further behind the large/tall ones, until they die (Black 1958). The resulting size distribution may pass through a bimodal phase, as the size asymmetry develops and before all of the small/short individuals have been excluded (Turley and Ford 2011). When competition is below-ground no such effect will be present (Wilson 1988b) so self-thinning should be much less prominent and the −3/2 (etc.) relation may be absent.

---

[3] Anyway, plants do not compete for space: see Section 2.2.1.1.

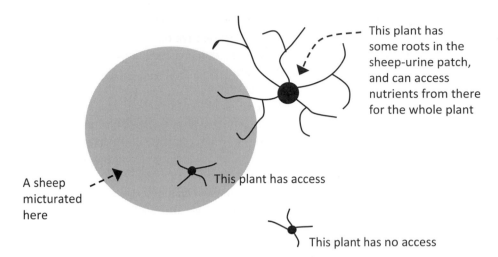

Figure 2.2 A possible mechanism for size-asymmetric competition below-ground, when nutrient supplies become available patchily in time and space, e.g. from sheep micturition.

This has not been investigated, probably because it would be difficult in a dense population to impose realistic competition for nutrients or water without there being competition for light.

Self-thinning mortality has sometimes been seen in shoots generated vegetatively via rhizomes or stolons, e.g. in the culms of the *Gynerium sagitatum* (uva grass; de Kroon and Kalliola 1995), in the tufted regeneration of two *Sasa* spp. (bamboo; Makita 1996) and in shoots of *Urtica dioica* (nettle; Hara and Srutek 1995). This emphasises the plant as a population of growing points (Section 1.2.2). However, the effect will sometimes be eliminated by ramet support.

2.2.2   Allelopathy

Allelopathic interference involves, like competition, reaction on the environment, but via a toxin, not a 'shared resource in limited supply' (Section 2.2.1). The toxin, or a precursor that is broken down to a toxin perhaps by soil microbes, is either leached or volatilised from the living parts of plants (above- or below-ground), from standing dead plants, or from decomposing litter above-ground or below (Section 2.4.2.3). Since compounds are readily leached from most plants, allelopathy may be widespread and important (Inderjit et al. 1994), or since soil bacteria and colloids can neutralise their effect it may be rare (Harper 1977). Field experiments are difficult to perform since '[t]he suspected allelochemicals must be isolated, identified, then quantified on a rate release basis, purchased or synthesized, and finally reapplied simultaneously in the field at natural rates over a prolonged period of time' (Williamson 1990).

Most work has been with shoot elutants, but Thorpe et al. (2009) obtained good evidence for allelopathy via root-produced toxins from the invasive forb *Centaurea maculosa* (spotted knapweed), realistic in that the supposed toxin was applied to plants growing in the field and in what were, to the best of their knowledge, realistic concentrations. Besides complex chemicals, reactive oxygen species are also possibly belowground allelochemicals (Bais et al. 2006). Many compounds leak from roots: sugars, amino acids, proteins, organic acids, fatty acids, sterols, etc. (Pierik et al. 2013), and could reach other plants through the soil solution, though whether they affect neighbours negatively, positively or not at all is little known.

Allelopathy might start early in the plant's life. The interference that has been demonstrated between germinating seeds (Harries and Norrington-Davies 1977; Murray 1998) could be competition for water for imbibition and mucilaginous anchorage, but allelopathy is more likely. Other workers, on the other hand, have found faster germination when seeds are denser (Orrock and Christopher 2010).

Plant development can be modified by allelopathy. Mahall and Callaway (1991, 1992) conducted greenhouse experiments between *Larrea tridentata* (creosote bush) and *Ambrosia dumosa* (white bursage) that occur together in the Mojave Desert. Roots of *A. dumosa* were inhibited when they came near to roots of *L. tridentata*. This effect was reduced by activated carbon, so it was presumably via a chemical. The two species apparently have similar water extraction capabilities (Yoder and Nowak 1999b), so this allelopathic 'signalling' would cause plants to avoid each other and postpone the onset of competition. This may be a more common phenomenon than is realised at present. However, Nord et al. (2011) suggest, with evidence from a greenhouse experiment with *Phaseolus vulgaris* (French bean), that at least some such effects are due to simple competition for nutrients.

Ecologists are inclined to assume that allelopathic ability has evolved as an adaptation: species that make their locale unfavourable for the growth of other species will gain in selection. But can this evolve? If a mutant produces an allelochemical active only against other species this will be at some cost to itself (synthesis of many of the secondary compounds involved is energetically expensive), it will benefit non-mutant plants of the same species as much as itself,[4] and the trait will not evolve. It is then necessary to invoke group selection, a controversial and probably weak force in plants (Wilson 1987). This problem would not occur with selection for a plant to protect itself, and thus eventually its species, against its own allelochemical. Williamson (1990) commented that there would be equal selective pressure on the species' neighbours to develop such defences, but this ignores the issue that a species always grows with itself, but has a variety of neighbours against which it would have to protect itself.

A species' toxins can inhibit its own seed germination or growth: autoallelopathy (Singh et al. 1999). Newman and Rovira (1975) tested a set of British meadow herbs and found autoallelopathy via root leachate in *Hypochaeris radicata* (cat's ear) and *Plantago lanceolata* (ribwort plantain), which might explain why those species never form pure stands and individuals of them are short-lived in meadows, resulting in very small-grain cyclic succession (Section 4.4). Newman (1978) questioned whether selection was involved in allelopathy at all, finding in a literature survey that autotoxicity is

---

[4] Unless the mutation simultaneously causes production of the allelochemical and resistance to it, and it is not likely that a single mutation will do this.

as common as toxicity to other species, and argued that allelopathy is an incidental result of the production of secondary compounds as pest-defence compounds or as metabolic by-products.

Given the huge numbers of papers that have been published on allelopathy without any consensus having been reached on whether it is important or insignificant, and given the problems in demonstrating that it operates effectively in the field, there can be no hope for a resolution any time soon.

2.2.3   Spectral Interference (Red/Far-Red)

Plant canopies change the spectral quality of light that passes through or is reflected off them (reaction), notably reducing the red/far-red ratio (R:FR), and this could affect neighbours' seed germination, relative growth rate, tiller production, leaf area ratio, leaf chlorophyll content (and hence photosynthetic capacity) and leaf longevity (Barreiro et al. 1992; Dale and Causton 1992; Skinner and Simmons 1993; Rousseaux et al. 1996; Batlla et al. 2000). Species differ in their response to R:FR. For example, Leeflang (1999) found that only one of six stoloniferous forbs responded to R:FR ratio in terms of biomass accumulation, and Dale and Causton (1992) found differences between three *Veronica* species (speedwell), such that the species that grew naturally in deeper shade responded more to R:FR, e.g. in stem weight ratio. Such differences between species might be the basis for assembly rules.

2.2.4   Remote Signalling of Impending Interference

Alterations to the R:FR ratio may be less important for its direct effect on growth than as a signal to a plant that competition is impinging, triggering plastic changes, some of which would increase competitive ability for light – faster leaf elongation, more upright shoots and leaves, and a higher shoot:root ratio (Skinner and Simmons 1993; Vanhinsberg and Vantienderen 1997). Other responses, such as altered stem angle and longer internodes, can help the plant avoid competition by foraging for light gaps (Ballaré et al. 1995). Thus, some plants, whilst still at some distance from neighbours and before they are directly shaded, sense the changed spectrum of light that has been reflected onto them and then grow away (Schmitt and Wulff 1993). Manipulation of R:FR ratios and experimentation with pieces of plastic with the same spectral characteristics as plant leaves show this to be due to R:FR effects (Figure 2.3; Novoplansky et al. 1990). Germination can also be affected (Dyer et al. 2000). However, it can be difficult to separate effects of light intensity and spectrum, and intensity effects may be greater: Muth and Bazzaz (2002) found that visible (PAR) light was more important than R:FR in the gap-seeking of *Betula papyrifera* (birch) seedlings. Very little is known of the processes causing the three-dimensional positioning of plant modules in a mixed-species canopy, but R:FR effects may be involved here. Blue light depletion can also affect neighbours, e.g. causing petiole elongation (Pierik and de Wit 2014).

Another mechanism that might potentially signal the presence of a neighbour is via volatile compounds. Although volatiles have been implicated in plant–plant interactions

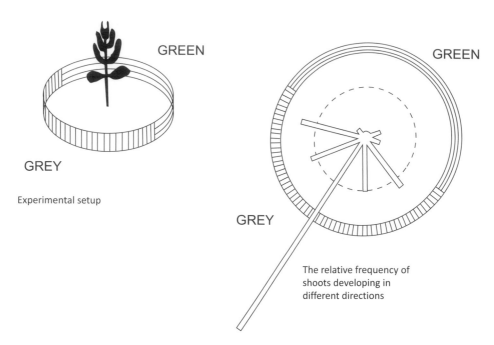

Figure 2.3 The effect of R:FR light, as seen in surrounds of different colours. Reprinted from Novoplansky et al. (1990), by permission from Springer Nature.

via heterotrophs, notably herbivores (Section 2.7.3.1 below), ethylene and many other volatile organic compounds from one species can affect the growth of another directly, e.g. altering its shoot:root ratio, perhaps thus giving advance warning of potential future interference (Pierik et al. 2013). It has also occasionally been suggested that plants might communicate by electrical signals (Shepherd 1999; Gruntman and Novoplansky 2004). Such signals may well operate within a plant (Yan et al. 2009), but there is no evidence they can between plants, and we doubt it.

2.3   Facilitation: Positive Effects between Plants

Reaction does not necessarily result in interference; it can also result in facilitation, i.e. a positive effect by one species on the survival and/or growth of another. Just as the plant community provides habitats for heterotrophic organisms, so facilitation can create new niches for other plant species, both horizontally and vertically (Section 3.2.3). Gleason (1936) gave facilitation ('favourable interference') equal weight with negative interference, but facilitation has been given appropriate attention only recently. We note that the existence of facilitation by Species **A** on Species **B** does not imply that they have coevolved. It is simply a corollary of reaction.

The terminology is confusing. 'Facilitation' was used by Connell and Slatyer (1977) for Clements' (1916) concept of the mechanism of succession, effectively 'altruistic facilitation' in Box 2.2, but 'facilitation' is now used more widely, extinguishing unambiguous

---

> **Box 2.2** Types of interspecific facilitation (positive effects of reaction on fitness)
>
> Commensalism: species **A** benefits species **B** with no benefit or cost to **A**
> Mutualism: **A** and **B** both benefit (each species benefits relative to its performance in single-species stands)
> Altruistic facilitation: **A** benefits **B**, but with a cost to **A** (relative to single-species stands)

terms such as 'beneficence' (Hunter and Aarssen 1988). There are many environmental/resource factors that can be modified by plants to the benefit of others, and in spite of our efforts we have probably failed to include them all below (see Callaway 1995) (we consider facilitation via litter in Section 2.4.2.1 and via heterotrophs in Section 2.7).

2.3.1   Water and Nutrient Redistribution and Availability

Plants intercept water from rainfall (Agnew et al. 1993a), though up to half of this can be lost by evaporation from the canopy (Cape et al. 1991), and plants in forests, shrublands and even herbaceous communities make the distribution of soil water patchy because of canopy gaps, leaf drip and stemflow. In a few ecosystems, input is increased by fog capture (Section 4.5.4.2). The plant-enhanced water input, over the whole community or in patches, facilitates the understorey, especially in arid areas. Patchy litter can increase local infiltration into the soil by preventing runoff (Tongway et al. 2001) and reducing evaporation.

Deep-rooted plants can also redirect water from below via hydraulic lift from deep layers in the soil, perhaps from an aquifer, to the surface layer where shallow-rooted species can benefit. This is likely to be important in arid climates and Facelli and Temby (2002) show that in arid South Australia two chenopods, *Atriplex vesicaria* and *Maireana sedifolia*, enhance the water relations of surrounding annuals, though they also compete with the annuals for light. The phenomenon is also known in temperate forest where Dawson (1993) and Emerman and Dawson (1996) have shown that *Acer saccharum* (sugar maple) uplifts considerable quantities of water, much of which may be taken up by understorey plants from near the exuding tree roots. In a nice variant of the story, the CAM photosynthesising *Yucca schidigera* in the Mojave Desert apparently redistributes soil water, creating a higher upper-soil water potential during the day rather than at night when it has open stomata generating a higher water demand (Yoder and Nowak 1999a). Surrounding non-CAM plants, with a high water demand, must benefit. Leaky hydraulic lift is commensalism or possibly altruistic facilitation, and may be much more common than formerly supposed (Caldwell et al. 1998); but most authors stress that soil particle size and structure can be critical in the process. Roots can also distribute water from the surface down to deeper soil layers (Ryel et al. 2003). Conversely, in waterlogged conditions water uptake by one species can dry the soil and benefit a less tolerant species (Berendse and Aerts 1984), or plants with aerenchyma could oxygenate the soil, potentially benefitting associated species (Luo et al. 2010).

Plants make mineral nutrients patchy in the same way. Stemflow can be an important source of nutrients for epiphytes, especially of N for which stemflow concentrations can exceed 10 times those in rainwater (Awasthi et al. 1995; Whitford et al. 1997). Nutrients such as N can be leached from canopy leaves and absorbed by plants below (Cappellato and Peters 1995). This is best categorised as commensalism; although one organism loses and the other gains, there would be leakage from the loser whether there were a recipient below or not. Nutrients can be moved from deeper soil layers towards the surface, either along with water in hydraulic lift, or by exchanging deep cations for organic ions in biological pumping (Wilson et al. 2007a). Organic nutrients can be transferred between plants, notably by roots excreting amino acids that can be taken up by other roots – an important source of nitrogen in arctic vegetation (Kielland 1994). Since this is a loss of C and N to one plant, it might be categorised as altruistic facilitation.

The transfer of water, nutrients and other substances between plants can be direct when there is root and stem grafting, though root grafts are not necessarily beneficial (Tarroux and DesRochers 2011).

It is also possible for a facilitator to increase the total availability of nutrients in the system. This can occur through trapping nutrient-rich wind-borne dust, by root exudates with organic acids changing the pH and increasing the availability of P and Fe or with phytosiderophores increasing Fe availability, and of course by nitrogen fixation. In this way neighbours can benefit as well as the species increasing nutrient supply (Li et al. 2007; Muler et al. 2014). Patchiness in microelements can be affected too (Schlesinger et al. 1996). As a final example, physical soil texture might be improved by one species, for example by the underground organs of *Cirsium arvense* (field thistle: Reintam et al. 2008), benefitting associates even as it interferes with them.

These effects are good examples of facilitation without there being any question of coevolution.

2.3.2   Shelter

Plants alter aspects of the physical environment around them: light, ambient temperature, evapotranspiration and air movement. Some such reactions give shelter from stress (Section 6.5.3): i.e. facilitation. Reduction in light often decreases growth, but by decreasing photoinhibition growth could also be facilitated (Semchenko et al. 2012). Canopy trees dampen diurnal and seasonal fluctuations in subcanopy air temperatures, which can be beneficial to subcanopy inhabitants because cells can be killed by heat load – daytime temperature in the field layer of a forest is generally 1–3°C less than above the canopy – or by frost (Stoutjesdijk and Barkman 1992; Chen et al. 1993). The amelioration of frost can interact with shading to prevent photoinhibition in plants in the subcanopy environment (Germino and Smith 1999). The canopy lowers windspeed, reducing transpirational load and the abrasion of photosynthetic organs, and maintaining humidity thus reducing water stress. Similar effects are seen in other community types: Liancourt et al. (2005) gave evidence of facilitation between species in a calcareous grassland due to amelioration of water stress. In extreme environments some plants survive only through facilitation from another species, for example in the alpine zone (Section 5.11).

Shelter from ultraviolet light (UV-B) is a possible type of facilitation; in some species, perhaps most, it damages cells and their photosynthetic apparatus, and UV-B shielding from above could be beneficial (Mazza et al. 2013). The transmittance of UV-B to lower strata does differ with the canopy structure of different species (Shulski et al. 2004), and species do react differently in growth rates to UV-B (Furness and Upadhyaya 2002), so all the components for facilitation seem to exist, but we can find no evidence that it does so in nature. In contrast, in some species, in some field light environments, UV-A damages photosystem II less than UV-B, and can contribute to photosynthesis (Turnbull and Yates 1993, using the shrub *Pimelea ligustrina*, tall rice-flower, in subalpine Australia). Facilitation via shelter is also unlikely to be the result of coevolution.

2.3.2.1 Shelter in Particular Habitats

Near coasts there could be plant-to-plant shelter from salt spray deposition (Malloch 1997) but this effect does not seem to have been demonstrated. However, soil salinity facilitation has: Bertness and Hacker (1994) removed *Juncus gerardii* (a rush) from a New England, USA, saltmarsh and showed that its presence had raised the redox potential of the soil and, by reducing evapotranspiration, kept down the soil salinity. Without this benefit, the associated *Iva frutescens* (marsh elder) died. Transplants of *J. gerardii* to the low- and mid-zones of the marsh also benefitted from the presence of neighbours of *J. gerardii* and/or *I. frutescens* because of lower salinities. However, transplants into higher zones in the marsh were suppressed by the *I. frutescens* there. This is commensalism or altruistic facilitation, rather than mutualism.

In arctic and alpine/montane habitats, krummholz trees can shelter shorter species from wind desiccation, storm damage and wind-borne ice particles (Marr 1977; Callaway 1998). Similar facilitation can occur among alpine herbs, possibly as a mutualism (e.g. Cavieres et al. 2007). Shelter from frost heave was implicated when Ryser (1993) found that, of six species examined in limestone grassland in Switzerland, two appeared to require the relief from both frost heave and drought provided by neighbouring plants. Shelter from UV-B would be especially important in alpine communities where, in clear weather, high UV-B radiation is received.

Facilitation might be especially important in arid and semiarid areas, where isolated 'nurse' shrubs and trees have a herbaceous flora on the fine soil and organic matter accumulated beneath them from litterfall, from trapping of wind-borne material, and even with nutrients from animals that come under shrub/tree canopies to defaecate or die. Wind-blown seed of subordinates can be trapped too. The association between nurse plants and their beneficiaries is often so clear that no attempt is made to determine the precise environmental variables responsible. It is sometimes assumed to be temperature combined with humidity (e.g. Brittingham and Walker 2000), and at other times to be both water and nutrients (e.g. Muro-Pérez et al. 2012). Protection from herbivory (see Section 2.7.3) can contribute too, and just possibly pollination facilitation (see Section 2.7.2). It has even been suggested that the nurse species could benefit somehow. We must be careful, because the vast majority of reports of nurse effects are correlative rather than experimental. Temperature differentials have frequently been documented, e.g. by Muro-Pérez et al. (2012) beneath *Prosopis laevigata* (mesquite) nurse trees in the Chihuahuan Desert, though there were soil differences there too. In hot deserts with summer rainfall, shelter from heat load can be critical in cell survival, especially for succulents, which cannot lose as much heat by convection as thin-leaved plants. The canopy shade also reduces evaporative load for the understorey, though below-ground there will be bidirectional competition for water between nurse and understorey. A study by Gómez-Aparicio et al. (2005) in dry Spanish shrublands showed that most of the nurse effect there was due to canopy shade rather than water or nutrients. Facilitation can be indirect: Cuesta et al. (2010) demonstrated that in a Mediterranean shrubland the shrub *Retama sphaerocarpa* facilitated the establishment of *Quercus ilex* (holm oak) seedlings by suppressing competing herbs. Nurse effects can also be found in alpine environments. In the high Andes of Central Chile Badano et al. (2015) observed increased performance of the invasive herb *Cerastium arvense* within native cushion patches compared to surrounding open areas, suggesting a potential role for nurse plants in modulating invasibility in harsh environments.

The balance of facilitation and interference between a nurse plant and its understorey is delicate. Walker et al. (2001) demonstrated by plant manipulations in a desert scrub community that, whilst 'islands of fertility' around woody plants had increased both water and nutrient availability to the understorey, the net effect of 'nurse' plants was negative due to shading and/or root competition. Belsky et al. (1993) meticulously described complex interactions in East Africa, whilst Holzapfel and Mahall (1999) carefully separated facilitation from interference between shrubs of *Ambrosia dumosa* (white bursage) and annuals in the Mojave Desert. In a fascinating series of facilitations in western Texas, a *Prosopis velutina* (mesquite) canopy allows the establishment of *Juniperus pinchotii* (juniper), whilst *J. pinchotii* helps the establishment of three other woody species with a more mesic habitat (Armentrout and Pieper 1988; McPherson et al. 1988).

2.3.3   Facilitation Conclusion

One of the categories of facilitation is mutualism (Box 2.2), and examples are well known between plants and heterotrophs, for example in mycorrhizae. However, we have been able to discover hardly any examples of demonstrated mutualisms between two embryophytes. This suggests that facilitations may be accidents of evolution. The evolution of facilitation, and especially altruistic facilitation, is a problem: benefitting another species is not selectively advantageous. Plants do not rise to the top of the interaction pecking order by being nice to others. There is evidence that under environmental stress, facilitation becomes more important than interference, though this is contentious (Section 6.8). Competition will be stronger between functionally similar species, tending to overwhelm facilitation; Valiente-Banuet and Verdú (2008) gave evidence for this in Mexican semiarid communities, using 'phylogeny' as a proxy for function and using spatial associations as evidence for species interactions. Some altruistic facilitation clearly occurs during succession (Section 4.3.2), probably as a combination of mechanisms, for example in an N-fixing and litter-accumulating pioneer that is shaded out by subsequent species.

There is considerable current discussion on facilitation, almost overwhelming that on interference. Seifan et al. (2010) declared that 'positive interactions among plants (facilitation) are common in nature'. They offered no evidence. Field removal experiments, trenching experiments and 'constant final yield' experiments make it clear that almost everywhere interference is a stronger force in plant communities than facilitation. We know this from weeding our own gardens. That is not to deny that facilitation occurs in nature alongside interference. In some special situations, alpine or desert, it may predominate. Even here, though, we cannot assume it: Onipchenko et al. (2009) in a site in alpine Russia, with a mean annual temperature of $-1.2°C$, could find only positive or neutral effects of plant removal. Careful evaluation is needed, rather than following fashionable research lines and assumptions.

2.4   Litter: The Necessary Product

We described plants in Chapter 1 as largely composed of colonies of modules. The discard of old modules and tissue comprises litter. Most organs are shed when dead or dying, but some living material is lost from a plant through physical damage or more uncommonly through self-pruning under either light or water stress. All types of plant organ are included in litter, especially leaves, but also twigs, coarse woody debris (CWD, i.e. branches and whole trees), bark and reproductive parts. Roots and underground stems also have limited life and die below-ground, but little is known of the nature and fate of root litter (Birouste et al. 2012), save for a little on root litter allelopathy, and rates of fine root turnover. Our focus is therefore above-ground litter. Considerable amounts of litter can be deposited in terms of the number of modules, e.g. 37,000 plant fragments from one *Quercus robur* (oak) tree in a year (mostly short shoots up to two years old (Buck-Sorlin and Bell 1998)), and mass, e.g. 11.4 tonnes ha^{-1} $year^{-1}$ of leaves and small branches in a typhoon-prone hardwood forest in Taiwan (Wang et al. 2013). Amounts per area are naturally less in more open forests in drier areas, for example c. 1–2 tonnes ha^{-1} $year^{-1}$ (total; mostly leaves) for two *Eucalyptus* savannahs in Queensland, Australia (Grigg and Mulligan 1999).

Litter production is a mandatory process in the life of plants and the ultimate fate of most plant growth. Most often, litter is examined for its role in nutrient cycling (e.g. Orians et al. 1996), but it can play a major part in all aspects of community structure and ecosystem function. It must change the plant's immediate environment, i.e. cause Clementsian reaction, and in this book we discuss evidence for a range of its effects on species–species interactions, and hence for the community processes of succession-to-climax, cyclic succession and switches (Chapter 4) and even assembly rules (Chapter 5).

Litter produced above-ground can be removed from its site of deposition by wind, floodwater or animals, and it may be buried by river gravels or wind-blown loess. In its final location, if it breaks down more slowly than it is deposited, in moist conditions it becomes peat (see Section 2.4.3), or in a dry climate or on a dry substratum it can be removed by fire (see Section 2.4.3.2). Most generally, the break-down of litter directly contributes to soil formation via the build-up of soil organic matter. These processes are indirect effects of litter production.

Here, we discuss the phenology/morphology (Section 2.4.1), effects (Section 2.4.2) and fate (Section 2.4.3) of this important material, and finally whether litter production or its effects can be susceptible to Darwinian selective processes (Section 2.4.4).

2.4.1   The Timing and Type of Litter Production

Dead plant modules are a necessary concomitant to growth, including: (a) standing dead litter retained on the plant, (b) programmed litter discarded in response to seasonal cues and (c) stochastic litter shed by accidental, environmental or herbivore damage. We distinguish them because we think that selection can act more easily on predictable litter deposition.

2.4.1.1   Standing Dead Litter

This comprises litter retained on the plant for some period of time. It may be actively decaying, and includes not only whole dead trees but also dead wood, partially detached bark, flowering stems, and erect or drooping attached leaves. Dead trees can persist for several centuries (Mukhortova et al. 2009), and in many perennial herbs the vegetative shoots collapse in winter but the dead flowering stems remain erect. Any standing dead material may have effects such as channelling rainfall or acting as bird perches, with consequent input of nutrients and disseminules. Many annuals die by programmed senescence. How such plant suicide evolves is something of a puzzle (Wilson 1997), but anyway it can result in plant–plant interactions. Bergelson (1990) demonstrated in a field experiment that the seedling emergence of the annuals *Senecio vulgaris* (groundsel) and *Capsella bursa-pastoris* (shepherd's purse) was very sparse where there was a high density of dead *Poa annua* plants.

2.4.1.2   Programmed Litter

We define litter as programmed when it is disconnected by abscission, the result of a physiological process endogenous to the plant and initiated some time before the litter actually falls (Lewis et al. 2006). It is almost always pulsed by season (e.g. Arianoutsou 1989; Enright 1999), exceptions being tropical figs (Nason et al. 1998) and the whole-plant senescence of short-lived ruderals in oceanic climates. There can be several peaks of programmed litterfall during the year, often of different types of a species' replaceable modules (leaves, twigs, flowers, fruit, etc.) all with their special phenology (Figure 2.4). The signal for abscission can be exogenous (e.g. daylength) or endogenous (e.g. the mutual shading of crowded twigs) (Buck-Sorlin and Bell 1998).

Programmed litter is usually lightweight and so can be redistributed by wind and rain into deep patches, perhaps having its greatest local effect some time after it first falls

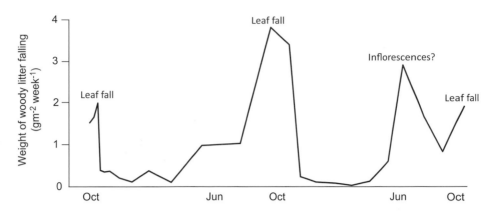

Figure 2.4 The seasonal pattern of litterfall in a *Quercus robur* (oak) wood.
Reprinted with permission from John Wiley and Sons Inc., from Christensen (1975).

(Fahnestock et al. 2000) (this applies to lightweight stochastic litter too). Animals can disturb such litter, for example bird scratching can remove the litter load locally and permit germination of surface seeds (Theimer and Gehring 1999). Such processes have the potential to alter plant populations and to select for certain seed-producing species, which could indirectly affect plant interactions.

In some species, twig fall and branch fall (= branch shedding = limbfall = cladoptosis) are programmed with well-defined cleavage zones, across a cork band produced from a phellogen (Christensen 1975). Large branches can be abscissed, especially by conifers (e.g. up to 17 cm mid-branch diameter in the conifer *Agathis australis*: Wilson et al. 1998). The peak of programmed wood fall is generally in autumn, but in a few species in spring (Millington and Chaney 1973). The effect of this process on species–species interactions appears to have been overlooked.

2.4.1.3   Stochastic Litterfall

Litter shed in response to allogenic disturbances is stochastic, i.e. not programmed, not initiated by the plant, but caused or at least triggered by storms, frosts, droughts, floods, geomorphological change, gravitational overload, herbivore destruction, pathogenic dieback, etc. Some of these stimuli occur irregularly, though others are seasonal, for example equinoctial storms. Falling coarse woody debris (CWD) adds to the direct effects of the disturbance.

Stochastic litter can be dead or alive. Leaves are a major component, 68 per cent in the upland forest studied by Shure and Phillips (1987), but twigs and CWD are often prominent, e.g. 46 per cent of the litter recorded in an Amazonian rain forest (Chambers et al. 2000). Falling or breaking dead trees can also contribute, e.g. 7 per cent in a boreal forest in Finland (Siitonen et al. 2000). In many climates, epiphytes add appreciably to the litter fall, being 5–10 per cent of the litter in old tropical cloud forest in Costa Rica and c. 5 per cent (all lichen) in *Picea abies* (Norway spruce) old-growth forests in Sweden (Esseen 1985; Nadkarni and Matelson 1992).

2.4.2   Effects of Litter

Fallen litter must have both physical and chemical effects, which can change community composition because of its differential effects on different species. It can be shown experimentally to affect interactions between species (Sayer 2006). It can speed or redirect succession, and deep persistent leaf litter can inhibit succession by preventing the establishment of later successional species (Crawley 1997). Low species richness in communities dominated by fast-growing, tall species has usually been seen as due to competition, but it could additionally be due to litter effects (Berendse 1999; Grime 2001). It is often impossible to tell the precise ways in which litter is modifying plant interactions, but we can group mechanisms into three classes:

1.   Physical effects, i.e. alteration in physical environment (Section 2.4.2.1), or direct damage, predominantly by CWD (autogenic disturbance, Section 2.5.2).
2.   Chemical effects, i.e. alteration in chemical environment (Section 2.4.2.1) or control of nutrient cycling (Section 2.4.2.2). This is predominant for finer litter.
3.   Complex environmental interactions, e.g. in peat formation (Section 2.4.3.1), with fire (Section 2.4.3.2) and with herbivores (Section 2.8).

2.4.2.1   Reaction Caused by Litter

Since leaf litter covers the soil surface, clear effects would be expected on germination and establishment, perhaps especially of species with small seeds giving rise to seedlings unable to penetrate a litter layer. Facelli and coworkers (e.g. Facelli 1994; Facelli et al. 1999) have shown that in succession from oldfields to Australian *Eucalyptus* forest, litter can affect seed germination, and seedling establishment, predation, growth and shoot:root ratio.

Litter causes interference via reaction on physical factors. The growth of low-growing plants must be suppressed by the low light under litter (e.g. Holdredge and Bertness 2011), though this has been much less documented than effects on germination and seedling establishment. Red/far-red (R:FR) ratios, already reduced by a live canopy, can be halved again by the litter (Vázquez-Yanes et al. 1990). Litter might intercept rain and allow it to quickly evaporate, reducing soil water content (Myers and Talsma 1992), though it can also reduce evaporation from the soil (Eckstein and Donath 2005) and increase infiltration. Temperature fluctuations at the soil surface can also be dampened.

A simple effect, probably very common, is that carbon added to the soil by litter can stimulate microbial growth and therefore N absorption, reducing the N available for plant growth (Meier et al. 2009). Chemical interference (allelopathy) is also possible. For example, Bosy and Reader (1995) found that cover by grass litter in an oldfield markedly reduced the seed germination of at least three of four forb species, and for two species an appreciable component of the suppression could be explained by allelopathic effects of litter leachate. Chapin et al. (1994) suggest that in Glacier Bay successions *Alnus viridis* subsp. *sinuata* (sitka alder) suppresses the creeping *Dryas drummondii* partly by the allelopathic effects of its litter, though as part of complex replacement interactions. However, experiments are easier in vitro than in the field, and only a few clearly demonstrate that litter allelopathy has real effects in nature. The study of Welbank (1963) is an impressive example, with the root/rhizome litter of *Elytrigia repens* (≡ *Agropyron repens*, couch grass) producing toxins when decomposing in anaerobic conditions.

Facilitation via litter can be very significant for particular species. It can increase germination, probably by increasing humidity at the soil surface (Fowler 1986b). In temperate rain forests, CWD can be an essential substratum for tree seedling regeneration (McKee et al. 1982; Agnew et al. 1993a; Duncan 1993). CWD can also interact with herbivory: Long et al. (1998) showed that in a Pennsylvania forest, treefall mounds provided a refuge from herbivory for *Tsuga canadensis* (hemlock) regeneration, representing more favourable sites for regeneration than the forest floor in spite of the adverse physical environment. Although bryophytes are often suppressed by litter cover (Barkman et al. 1977), especially by the combination of litter from woody species and from forest herbs, During and Willems (1986) named five bryophyte species that they suggested 'may thrive on litter'. It is possible that nutrients released from litter are taken up by bryophytes (Tamm 1964). CWD often harbours a specialised bryophyte flora and in some Scandinavian forests the protonemata of many bryophytes are energetically dependent on it (Siitonen 2001).

Reaction by litter can cause switches (Section 4.5), importantly including acidification (Section 2.4.3.1). There can be effects on community dynamics. For example in forests of Michigan and Wisconsin, USA, *Tsuga canadensis* (hemlock) cannot enter mid-successional stands dominated by hardwoods such as *Quercus rubra* (northern red oak) because its seedlings cannot penetrate the litter produced, whereas *Acer saccharum* (sugar maple) seedlings can tolerate it, and invade (Rejmánek 1999).

2.4.2.2 Control by Litter of Nutrient Cycling

The process in which litter returns nutrients to the soil as mineral compounds, or as organic compounds which are then broken down to mineral ones, differs between species, and such differences have the potential to drive succession. Berendse (1994) suggested that in fast-growing species (i.e. with high RGR_{max}), which are often dominant in N-rich environments, less N is retranslocated to the living parts before the leaves fall, and the resulting litter breaks down quickly. He gave the example of the deciduous *Molinia caerulea* (purple moor grass), which in one study lost 63 per cent of the N present in its above-ground standing biomass in a year and 34 per cent of its P (Berendse et al. 1987). In contrast, species with a low RGR_{max}, often found early in primary succession when N is in short supply, retranslocate a greater proportion of their leaf N, leaving less in the litter, Berendse's example being the evergreen shrub *Erica tetralix* (heath) where comparable losses were 27 per cent of N and 31 per cent of P. This suggests that under nutrient-poor conditions the nutrient-conserving *E. tetralix* will be the dominant species, and that if nutrient availability increases, *M. caerulea* will replace it, possibly comprising a litter-mediated switch (Section 4.5.4.5). There is some empirical support: in an experiment in a similar system, addition of nutrient solution caused *Calluna vulgaris* (a species ecologically similar to *E. tetralix*) to decline and cover of four grass species to increase (Berendse et al. 1994).

2.4.2.3   Litter Decomposition

The rate of litter decomposition sets the period during which litter can affect the community. Break-down occurs by a succession of organisms: invertebrate shredders, commutators, fungi and bacteria (Cadisch and Giller 1997). The rate varies with the climate, but also with the particular characteristics of the litter: mechanical, nutrient content, polyphenol content, cellulose:lignin ratio, etc. (Cornelissen and Thompson 1997; García-Palacios et al. 2016). Sometimes mixtures of leaf litter from several species decompose faster than single-species litter. The reduced litter residence times could be important in species coexistence, giving multispecies communities different properties from monocultures. For example, Kaneko and Salamanca (1999) found that in litter bags containing species collected from a Japanese *Quercus serrata* (oak)/*Pinus densiflora* (red pine) forest, mixtures containing *Sasa densiflora* lost mass faster than expected from the decomposition rate of their components. Gartner and Cardon's (2004) review of the literature showed enhanced decomposition in mixtures in 47 per cent of cases reported. Such effects might operate by mixtures enhancing the commutating fauna or by the high N content of some species speeding the decomposition of their lower-N associates (Vos et al. 2013). Other experiments have shown that, as expected, mixture effects are dependent on the types of litter or the species identity being used.

2.4.3   Ultimate Fate of Litter

Decomposition and the incorporation of residues into the soil matrix is the most important fate of litter globally, providing the raw material for organic soil formation and significantly influencing the biogeochemical cycling of carbon and nutrients (Section 2.4.2). There are, however, at least two other pathways of litter cycling that have the potential to impact plant–plant interactions, and ultimately community development: peat formation (a subset of soil formation more generally) and fire.

2.4.3.1   Peat

On wet substrates, if litter breaks down at a slower rate than it is produced, there is an accumulation of peat, i.e. plant remains partially hydrolysed via microorganisms. This is the process of humification. There are several modes of peat formation. In eutrophic or mesotrophic conditions, litter production and decomposition are both fast. Only in still lakes, where redox is low, will peat accumulate in the sediment, infilling them to form a fen or carr, and perhaps eventually dry land. This is a classic succession story (Walker 1970).

Other types of peat formation represent a switch, i.e. positive feedback between plant community and substrate (Section 4.5). Where there is high rainfall and/or a substrate of free-draining siliceous sands, nutrients are lost through leaching and peat forms. This process, which can occur on a range of landforms, is paludification, and it forms an oligotrophic mire. Most of the species of such low-nutrient mires bear unpalatable evergreen leaves, giving them longer to amortise the cost of their production. They usually have a high polyphenol content that slows decomposition of their litter, further contributing to peat accumulation (Inderjit and Mallik 1997; Northup et al. 1998; Figure 2.5a). This is a widespread phenomenon, which includes heaths, heath forests, forests on podsols, and slightly-sloping aapa mires with their string/flark structure.

The final type of mire is a bog, ombrotrophic in that it obtains nutrients only from rain. It can develop from a fen, or by paludification, or from moss mounds in a wet valley. A raised bog is raised above its surroundings and therefore isolated from any substrate or mineral-bearing aquifer (Anderson et al. 2003), but blanket bogs are quite flat, receiving almost all their water and nutrient input from rain because the rainfall is high, evaporation low, and the slope of the land shallow. The species that can survive, and even cause, these conditions are mostly bryophytes (particularly *Sphagnum* spp.), but in many New Zealand bogs the restiad *Empodisma minus* (wire rush) is the driver, an odd flowering plant with a partly negatively geotropic root weft with persistent root hairs (Agnew et al. 1993b). A complex set of environmental changes decreases litter production, but decreases its decomposition even more. The cellulose cell walls of *Sphagnum* spp. and *E. minus* hydrolyse to weak uronic acids that can scavenge cations from precipitation by exchanging them for hydrogen ions, thus yielding sufficient nutrients for growth and maintaining a low pH in the interstitial water (Clymo and Hayward 1982; Kooijman and Bakker 1994; Figure 2.5b). The low pH reduces the availability of most nutrients, inhibits nitrification and increases the availability of Al, Mn and Fe to toxic levels. The rate of percolation is slowed by the humic colloids present (i.e. there is waterlogging), reducing the redox potential. The low nutrients, low pH, high toxicity and low redox potential favour the species that cause them, i.e. there is positive feedback constituting a switch. Peat decomposition is decreased, increasing the isolation from ground water: further positive feedback. Malmer et al. (1994, 2003) suggest that since *Sphagnum* spp. can displace vascular plants when ombrotrophic conditions arise, bog conditions/vegetation may arise rather suddenly, and this has been observed where rapid *Sphagnum* spp. expansion has displaced a brown moss fen (Kuhry et al. 1993), or invaded and destroyed forest (Anderson et al. 2003). This is precisely the state change that is expected when a switch is operating (Section 4.6.5). The state change can be completely autogenic, or initiated by climatic change to colder, wetter conditions with the switch hastening the change (Korhola 1996). In other situations/conditions ombrotrophic and rheotrophic communities might be alternative stable states (ASS; Section 4.6). These are complex processes (Figure 2.5), and although several links are well established, a complete understanding is yet to be developed for any system.

2.4.3.2 Fire

Sometimes the autogenic disturbance of litter (Section 2.5) interacts with allogenic disturbance. Fire is perhaps the best example. If litter accumulates faster than it decomposes, and local conditions of climate and soil are too dry to allow peat accumulation, its ultimate fate on a landscape scale is destruction by fire. This is basically an autogenic effect because there can be no fire without fuel, and the initiating fuel of a fire is usually a build-up of standing dead, programmed and stochastic litter. However, an

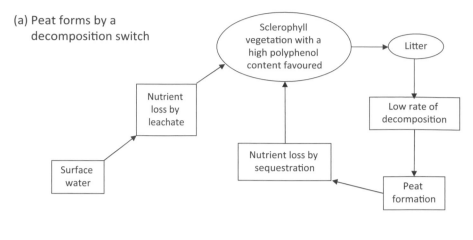

(a) Peat forms by a decomposition switch

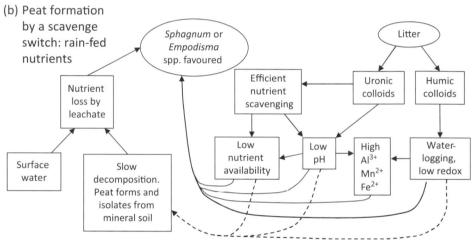

(b) Peat formation by a scavenge switch: rain-fed nutrients

Figure 2.5 Two pathways (a and b) of peat formation in which plant litter plays a decisive role through a switch (i.e. positive feedback loop) by changes in the ambient soil water.

allogenic process has to start the fire. This process is intrinsic to heathlands, savannah and dry forests worldwide.

In many vegetation types, fine fuel components (leaf and twig litter) and branch fall provides the fuel build-up, and because CWD is long-lived, it often persists until burnt. Treefall also provides fuel (Whelan 1995). The spread of fire into the canopy is facilitated by the retention of dead limbs (Johnson 1992), so branch fall will help to confine fire to the ground flora, sparing the canopy stratum. Lianas can have a similar effect: Putz (1991) observed flames travelling up lianas, with the result that *Pinus* spp. (pine) trees with lianas (especially *Smilax* spp., greenbriar, etc.) were more likely to suffer crown scorch than liana-free trees. Fire is especially prevalent in communities of conifers (particularly *Pinus*) and *Eucalyptus* (gum) apparently because of their flammable resin/oil content (Whelan 1995). At the other extreme, some types of vegetation are very fire avoiding. For example, in semi-arid Australia, where most vegetation is fire-prone, *Acacia harpophylla* (a wattle) rarely burns because of several features: the chemistry of its phyllodes makes them barely flammable; the simple phyllode shape results in flat, poorly aerated litter fuelbeds; phyllode fall is keyed to intermittent rains when fire is less likely; bark is not shed; and finally the microclimate of the forest floor discourages drying and promotes decomposition (Pyne 1991). Thus, fires affect community composition by having greater impact on the less fire-avoiding and more fire-intolerant species.

The products of fire – charcoal and smoke – also affect community composition, especially since the fire products from one species can trigger the germination of other species (Keeley and Fotheringham 1997).

2.4.4   Can Reaction via Litter Evolve?

Evolution is a problem: are the litter processes within plant communities simply by-products of growth, or could they have evolved? It is beyond doubt that interference can evolve. As Clements (1904) pointed out, all plants react on their environment and the extent and type of reaction will be genetically controlled. A particular reaction will evolve if it enhances the fitness of the individual plant (i.e. genet) more than it enhances the fitness of its conspecific neighbours, after taking into account any cost of reaction.

For reaction via litter there is a problem. Each species must produce litter that is unique in its abundance, form, chemistry and phenology. However, litter can benefit neighbours as well as increasing the fitness of the plant that produces it. Moreover, the effects of litter sometimes appear too late to affect whether the genes that caused them are transmitted, i.e. when the litter is produced as the plant dies or has its effect after the plant's death via paludification, control of nutrient cycling or being fuel for fire. In this case there is no possibility of plants enhancing their own fitness, in survival or fecundity, only that of their offspring. An explanation in terms of kin selection is therefore needed, with the mechanism that a plant benefits especially its neighbours, which are especially likely to carry the same alleles (Wilson 1987; Dudley et al. 2013). Another way of expressing this is that selection occurs through the extended phenotype. Northup et al. (1998) commented that 'decomposing leaf litter is not generally considered to be part of the plant's phenotype' so that litter characters must evolve through selection on the extended phenotype. Theoretical analyses by Odling-Smee et al. (1996) and Laland et al. (1996, 1999) indicate that selection for characters of the extended phenotype, i.e. 'secondary' characters, is possible. A different solution was proposed by Whitham et al. (2003); that traits of the extended phenotype could evolve because they affect the whole community and there could be selection between communities. In the present context, if the community is a mosaic of patches caused by reaction (Section 7.3.2), a patch type might be favoured due to a litter effect of its predominant species, which could favour it and the few individuals of the species that engineered the patch. This is the subpopulation group selection of Wilson (1997). However, many ecologists find the concept hard to swallow if the evolving trait is altruistic, because non-altruistic genotypes can invade (see Wilson 1987). Most plants do produce litter during their life and have an opportunity for that litter to affect their fitness, and for this the argument of

Northup et al. (1998) does not apply, but it is still a problem that litter affects all the plants and species in a community, not just the litter-producing genet. Similarly, its effect needs to be very local for kin selection to operate (Wilson 1987).

Perhaps litter characteristics are just a by-product of growth, so no evolutionary explanation is necessary. The evolution of reaction caused by litter needs further theoretical consideration.

2.5   Autogenic Disturbance: Plants As Disturbers

Disturbances – marked changes in the environment for a limited period often with the removal of plant material – are significant influences on plant communities (White 1979). Most consideration has been of allogenic disturbance, i.e. caused by external factors such as animals or extreme weather events. For example, Bazzaz (1983) described disturbance as continuous and necessary in all ecosystems, but discussed only allogenic causes. No less important, we suggest, are the minor disturbances within the community that inevitably result from the growth and death of plant parts, the 'endogenous disturbance' of White (1979); the 'autogenic disturbance' of Attiwill (1994). Autogenic disturbance occurs by plant contact, litter, treefall, etc. (though there seems little opportunity for autogenic disturbance by subterranean organs). Vegetation, by its nature, disturbs itself: as we pointed out in Chapter 1: 'plants move; whereas animals stay within their adult body', yet this is an underresearched and poorly understood topic. Autogenic disturbance has potentially positive and negative effects on population fitness and density, yet stands apart from the general consider- ations of interference and facilitation described above. There is a wide range in the intensity of autogenic disturbances, from delicate adjustments resulting from leaf growth to the fall of a group of trees, but plants must tolerate all these and they may profoundly affect vegetation dynamics. Autogenic disturbance may be more predict- able than allogenic disturbances such as storms and earthquakes, and if so, selection for adaptation to it is more likely to occur. However, autogenic and allogenic disturbance can interact.

2.5.1   Movement, Contact and Sound

Even gentle contact between plants can produce effects such as reduced growth, reduced soluble carbohydrates and increased transpiration (van Gardingen and Grace 1991; Keller and Steffen 1995). For example, when Biddington and Dearman (1985) brushed *Brassica oleracea* var. *botrytis* (cauliflower) seedlings with typing paper for 1.5 min day^{-1} it reduced their shoot dry weight (Figure 2.6). Experimentally shaking the plant can have an even greater effect, e.g. reduced shoot extension, shorter and wider stems, lower shoot:root ratios, altered root development, shorter petioles, shorter flowering culms, reduced reproductive effort, induction of dormancy and changed gene expression (e.g. Niklas 1998; Wang et al. 2008). The changes are induced by the up- regulation of a large number of genes (Braam and Davis 1990). Whilst brushing with

Control   Brushed

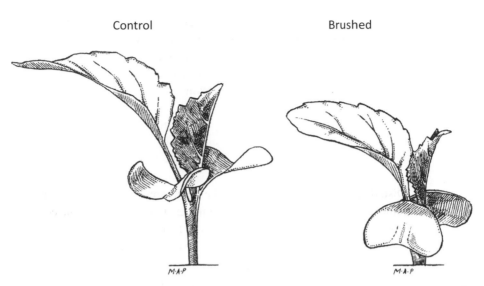

Figure 2.6 The effect of brushing seedlings of *Brassica oleracea* (cauliflower) for 1.5 min day^{-1}. After Biddington and Dearman (1985).

typing paper and shaking by hand are artificial treatments, they are very similar to the leaves of adjacent plants brushing each other, or close trees rocking each other. These processes probably affect plant–plant and species–species interactions in nature, but this is yet to be investigated.

Shoots can thrust each other aside physically as they grow, a more sustained contact effect (Campbell et al. 1992). For example, the leaves of *Juncus squarrosus* are vertical when they first emerge, but then become horizontal, form a rosette and press down surrounding plants (authors, pers. obs.) as do leaves of *Plantago* species (Campbell et al. 1992). Surprisingly little is known of this process. Circumferential tree trunk and branch growth produces new sites for the establishment of epiphytes. Circumferential growth of tree roots near the soil surface creates shallow soils and diversifies the forest floor habitat (authors, pers. obs.).

Crown shyness is the phenomenon where there are canopy gaps between adjacent trees, probably caused by abrasion in the canopy. Franco (1986) observed the process, reporting that when branches of *Picea sitchensis* (Sitka spruce) and *Larix kaempferi* (Japanese larch) met, physical damage occurred by abrasion, causing the death of the leading shoots. New shoots arising from lateral buds were also killed. Long and Smith (1992) demonstrated the mechanism by fixing wooden pickets into the canopy as artificial branches: they were broken back to the edge of the crown, due to wind-rock and consequential abrasion. Tall, slender trees were subject to greater wind-rock and thus greater frequency of impact with neighbouring crowns, leading to greater crown shyness. Putz et al. (1984) found that branches of *Avicenna germinans* (mangrove) bordering crown shyness gaps generally had broken twigs and few leaves. As in Long and Smith's (1992) observations, flexible crowns were more widely spaced than still crowns. These effects are the result of the compulsory movement imposed by the modular construction of plants. It is possible that red:far-red effects and even shading play a role, but there seems to be no evidence of this.

Plants can also change position as a result of their own growth (Evans and Barkham 1992), for example *Salix* (willow) trunks collapse due to their own weight in wet carr forest (authors, pers. obs), and shrub stems can be bent by their own weight and fall to the ground (Wilson 1997).

The reader may like to take the 'pers. obs.' in this section as confirmation that autogenic disturbance has not received due attention in ecological research.

2.5.2   Coarse Woody Debris Disturbance

Coarse woody debris disturbance clearly has the potential to be very destructive to the forest floor flora.

The most destructive type of stochastic litter is tree fall, which is the autogenic disturbance par excellence and implicated in all demographic processes in forests of mixed canopy species. It is a normal event in many undisturbed stands, and can be measured and built into models of species and biomass turnover in boreal (e.g. Hofgaard 1993), tropical evergreen (e.g. Chandrashekara and Ramakrishnan 1994) and tropical rain forests (e.g. Kellman and Tackaberry 1993).

In a forest the upper canopy trees are more likely to fall, having reached their greatest potential and being too heavy for their root support or too loaded with lianas/epiphytes (Strong 1977; Putz 1995). Tree fall is associated with storm damage, particularly in mid-latitudes (e.g. Busing 1996; Allen et al. 1997). Understorey trees are more likely to die and remain standing due to light reduction by neighbours (self-thinning). Moreover, understorey trees are not as exposed to storm damage as are the canopy species. An exception occurs in the dry Knysna (South Africa) forest, where most canopy trees die standing due to 'competition, senescence and secondary pathogens' and gradually break up (Midgley et al. 1995). There is not enough evidence to suggest this as a general feature of any forest types.

In old-growth forest, tree fall is the usual process that forms canopy gaps, resulting in several sudden environmental changes. The most important is that the forest floor is brightly lit with light of a very different spectral quality, promoting or modifying growth, or causing photoinhibition of shade-tolerant juveniles (Mitamura et al. 2009). Another, often overlooked, factor is that a full canopied forest evaporates up to 40 per cent of incoming rainfall so the soil is seldom at field capacity, but gaps receive the rain almost uninterrupted. The temperature regime is changed and the soil is disturbed, with a root pit and adjacent root-plate mound. All these changes allow plants of pioneer species to establish (e.g. Vranckx and Vandelook 2012, in forests of northern Europe).

Branches can also cause severe impact when they fall. For example, in the tropical liana *Connarus turczaninowii* in Panama, 20–45 per cent of the mortality of young plants is caused by branch fall (Aide 1987); in a tropical rain forest in Amazonia, Uhl (1982) found the most common (38 per cent) cause of death in small (1–10 cm diameter at breast height [dbh]) trees to be a branch fall or treefall; and in Costa Rican rainforest, Vandermeer (1977) found 46 per cent of seedling deaths were due to falling leaves or branches, mostly falling palm fronds. In a review of seedling mortality resulting from litterfall, Gillman (2016) identified a latitudinal trend, with mortality rates rising steeply with decreasing latitude, reaching a peak of 30–50per cent in tropical forests.

In general, however, the existence of CWD is much better known than its effects, especially its mechanical effects.

2.5.3   Lianas and Epiphytes As Disturbers

Autogenic treefall, with its associated disturbance to the forest floor, can be generated by the weight of adherent lianas (e.g. *Hedera helix*, ivy) and by free-swinging lianas (e.g. *Ripogonum scandens,* supplejack). Competition adds to the stresses: lianas can be so abundant in the canopy of tropical and subtropical forests that they compete with the emergent trees, below- and above-ground (Schnitzer et al. 2005). Phillips et al. (2005) found that in an upper-Amazon tropical rain forest, large trees were three times as likely to die if they were infested by lianas. The resulting canopy gap can benefit lianas; needing less investment in support they can extend fast and keep pace with the fast-growing secondary species filling the gap.

Twining or tendril lianas can constrict the host stem (Lutz 1943; Clark and Clark 1991; Matista and Silk 1997). Usually there is little active strangulation, but the liana stays put and the tree grows, which can produce deformations in the trunk/branch and lead to changes in the vascular tissue, including the wood fibres (Falcone et al. 1986; Putz 1991; Reuschel et al. 1998). Constriction can inhibit or even stop downward translocation of organic solutes, though sometimes new parallel conducting tissue is formed (Lutz 1943; Hegarty 1991; Putz 1991). The constriction can also be a point of weakness, at which a branch or young stem is more likely to break (Lutz 1943; Uhl 1982; Putz 1991). Deformed trees are also more susceptible to pathogens (Putz 1991). Not infrequently the tree dies (Lutz 1943; Uhl 1982; Clark and Clark 1991). These effects have been observed, but their effect on community structure is not known.

The interactions between support and liana are complex. Lianas often climb over more than one tree (with a record of 27 trees for one liana on Barro Colorado Island, Panama: Putz 1984). This causes interdependence between trees that may stabilise canopy trees adjacent to treefalls (Putz 1984; Hegarty 1991). However, a corollary is that if such a connected tree falls, it damages more trees: Putz (1984) found on Barro Colorado Island that 2.3 trees fell with each gapmaker, often ones that shared lianas. Liana-laden trees can also pull branches off neighbouring trees as they fall (Appanah and Putz 1984).

We have found no record of an epiphyte causing treefall, though they can cause the fall of branches that already had some structural flaw (F.E. Putz, pers. comm.). Stranglers and hemi-epiphytes certainly damage trees (Richards 1996). The strangler species involved are evergreen; many are in the genus *Ficus* (fig) and curiously revered by the indigenous people in some areas (*F. benghalensis* in Hindustan, *F. natalensis* in Kenya), feared in others. Such stranglers begin life as epiphytes, gradually extend aerial roots downwards, eventually root in the soil, then grow around their host as a tree. The host usually dies, and does so standing, though the cause of the host's death is not clear

(Richards 1996, p. 134). True epiphytes also seem to reduce their host's fitness; there are many possible mechanisms and it is not clear which are significant (Richards 1996, pp. 144–146).

2.6   Parasitism

Shoot parasites can be partial (Loranthaceae, Santalaceae) or total (e.g. *Cuscuta*, *Cassytha*). Root parasites can also be partial (e.g. *Rhinanthus* spp.) or total (e.g. Balanophoraceae, Orobanchaceae, Rafflesiaceae, *Striga gesnerioides*). These parasites can reduce the growth of their hosts considerably, by 70 per cent with some root parasites of crops (Graves et al. 1992; Press 1998), and apparently cause death (Prider et al. 2009). We are concerned here with impact on the community, and the result is often an increase in species diversity (Press 1998; Bardgett et al. 2006). It is not clear why. Perhaps the parasite's most common hosts are the competitively dominant species, and a reduction in their productivity allows subordinate species to flourish, possibly also reducing cumulative competition (Section 2.2.1.2), but it is not clear why parasites should infect particularly the dominants. Perhaps the effects are noticed more when they do. Effects of parasites on diversity have so far been shown with annual parasites, which might be able to maintain populations only when their hosts are abundant. The production of a parasite+host will be less than that of an unparasitised host because the parasite utilises carbon but does not fix it (Hautier et al. 2010), and if this reduces whole-community productivity from high to medium, more species will be able to coexist according to the humped-back theory of Grime (1973), again because of a reduction in cumulative interference.

2.7   Plant–Plant Interactions Mediated by Heterotrophs

The populations of organisms using a macrophyte as a carbon energy source – herbivores, parasites and pathogens – are all capable of affecting plant fitness. We discuss here only ways in which this modifies a plant's impact on its neighbours, though surely there are many mechanisms that we have missed. In most cases these interactions occur on a particular spatial scale – a match between the patchiness of the plant and the behaviour of the heterotroph.

2.7.1   Below-Ground Interactions via Heterotrophs

It is generally assumed that species with nitrogen-fixing microorganisms in root nodules enhance the nitrogen economy of the whole community, benefitting associated non-nodulating plants, though documentation is best for agricultural pastures (e.g. Dahlin and Stenberg 2010). Kohls et al. (2003) estimated (using $\Delta^{15}N$ values) that in the well-studied site of Glacier Bay, Alaska, nodulated plants account for most of the fixed N in soils and plants for up to 40 years of succession, the majority fixed by *Alnus viridis* (green alder). Such fixed N can be passed to other species via above-ground litter, below-ground litter or root exudates, or N-fixers can simply spare soil N for non-fixers (Roggy et al. 2004).

Nutrient transfer between species can occur also through an underground common mycorrhizal network (CMN). In all three major types of mycorrhiza – arbuscular, ectotrophic and ericoid – most of the fungal species involved can infect more than one plant species, though there are always quantitative preferences (Simard and Durall 2004). The arbuscular and ectotrophic ones are known to establish CMNs connecting different species, and this may be true for the ericoid type too (Cairney and Ashford 2002). There is evidence for greater nitrogen transfer between plants when mycorrhizae are present, but it is not clear whether the transfer is through a CMN or whether the fungus increases release of N from one plant into the soil and/or uptake by the other plant from the soil. Transfers via CMNs could also operate for P. However, Newman and Ritz (1986) and Newman and Eason (1993) carefully monitored the transfer of ^{32}P and concluded that any direct transfer of phosphorus via CMNs was very minor, and too slow to substantially influence the nutrient status of the recipient plant.

Moreover, nutrient transfers do not always help the species expected. He et al. (2004) found that there was net CMN transfer of N (be it at very low rates) from non-fixing *Eucalyptus maculata* (gum) to N-fixing *Casuarina cunninghamiana* (sheoak), the latter with close to double the N concentration in its tissues. Mycorrhizal transfer of nutrients may not support a suppressed plant as often expected: Fellbaum et al. (2014) found that the fungus supplied more N and P to a non-shaded *Medicago truncatula* (barrel medick) host than to a shaded one.

It has been suggested that CMNs can also affect the performance of plants via carbon transfer (e.g. Booth 2004). However, it is unclear whether plant-tissue → hyphae → plant-tissue transfer of carbon actually occurs (Simard and Durall 2004). There has never been a convincing demonstration of it in ectomycorrhizae and especially not in arbuscular ones, at least not to a degree sufficient to benefit the recipient plant (Teste et al. 2010). There is the possibility that observed 'transfer' comprises labelled carbon retained in the fungal hyphae within the recipient's roots, not in the plant tissues, or is respired and finds its way into the higher plant via photosynthesis. The one clear case of transfer is into achlorophyllous plants, 'saprophytes'.

Roots and microbial activity in the soil are intimately linked, since roots provide most of the carbon source below-ground. Their exudates can increase microbial populations, and some bacteria and fungi can increase P and N availability, in the case of P by mineralising/solubilising organic or mineral forms and in the case of N by mineralising organic forms or by stimulating free-living bacterial N-fixers (Richardson et al. 2009). These effects differ between plant species, e.g. Innes et al. (2004) found that in soils of higher fertility the grasses *Holcus lanatus* (Yorkshire fog) and *Anthoxanthum odoratum* (sweet vernal) tended to stimulate microbial activity, whereas two forb species depressed it. It would be possible for one species to benefit an associated species in this way, and such effects require investigation. However, root exudation of bacterial substrates is not always beneficial: it can reduce N availability in the soil in the same way that litter can, reducing soil N availability to another plant (Cipollini et al. 2012).

2.7.2   Pollination and Dispersal

When fruit set is limited by pollen, plants of the same or different species can interfere with each other's reproduction by competing for the service of pollinating animals (Thijs et al. 2012). When pollen from another species does land on a stigma it may interfere with the germination or development of the species' own pollen by occlusion or allelopathy, reducing seed set and perhaps increasing self-fertilisation (Arceo-Gómez and Ashman 2014).

There can be facilitation too, if one species attracts pollinators to the vicinity of another by mimicry, by being more attractive (a 'magnet species') or simply through the adjacent mass of flowers. For example, Moeller (2004) found that among populations of *Clarkia xantiana* (an annual in the Onagraceae) in California there were more pollinating bees and less pollination limitation in populations with a greater number of congeners, implying that the mass of plants of the same pollination type attracted pollinators. This could be a very rare case of a mutualism between embryophytes. If species with shared pollinators have flowering times staggered but with some overlap, there will probably be some competition for pollinators, but the continuous availability of flower rewards over the period may maintain pollinator populations. Waser and Real (1979) gave evidence from flower numbers, fecundities and pollinating hummingbird numbers across four years at a site in the Rocky Mountains, Colorado, that the earlier-flowering *Delphinium nelsonii* was benefitting the later-flowering *Ipomopsis aggregata* in this way. Considering the maintenance of hummingbird populations from one year to the next, *I. aggregata* could be benefitting *D. nelsonii* in a similar way, forming a genuine mutualism of an almost-undocumented type.

There is evidence that species can also compete for the services of animal dispersers (Wheelwright 1985). Facilitation would be possible too, analogous to that with pollination. The Janzen–Connell effect assumes that fruit/seed predators are attracted to higher densities of food. The same is said to occur after masting and there is some evidence that this can occur due to the synchronous masting of different species (Curran and Leighton 2000). Moreover, dispersal is linked with fruit/seed predation in many species. However, we know of no study that links all these effects to demonstrate facilitation between plant species due to their joint mass of fruits/seeds attracting dispersers.

All these interactions have the potential to affect species' coexistence patterns, and we shall return to them in Chapter 5 (Assembly Rules).

2.7.3   Pests

2.7.3.1   **Herbivores**

Both invertebrate and vertebrate herbivores could mediate interactions between plant species. Invertebrate grazing pressure can be as great below-ground as above-ground (Brown 1993; Wardle 2002). Below-ground vertebrate herbivory is less common, but it is found in some birds, pigs, cows, sheep and rodents, some species of the latter being root-feeding specialists (Andersen 1987). Since they are small, invertebrate herbivore generation times can be short, allowing efficient selection to overcome plant defences and to specialise on particular plant species, but like vertebrates they have parasites and carnivores that control them. There are therefore complex systems affecting plant population fitness through invertebrate herbivory, but since the observation of both the herbivory and the effect on the plant are difficult, studies of the effects of invertebrates on plant community structure, such as that by Brown and Gange (1989a), are rare.

Since herbivores can remove large amounts of all types of plant material, and since herbivores are always selective to some degree, they can effect positive or negative interactions between plant species. There are several mechanisms (Table 2.1).

2.7.3.2 Diseases

Many of the interactions listed in Table 2.1 for herbivory apply also to pathogens and their resulting diseases, so we shall discuss them in parallel where possible. The impact of disease is obvious when it appears suddenly, as in the case of *Cryphonectria parasitica* (chestnut blight) in eastern USA (Redmond et al. 2012), but there have surely been disease epidemics of this magnitude in the past and occasionally there is circumstantial evidence for them (e.g. decline in *Tsuga canadensis*, hemlock, 6000–5000 years BP: Fuller 1998). The ghosts of these diseases past, and continuing, must be a major factor in shaping the plant communities that are around today. This is seen in the few experiments that have applied fungicide in the field. For example, Allan et al. (2010) found that applying foliar fungicide in a Silwood Park, UK, grassland not only increased plant community biomass but also decreased species richness and changed the species composition, increasing the abundance of several species and decreasing that of a couple of dominants. Because abundance was represented as estimated cover we cannot be sure of the details, but it is clear that fungi were having considerable impact on the control community.

2.7.3.3 Types of Interaction

(1) **+ve: Species *A* Repels the Herbivore/Pathogen, Reducing the Pest Load on *B* ('Associational Defence')**

A spiny shrub can deter a medium-large grazer/browser from consuming other species within its canopy: 'associational defence' or less accurately 'associational resistance'. For example Baraza et al. (2006) found that in two Mediterranean woodlands the protection from mammal grazing that shrubs afforded to tree saplings was proportional to their spininess and unpalatability. A recognisably unpalatable species in a grass sward deters a herbivore from taking its bite at that point, for example Suzuki and Suzuki (2011) found that in a Japanese deer park unpalatable *Urtica thunbergiana* (nettle) protected a palatable annual from deer grazing, the effect decreasing away from the centre of the nettle patch. The avoidance of the deterrent patch would be part of an 'extended phenotype' of the nettle species, but non-specific, benefitting all the susceptible plants in that neighbourhood.

The herbivore repulsion can be complete or partial. Active repulsion can be mechanical (e.g. spines, which has been called 'associational refuge'), textural (e.g. silica

Table 2.1 Possible mechanisms of herbivore- and pathogen-mediated plant–plant interaction, the effect of Species **A** on Species **B**. Effects include: 'mitig' = species **A** mitigates the negative effect of the pest on species **B**; '+ve' = the species has higher fitness in the presence of the pest; '–ve' = the species has lower fitness in the presence of the pest; '+ve via **B**' = species **A** has higher fitness due to reduced interference from **B**. 'Oper' indicates whether the mechanism could operate with vertebrate herbivores (V), invertebrate herbivores (I) and/or pathogens (P). 'Common ?' is our estimate of whether this mechanism is common in the real world

+ve/–ve		Mechanism	Effect on B	Effect on A	Oper	Common ?
Positive (species **B** increases in fitness)	1	Species **A** repels the herbivore/pathogen, reducing the pest load on **B** (*'associational defence'*)	mitig	none / -ve	V I P	Common
	2	Species **A** attracts the herbivore/ pathogen's enemy, reducing the pest load on **B** (*'my enemy's enemy is my neighbour's friend'*)	mitig	–ve	I P	Occasional
	3	Herbivory/disease induce defences in species **A**, and signals received by neighbour **B** induce defences in it (*'talking trees'*)	mitig	–ve	V I P	Small effects
	4	Species **A** attracts the herbivore/pathogen, diverting the pest load from **B** (*'attractant decoy'*)	mitig	–ve	V I P	Not long term
	5	The herbivore/pathogen weakens **A**, which is a more palatable/susceptible, interfering neighbour of **B**	+ve	–ve	V I P	Yes (but little evidence)
	6	Species **A** suffers from a herbivore/ pathogen, creating open ground that is the small-scale beta niche of **B**	+ve	–ve	V	Occasional
Negative (species **B** decreases in fitness)	7	A mutualist of **A** is a herbivore of neighbour **B**	–ve	+ve via **B**	I	Occasional
	8	Unpalatable/non-susceptible **A** diverts the herbivore/pathogen to **B**	–ve	+ve via **B**	V I P	Not long term
	9	The herbivore/pathogen requires alternate hosts **A** and **B** to complete its life cycle	–ve	none / –ve	V I P	Not common
	10	Species **A** attracts the herbivore/pathogen, increasing the pest load on **B** (*'magnet species'*)	–ve	–ve	V I P	Probably common
	11	The herbivore/pathogen of **A** removes a third palatable/susceptible species **C** that had facilitated **B**	–ve	–ve / +ve via **B**	V I P	Uncommon

content), by taste or olfactory (the latter probably a volatile through the air, but potentially through the soil or through a mycorrhizal network). Below-ground, or in its litter, a plant might release a compound toxic to invertebrates such as soil nematodes. In passive effects, a plant might mask, via scent, via colour or as a visual block, the presence of a palatable neighbour from a herbivore (Agrawal et al. 2006), making herbivore foraging less efficient.

The repelling of herbivores by a neighbour affects plant communities at the scale at which the repulsion occurs. Especially, the effectiveness of olfactory associational defence depends on the distance over which the chemical diffuses and on the foraging behaviour of the herbivore, and would operate mainly with invertebrate herbivores. The mechanism is widely discussed as a goal of organic gardeners ('companion planting'), but supporting evidence is rare. In an experiment by Tahvanainen and Root (1972), planting *Lycopersicon esculentum* (tomato) amid *Brassica oleracea* (collards) plants decreased flea beetle populations on *B. oleracea* and increased its plant weight. In a choice experiment, adult flea beetles (*Phyllotreta cruciferae*) preferred excised *B. oleraceus* leaves with no tomato mixed in with them, indicating that the repellent was chemical. Himanen et al. (2010) suggested that the sesquiterpene ledene was released by *Rhododendron tomentosum* and adsorbed onto leaves of *Betula pendula*, reducing the palatability of the latter. It is widely suggested that companion planting of *Tagetes* (African marigold) or *Calendula* (pot marigold) species repels insect pests and thus benefits the crop (Figure 2.7). There is a little evidence for this (McSorley and Dickson 1995). Nematodes, below-ground, may be mediating such an interaction (cf. van Ruijven et al. 2003), though apparently not due to any chemical coming from a *Tagetes* plant (Ploeg and Maris 1999).

With pathogens, species **A** might release a compound toxic to bacteria or to fungal spores (Ratnadass et al. 2012). In an indirect effect, they could repel viral vectors such as aphids (Birkett et al. 2000), actively or passively as discussed above, reducing the disease load on **B**.

Tuomi et al. (1994) suggested that the evolutionarily stable strategy was for any associational defence of conspecific neighbours to be limited, otherwise an undefended 'selfish' genotype would too readily invade.

(2)   *+ve: Species **A** Attracts the Herbivore/Pathogen's Enemy, Reducing the Pest Load on **B** ('My Enemy's Enemy Is My Neighbour's Friend')*
The attractant of the enemy of an invertebrate (probably a carnivore) can be food or a volatile chemical. Either the predator of **A**'s herbivore can spill over onto **B**, or the predation occurring on **A** can reduce the local population of the herbivore, benefitting **B**. White et al. (1995) found that planting *Phacelia tanacetifolia*, which was attractive to hover flies around *Brassica oleracea* (cabbage) patches (the control was bare ground around the patches), reduced the density of *Brevicoryne brassicae* and *Myzus persicae* (both aphids) on *B. oleracea*. The predator could have been attracted directly to prey on the herbivore, but in this case the two hover fly species were apparently attracted to *P. tanacetifolia* for its pollen, but their larvae fed on the aphids. A more complex situation is that *Lythrum salicaria* (purple loosestrife) is grazed by a chrysomelid beetle that is parasitized by the 'bodyguard' hymenopteran *Asecodes mento*, which preferentially parasitizes another chrysomelid that grazes *Filipendula ulmaria* (meadowsweet). In mixed populations of the two plant species this reduces the herbivory of *F. ulmaria* and increases its fitness, i.e. *L. salicaria* → herbivore → parasitoid → herbivore → *F. ulmaria* (Stenberg et al. 2007). The 'predator' could also be a bacterium, for example one that infects nematodes (Ratnadass et al. 2012).

Figure 2.7 Most of the apprentices' gardens at Kew Royal Botanic Garden include companion *Tagetes* or *Calendula*.

Surely species **A** could not attract an enemy of a pathogen, but it could attract an enemy of a disease vector, as in the study by White et al. (1995) where two aphids, which are known virus vectors (Schliephake et al. 2000), were reduced.

(3)   *+ve: Herbivory/Disease Induce Defences in Species **A**, and Signals Received by Neighbour **B** Induce Defences in It ('Talking Trees')*
Attack by either vertebrate or invertebrate herbivores can induce plant chemical defences such as alkaloids, nicotine, phenolics and tannins (e.g. Boege 2004) and even morphological defences (e.g. Young et al. 2003). Perhaps an attacked plant can send a chemical signal such as methyl jasmonate through the air to neighbouring plants, inducing the same defence in them: 'talking trees'. A signal could also be transmitted through common mycorrhizal networks (Babikova et al. 2013), in root exudates, or in vectors such as aphids. Induced defences are usually partial, so **A** will still suffer from the herbivore.

Early work was plagued by problems such as pseudoreplication, but more careful work has validated the effect (Pearse et al. 2013), including in unmanipulated field situations (Himanen et al. 2010). The remaining puzzle is whether 'tree talk' can be a product of natural selection since it benefits the neighbours of the plant (genotype) producing the signal, not the plant itself. It could have evolved by kin selection (Wilson 1987), or between-plant signals could be an incidental result of within-plant signalling. However, Pearse et al. (2013) and others have produced some, albeit inconsistent, evidence of greater 'talking tree' effects between two genetically identical ramets than between those of different genotype, and Karban et al. (2013) a greater effect between kin. The mechanism is hard to envisage: can the volatile signals really be chemically specific to a plant genotype? If so, this solves the evolutionary problem.

A similar mechanism for plant pathogens could be envisaged (Yi et al. 2009), but it has not been investigated.

(4)   *+ve: Species **A** Attracts the Herbivore/Pathogen, Diverting the Pest Load from* ***B** ('Attractant Decoy')*

This mechanism is theoretically possible, and Atsatt and O'Dowd (1976) mention some possible examples. Mensah and Khan (1997) found that *Medicago sativa* (lucerne), which had a higher palatability in choice experiments, diverted *Creontiades dilutus* (a mirid insect) pests from a crop of *Gossypium hirsutum* (cotton). There was a dramatic effect over the four-month season. (Since the mechanism depends on relative palatability, this could be due to induced defences in **B**). The effect assumes that herbivore pressure is constant irrespective of food source, which is unlikely. It could operate temporarily until the herbivore population built up on **B,** probably resulting eventually in a higher population than before because of the 'magnet species' effect of **A** (mechanism '10 −ve'). This would not happen if **A** stimulated a dormant phase of the herbivore to emerge (e.g. cysts of a nematode to hatch), but failed to host it, leading to a lower level of infestation for **A**, especially if **A** is later germinating ('dead-end' trap species). This mechanism is used in some trap crops in horticulture (Scholte and Vos 2000). Such effects are also possible in the short term with vertebrate herbivores. Operation of the mechanism with vertebrate herbivores depends on the grain of species intermixing and on foraging behaviour: it cannot operate if the plant species are so intermixed that the herbivore will take several species in one bite.

The effect '4 +ve' could apply to pathogens if **A** had a strong 'fly-paper effect', trapping fungal spores and sparing **B**, or if it attracted disease vectors.

(5)   *+ve: The Herbivore/Pathogen Weakens **A**, Which Is a More Palatable/ Susceptible, Interfering Neighbour of **B***

If **A** is more palatable than **B**, vertebrate or invertebrate herbivory could reduce interference from it, allowing increased growth of **B**. This is another example of: 'my enemy's enemy is my friend', indirect mutualism. It is remarkably difficult to find evidence for this simple process. Part of the reason may be that estimates of relative palatability are not easy to obtain, but other reasons are confounding effects of the herbivore: e.g. gap creation, and species-to-species tradeoffs such as those between palatability and regrowth ability (Bullock et al. 2001). However, Hanley and Sykes (2009) grew seedlings of *Trifolium pratense* (red clover) and *T. repens* (white clover) in pots in a greenhouse, having established that *T. pratense* was considerably more palatable to snails (*Helix aspersa*). When there were no snails present, the community was dominated by *T. pratense*. However, when five snails were placed in each pot overnight, *T. pratense* was almost eliminated and *T. repen*s flourished. Carson and Root (2000), in a very careful experiment, applied insecticide experimentally in an oldfield dominated by *Solidago altissima* and *S. rugosa* (goldenrods). It became clear that in the untreated community specialist *Microrhopala* spp. and *Trirhabda* spp. (chrysomelid beetles) were suppressing the *Solidago* spp. and thus allowing increased growth of other forbs and of woody seedlings. As with other mechanisms involving release from interference, the herbivore must select between species at a fine scale.

Disease could also weaken or remove a more susceptible neighbour that normally interferes with **B**. For example, the introduction of the Asian pathogen *Cryphonectria parasitica* (= *Endothia parasitica*; chestnut blight fungus) to forests of northeastern USA led to almost complete death of the dominant *Castanea dentata* (chestnut). In one area this resulted within c. 14 years a doubling of the basal area of *Quercus rubra* (northern red oak) and a threefold increase in *Quercus montana* (Chestnut oak; Korstian and Stickel 1927). Almost all the other tree species increased slightly. In other areas different species took advantage of the released resources.

In the short term this mechanism can be an effect of '4' above: the herbivore is attracted to **A** which weakens it, and **B** benefits.

The induction of defences against herbivores and fungal pathogens can carry a cost (though the evidence for this is weak), again reducing the host's interference against **B**.

(6)   **+ve: Species A Suffers from a Herbivore/Pathogen, Creating Open Ground Which Is the Small-Scale Beta Niche of B**
Any type of disturbance by a vertebrate herbivore could reduce or eliminate species **A**, opening up ground. For example, mammalian herbivore hoof marks are the preferred microenvironment of some plant species (Csotonyi and Addicott 2004). This mechanism does not depend on the herbivore's having any selection between species or selecting at any particular spatial scale, though it is obviously more effective if **B** is not susceptible.

(7)   **−ve: A Mutualist of A Is a Herbivore of Neighbour B**
Ants are often associated with plants. For example, some *Macaranga* species (an old world genus in the Euphorbiaceae that is often conspicuous in secondary forest) bear extrafloral nectaries that attract ants, and have domatia hollows that shelter them. The relationship is close and complex. Fiala et al. (1989) found that on trees of *Macaranga* spp. in Malaya, ants considerably reduced the amount of herbivory damage on their host, mainly by forcing lepidopteran larvae (caterpillars) off the leaves, and possibly by harassing Coleoptera (beetles) and Acrididae (grasshoppers). The ants involved are generalist predators, often *Crematogaster* spp., yet they also seem to protect against lianas: *Macaranga* species that were myrmecophytic had far fewer lianas attached, and this difference was seen only with plants occupied by ants. The effect was caused by the ants biting the tips of invading liana, and Federle et al. (2002) showed that they also pruned adjacent tree species reducing canopy contact and presumably also interference.

It is hard to envisage analogues for vertebrate herbivores, or for fungal, bacterial or viral pathogens.

(8)   *−ve: Unpalatable/Non-susceptible **A** Diverts the Herbivore/Pathogen to **B***
This is the other side of the coin from mechanism '4 +ve' above, except that **B** does not have to be attractive to the herbivore/pathogen.

(9)   *−ve: The Herbivore/Pathogen Requires Alternate Hosts **A** and **B** to Complete Its Life Cycle*
In this situation (several examples in Dixon and Kundu 1994), a herbivore's abundance will depend on the presence of both species, though they need not be close neighbours. Species **A** might not be consumed, just provide shelter (Agrawal 2004). Some rusts require alternate hosts to complete their life cycle, allowing this mechanism. A less dramatic situation is where a herbivore, or just possibly a pathogen, is sustained through unfavourable seasons by using alternative food/hosts.

(10)   *−ve: Species **A** Attracts the Herbivore/Pathogen, Increasing the Pest Load on **B** ('Magnet Species')*
White and Whitham (2000) found that *Populus angustifolium* × *fremontii* (cottonwood) plants growing under the highly palatable *Acer negundo* (box elder) suffered much more herbivory from *Alsophila pometaria* (fall cankerworm) than those under their own species or in the open. Orrock et al. (2008) provided evidence that in Californian grassland exotic *Brassica nigra* (black mustard) plants provided refuge for small seed-eating mammals, reducing the density of *Nasella pulchra* grass tussocks. The magnet species mechanism can operate at any spatial scale and the vertebrate or invertebrate herbivore does not need to be selective so long as it is attracted to the patch by one species. It seems to be the situation to which the strange term 'apparent competition' has been applied, but the term 'magnet species' is more helpful. A variant is where a plant protects a herbivore, for example shading it from UV damage (Mazza et al. 2013). A more palatable neighbour, having initially benefitted **B** as a decoy ('4 +ve'), might soon become a magnet.

Effects are also quite possible with pathogens. A neighbour could be susceptible to the disease and, whilst not quite being a magnet, it could spread spores or vectors and increase the pathogen load on **B**. For example, there is evidence that the successful invasion of the exotic *Avena fatua* (wild oat) into Californian grasslands has been due to its being highly infectable by a particular generalist, aphid-vectored virus (Malmstrom et al. 2005). The virus has spread to native tussock grasses, severely reducing their growth. Power and Mitchell (2004) found in a field experiment with plots of 1–6 grass species that plots containing *A. fatua* were over 10× more heavily infected by the virus than communities lacking *A. fatua*. The process is termed 'pathogen spillover', but it is effectively our situation '10 −ve'. For those wishing to use the term 'apparent competition', this is the disease equivalent. The effect may be indirect, as in *A. fatua*'s increasing the abundance of the aphids that are vectors of the virus. In this mechanism, species **A** does not have to suffer from the disease; it can be just a carrier. For example, van den Bergh and Elberse (1962) concluded that *Anthoxanthum odoratum* (sweet vernal grass) was a symptomless carrier of a virus that suppressed *Lolium perenne* (ryegrass), even in conditions in which the latter was the stronger interferer (it is a pity that today's molecular or immunological techniques were not available then to back up their symptom-based virus identification). Similarly, preliminary evidence suggests that *Umbellularia californica* (Californian laurel) is infected with the fungus *Phytophthora ramorum*, but its growth is little affected. However, many spores are produced and these cause the death of nearby *Quercus* spp. (oaks) and *Notholithocarpus densiflorus* (≡ *Lithocarpus densiflorus*, tanoak) (Mitchell and Power 2006).

(11)   *−ve: The Herbivore/Pathogen of **A** Removes a Third Palatable/Susceptible Species **C** That Had Facilitated **B***

B could be either susceptible to the herbivore/disease, or not. As with other mechanisms involving release from interference, if the mechanism is via herbivory, the herbivore must select at a fine scale. The net effect of the pest on **A** could be negative because of its direct impact, or positive via reduced interference from **B**. This mechanism is a theoretical possibility, but we know of no example.

2.8   Combined Effects

Grazing, especially by mega-herbivores, can, but does not necessarily, result in:

1.   less litter production, because: (a) plant vigour is reduced so fewer leaves (etc.) are produced, and (b) leaves have a lower probability of reaching senescence, or
2.   greater litter production, comprising: (a) leaf litter, including necrotic tissue, due to collateral damage in herbivore movement and grazing, and (b) woody litter including CWD caused by tree canopy browsers (notably by elephants: Plumptre 1994), and/or
3.   more autogenic shoot-shedding around the point of damage, as Buck-Sorlin and Bell (2000) found after spring defoliation in *Quercus* spp. (oak) trees in Wales, and/or
4.   more litter from subcanopy species, when herbivores open the canopy and there is more light in the understorey, and/or
5.   redistribution of lying litter by trampling and habitat use, speeding decomposition (Eldridge and Rath 2002).

The complexity is well illustrated by Long et al. (2003), who examined the interaction between the herbivorous beetle *Trirhabda virgata* and its food plant *Solidago altissima* (goldenrod). Beetle populations were highest in the densest *S. altissima* patches, leading to reduced plant vigour and lower litter production. Litter, when present in quantity, increased above-ground biomass and reduced species richness. The net effect of the beetle was to enhance plant species richness.

In heavily grazed vegetation there is little litter, so litter effects are minimised and species will interact in other ways, such as via competition, contrary to the usual assumption that competition will be less intense under heavy grazing. If only living plant presence is recorded and litter is ignored, only growth interactions can be assessed.

This may explain why some of the strongest evidence for assembly rules has been found in heavily grazed (i.e. mown) lawns (e.g. Wilson and Roxburgh 1994), a heavily grazed saltmarsh (Wilson and Whittaker 1995) and a sand dune (Stubbs and Wilson 2004), in all of which there is very little accumulation of litter, and interactions due to growth are not obscured by the litter effect. We suggest that assembly rules should be sought that include the litter as well as the living components, as implied by Grime (2001) in his interpretation of the productivity–richness humpback curve.

There are often higher-order interactions between grazing, litter and fire. For example, in North American *Pseudotsuga menziesii* (Douglas fir) forest, grazing by livestock (including cattle, sheep and deer) reduces the herb layer and hastens the decay of litter by trampling, and the unpalatable herbs left tend to be less flammable (Zimmerman and Neuenschwander 1984). This reduces fire frequency, but allows the build-up of CWD so that when fires occur they are more severe and may reach the canopy. Similarly, grazing reduces the fuel load and hence fire frequency in the African savannah (Figure 2.8; Roques et al. 2001). In systems with both a grass and a woody component (savannah or ecotone) a number of processes can occur, triggered by herbivory and fire, which alter the vegetation.

There are many complexities in the dry African savannah system. The whole could converge to an equilibrium, it could operate as a cyclic succession (Section 4.4), be delayed or diverted by switches in three places (Section 4.5), or the switches could give alternative stable states (Figure 2.8; Dublin 1995; Touchan et al. 1995; Brown and Sieg 1999). In the 1960s and 1970s, fire caused a decline in Serengeti forest because higher rainfall led to greater grass growth (Dublin et al. 1990). *Loxodonta africana* (African elephant) inhibited the recovery and kept it as grassland, and because of their grazing the grass crop was not enough to support hot fires. On the other hand, elephants avoid the fire-resistant thickets, which are therefore stable as an alternative stable state. There are both negative and positive feedback processes here.

2.9   Conclusions

We have described in this chapter many processes that can be involved in the development of mixed-species stands, but though we tried to make the list exhaustive it surely is not.

Most species–species interactions are based on the universal principle of reaction: that organisms in general, and plants in particular, modify their environment. All such reactions have effects on neighbours, and so are very relevant for our exploration of the construction and structure of plant communities. F. E. Clements envisaged reactions as being on the physical environment, and these are indeed the basis of interference, facilitation, some litter effects and arguably the 'talking trees' mechanism. Other reactions via litter are indirect, such as the build-up of flammable material that promotes fire and thus affects other species.

We have expanded the term 'reaction' here to include biotic reaction. Pollination was a process well known to F. E. Clements, as in his monograph *Experimental Pollination:*

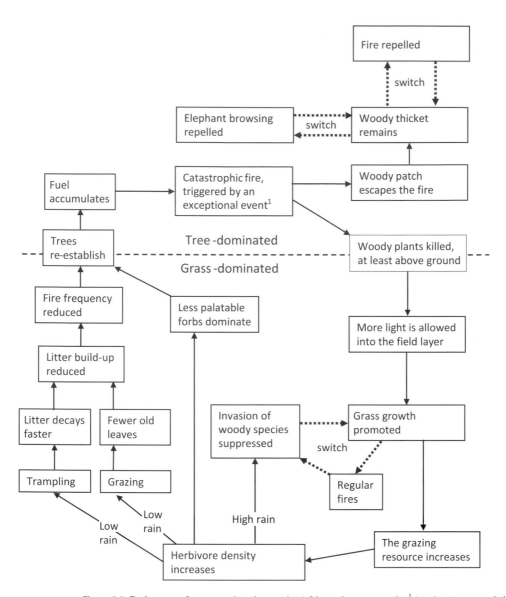

Figure 2.8 Pathways of community change in African dry savannah. [1]A rainy season giving a heavy crop of grass followed by a dry season; or a dry season making savannah burnable; or high browser density causing CWD accumulation.

an Outline of the Ecology of Flowers and Insects, but species–species interactions via pollination have been known more recently. The subtleties of interactions via herbivores have been considered in recent decades. There is also much more understanding of the role of microorganisms in species–species interactions, via diseases and below-ground processes. Surely many more interactions via microorganisms remain to be discovered.

Some biotic reactions depend on prior reaction on physical factors, e.g. the release of a bacterial substrate that indirectly reduces the availability of N to another species, or the release of a volatile that affects a herbivore or the herbivore's predator. These are basically biotic reactions, even if initiated by a change in the local physical environment. Some pollination effects depend on the release of a volatile scent. Other types of biotic reaction do not involve change to the physical environment, even locally, e.g. those via physical defences such as spines, those that depend on chemical unpalatability within the leaf, and most reaction via pollinators. All these complications make it impossible to draw an exact distinction between physical and biotic reaction.

Interactions that do not depend on reaction include autogenic disturbance, parasitism, CMNs and most of those involving pollination.

Plants must be seen within the framework of the ecosystem. Autotrophs make up most of the stored living carbon of terrestrial ecosystems, i.e. its biomass, and a plant's biomass is also a necessary part of its work in foraging for light and soil resources. But phytomass is never alone. Operators from other trophic levels are always present and can have profound influences on plant species presence and abundance, adding to the complexity. Paine's (1969) original usage of 'keystone species' is a clear expression of these. The plant community itself comprises continual renewal: disturbance, death, establishment, growth and both sexual and vegetative reproduction of great numbers of offspring: greater than the habitat can support. The negative and positive interactions that we have listed moderate those processes. These are the basis for the mechanisms of coexistence that we list in Chapter 3. In Chapter 4 we consider how those mechanisms operate at the whole-community level to determine successional changes after allogenic or autogenic disturbance, and to affect the spatial pattern of vegetation.

3   Mechanisms of Coexistence

3.1   Introduction

Most plant communities comprise persisting populations of several species. Populations may increase or decrease through neutral drift or climatic fluctuations, and species can immigrate to them or disappear, but long-term studies such as the Park Grass experiment (Silvertown 1987) and Bibury study (Dunnett et al. 1998) show that the basic tendency is persistence. Outbreaks are followed by a decrease back to the original abundance. This coexistence is the fundamental statement to be made about plant communities, and how it is achieved is the fundamental problem addressed in this book. We conclude that there are only five mechanisms of coexistence of importance.

Interference between species is demonstrable in all types of habitat, except perhaps immediately after disturbance, and since interference abilities can never be exactly equal, the result should be, by the competitive exclusion principle, the exclusion by interference of all but one species[1] (Gause 1934). The amazing thing about multispecies plant communities is that they exist at all. Hutchinson (1941, 1961) asked: 'How [is it] possible for a number of species to coexist in a relatively isotrophic or unstructured environment, all competing for the same sorts of materials?'. He called it the 'Paradox of the Plankton', but the question arises in all species-rich communities such as tropical rainforests, and indeed almost everywhere.

Monospecific stands of vegetation, i.e. with only one vascular plant species, do exist. Table 3.1 lists those that we have seen ourselves. They are often at land/water ecotones, in wet places and especially in open water or extreme saline environments. In arid countries, a monotonous vegetation of one halophytic species can dominate the landscape (Zohary 1973). We could generalise that these are habitats where only one species can survive due to a harsh environment, or where the exuberance of one species excludes others by interference. However, in some cases it is hard to know whether to credit the extreme habitat or the high interference, e.g. *Phragmites communis* reedswamps. Almost as notable are monospecific strata: Peh et al. (2011) list a number of species traits that could cause the relatively rare situation of a monospecific tree stratum in a tropical rainforest.

The other extreme is vegetation with extraordinarily high species diversity at various spatial grains (Wilson et al. 2012). Balslev et al. (1998) found 942 vascular species in

---

[1] Often loosely referred to as 'competitive exclusion'.

Table 3.1 Some examples of monospecific stands; we exclude here monospecificity in a single stratum or guild of vegetation, such as a tree species or understorey species

Habitat	Climatic zone	Exemplar taxa	Reference
Arid saline	Subtropical	*Halocnemum strobilaceum*	Zohary (1973)
Sand dunes	Temperate	*Zygophyllum dumosum,* *Ammophila arenaria*	Zohary (1973)
Marine submerged	Mediterranean Temperate	*Posidonia oceanica* (also other Helobieae) *Zostera marina*	Den Hartog (1970) Den Hartog (1970)
Freshwater submerged	Temperate Tropical	*Sagittaria sagittifolia* *Podostemon* spp.	Pieterse and Murphy (1993) Meijer (1976)
Freshwater floating	Temperate Tropical	*Lemna minor* *Azolla filiculoides* *Eichornia crassipes*	Scunthorpe (1967) Scunthorpe (1967) Pieterse and Murphy (1993)
Freshwater edge	Temperate Tropical	*Typha* spp. *Cladium mariscus* *Cyperus papyrus*	Weisner (1993) Tansley (1939) Lind and Morrison (1974)
Tidal/brackish edge	Temperate Subtropical Tropical	*Salicornia* spp. *Avicennia marina* *Rhizophora mangle*	Tansley (1939) Batanouny (1981) Gilmore and Snedaker (1993)
River edge	Tropical	*Pandanus* spp.	van Steenis (1981)

1 ha of Ecuadorian tropical rain forest and Steel et al. (2004) found 19 bryophyte species in 0.01 m^2 in the per-humid West Cape in New Zealand.

Spatial focus enters into any question about species coexistence. The rainforest tree *Swietenia mahagoni* (mahogany) occurs in the tropics and *Colobanthus quitensis* occurs in Antarctica, *Salicornia* spp. (glasswort) occur on low-altitude saltmarshes and *Androsace* spp. in the alpine region. This is not coexistence: they are growing in different beta niches. We are *not* concerned in this chapter with Warming's (1909) 'easy task' of explaining that different species grow in different beta niches, the 'Regional coexistence' mechanism of Barot and Gignoux (2004). No absolute spatial focus can be specified for the question of how species coexist, because environmental heterogeneity, both allogenic and autogenic, occurs down to the very finest scales. Our clue comes from Hutchinson's definition of the Paradox of the Plankton, asking how coexistence can occur in a 'relatively isotrophic or unstructured environment': the relevant scale is the two-dimensional one at which beta-niche explanations fail and the paradox exists.

It has been suggested that there are at least 120 mechanisms of coexistence (Palmer 1994; Vellend 2010). We believe most are synonyms, and can see only 12 distinct ones (cf. Wilson 1990), which we put into four categories (Box 3.1). The first basic type are

---

Box 3.1 Mechanisms of coexistence

Stabilising mechanisms

Niche differentiation
1. Alpha-niche differentiation (resource, heterotroph-imposed and constructed)
2. Environmental fluctuation (hourly to decadal change)

Balances
3. Pest pressure (heterotroph challenges)
4. Circular interference networks

Escape through movement
5. Allogenic disturbance (disrupting growth, mainly mechanically)
6. Interference/dispersal tradeoff
7. Initial patch composition
8. Cyclic succession (movement of community phases)

Equalising mechanisms

9. Equal chance (neutrality)
10. Spatial mass effect
11. Inertia (temporal and spatial)
12. Coevolution of similar interference ability

---

stabilising mechanisms (Chesson 2000a; mechanisms 1–8 in Box 3.1). In the longer term, population growth, r, is the critical question (we shall use RGR = relative growth rate, subsuming r, to cover all types of growth: vegetative RGR in the short term and in species that reproduce vegetatively, plus increase via propagules [Section 1.2.2.2]). Species abundances always fluctuate, so the question is not absolute stability at an equilibrium but the existence of an increase-when-rare process (see also Section 4.8.1.2). Increase-when-rare, i.e. negative abundance-dependence,[2] means that when a species is at lower biomass in the community its ramets have higher fitness, as seen in long-term RGR, than when it is at higher biomass (Chesson 2008). Such increase-when-rare[3] is a necessary and sufficient phenomenon for maintaining a species and thus stabilising a mixture. Johnson et al. (2012) found, analysing a forest-plot database, that conspecific increase-when-rare could be seen in most of the 151 species analysed, and little effect between species. Moreover, sparser species showed a greater effect, as if decreased RGR at higher abundance were limiting them.

---

[2] For animals, 'density-dependence' is often used, but since the concepts of 'individual' and 'density' are difficult in most plants (Chapter 1) the more general 'abundance-dependence' should be used.

[3] 'Sparse' would be a better term than 'rare', but 'increase-when-rare' is ensconced in the literature, and so we use it here.

Some species will always be present at low abundance, but the issue is what happens when their abundance is reduced towards the point of elimination, or equivalently at the point of introduction. The strength of the increase-when-rare process necessary for coexistence depends on the degree of difference in interference ability. However, the allee effect may disrupt increase-when-rare in the real world (Section 3.2.2).

The second basic type – equalising mechanisms – do not contain an increase-when-rare mechanism and so cannot on their own give long-term coexistence (mechanisms 9–12 in Box 3.1). However, they allow species to persist for a longer time in unstable coexistence, slowing species replacement. If the difference between the interference abilities of species is too large for a stabilising mechanism to work, a concomitant equalising mechanism might reduce the difference in replacement rate and allow a stabilising mechanism to operate (Chesson 2000a).

3.2   Alpha-Niche Differentiation

The increase-when-rare element in alpha-niche differentiation is that when a species is rare, the resource that it particularly absorbs and requires will be present in greater abundance, i.e. its niche is not fully occupied (though luxury uptake of nutrients would complicate this process, cf. Revilla and Weissing 2008). The converse is zero RGR when there is full use of its particular resource. The degree of niche separation required for coexistence will increase with the difference in interference ability between the species; if one species interferes strongly, another species will be able to coexist only if it is occupying a completely different niche. Coexistence by alpha-niche differentiation is impossible to disprove: each species must occupy a different niche (Section 1.4.4), and by reaction uniquely construct part of its niche. The issue is how much separation is needed, but in spite of the calculations of MacArthur and Levins (1967) this remains unanswered for the real world. Even the existence of such niche limitation has been controversial and difficult to prove (Chapter 4). However, if redundancy really occurs, i.e. there are coexisting species that are almost identical in alpha niche, some of the 11 other mechanisms (Box 3.1) must account for their coexistence.

3.2.1   Resources (Type of Resource and Time of Availability)

Tilman (1976) demonstrated that coexistence was possible between two algal species limited by different nutrients, P and Si, and concluded that any number of species can coexist if each species is limited by a different resource. He confirmed this in modelling (Tilman 1977), and there appears to have been no contradiction. Vance (1984) claimed to show that two species can coexist on one limiting resource, but only 'if each species interferes less with resource acquisition by the other than with resource acquisition by itself', which with pure competition must mean niche differentiation (e.g. the one resource is water, but it is taken up from different soil depths).

The primary resource requirements of most embryophytes are similar (light, water, CO_2, N, P, K, minor elements, sometimes pollination and dispersal, maybe soil O_2 and radiant heat). The concept of niche differentiation along a resource gradient is simple for seed sizes as a resource for birds, but it applies less readily to plants, for which most of the resources are absolute requirements. For example, one species cannot require a low-concentration type of P and another a high-concentration type, and the two concentrations cannot co-occur simultaneously anyway. However, the separation and specialisation of species along a gradient is an important mechanism of coexistence, examples being soil resources at different depths and pollinator service during the season (Section 3.3).

The above-ground structure of a plant is a light-capturing apparatus, so gross characteristics of plant form have great relevance. Modelling indicates that niche differentiation in light utilisation between several strata in a forest can lead to stable coexistence, short species coexisting with taller ones (which are necessarily short when regenerating), given the right parameters of fecundity, speeds of recruitment and canopy shape (Akashi et al. 2003; Kohyama and Takada 2009). These are subtle and devious ways in which resource differentiation takes place in this famous biome. Stratification below-ground, i.e. in rooting depth, is another important niche gradient (Section 1.4.4), e.g. in nitrogen (Clarkson et al. 2009), or water, especially when there is both precipitation and an accessible water table (Nippert and Knapp 2007). Suggestive support for the role of rooting-depth differentiation in coexistence is the experiment of Dornbush and Wilsey (2010), where there was lower species richness in field plots where rooting depth was restricted.

Species can separate along niche axes of vegetative phenology. An excellent example is the vernal ground flora of deciduous forests. Fargione and Tilman (2005b) found evidence that vegetative phenological niche differentiation added to rooting-depth differences in facilitating the coexistence of minor species at Cedar Creek with the dominant grass *Schizachyrium scoparium* (bluestem). Two types of temporal resource gradient can be seen: (a) If growth is triggered by the resource itself, species can differ in their speed of response to variation in resource availability. Opportunistic species react fast to resource availability, for example succulents that have surface roots ready to capture rainfall, and *Grewia* spp. (cross berry) that have a flush of ephemeral leaves after every rain in African summer-deciduous bushland. (b) More commonly, seasonal separation of species' growth patterns is controlled not by the resource itself, but by signals such as daylength and temperature, causing consistent seasonal separation of species' activity. One mechanism of temporal niche separation is the storage effect (Section 3.3.2).

Golubski et al. (2010) suggested, via a model, that species with different-sized 'individuals' could coexist. A large individual of one species, averaging over patches with different resource availability, could be held back by depleting a resource-poor patch, and thus leave resources in a resource-richer patch that smaller individuals of another species could invade. As we noted in Chapter 1, 'Individual' is a tricky concept for plants. Golubski et al. here mean a physiologically integrated plant/clone, though there are degrees of integration due to sectorisation and to the physiological state of the ramets (Marshall 1996). Their model assumes that there is no root foraging – greater root growth in resource-rich patches (Drew 1975) – though they recognise this, and say that coexistence is still possible under a more limited range of conditions if root foraging exists.

3.2.2   Heterotroph-Imposed Niches

Pollination and dispersal can be switch mediators (Section 4.5.4), but also means of niche differentiation. Pollinators come in many sizes and specialisations: insects, reptiles, birds and mammals. Among insect pollinators there is huge variation in characteristics and their interplay with plants can involve rewards, guides, mimicry, robbery, defences against robbery and sexually transmitted disease. Differences in flowering times will reduce competition for pollinators, as will differences in the species of pollinators used, giving coexistence based on niche differentiation. However, the pollination niche is liable to disruption of the increase-when-rare mechanism by the Allee effect, which may cause decrease-when-rare below a critical density. In animals and in wind/water-pollinated plants the Allee effect operates directly on the species: isolated individuals are unable to find mates or receive pollen. In other plants, if the target plant provides a resource – pollen or nectar – for a specialist pollinator, then low plant densities can reduce vector populations, feeding back to even lower plant populations. However, we believe the Allee effect will be rare in plants because each individual is effectively a colony and perennials have an indefinite life span (Section 1.2.2). Any occurrences should be in species with obligate outcrossing and/or inbreeding depression, and/or scattered distributions, and/or monocarpic reproduction, and/or specialised pollen vectors. There are a few examples of the effect, e.g. in *Banksia goodii*, a shrub of Australian dry savannah that is probably highly outcrossing (Lamont et al. 1993) and in the annual *Clarkia concinna* in California, which occurs in isolated roadside populations and experiences severe inbreeding depression (Groom 1998). There is a problem examining natural populations, in that poor reproductive performance in small/isolated populations could be because such populations are in habitats marginal for the species. Experiments are more definitive, for example Firestone and Jasieniuk (2013) found a lower seed set in small experimental populations of *Lolium multiflorum* (Italian ryegrass). In fact, most attempts to find an allee effect in any type of organism have failed, perhaps because of ecological noise or perhaps because the effect is truly rare.

Dispersal tends to be less specialised, without an equivalent to the close relation between flower morphology and pollinator morphology seen with some insects and birds, but differences in fruiting times could reduce competition for dispersers: disperser-based niche differentiation. A dispersal allee effect, neutralising increase-when-rare, is possible if the population is too small to attract specialised dispersers.

Vascular plants could occupy different niches by associating with different mycorrhizal fungi. However, specificity within the main groups of fungi (arbuscular, ecto-, ericoid/epacrid, etc.) is quantitative, in terms of efficacy rather than in absolute ability to colonise the roots. Moreover, the effect on the higher plant with all types of mycorrhiza is on availability of soil nutrients (especially P) and water (which all plants need anyway), and the loss is in carbon. We conclude that niche differentiation through mycorrhizae is unlikely.

3.2.3   The Constructed Niche (Restricted, Extended and Shifted)

Alpha niches are not preexisting boxes into which species have to fit. All species modify their environment to a lesser or greater degree, imposing autogenic heterogeneity (Section 1.3). Most of this heterogeneity is in environmental conditions and not relevant to our consideration of local coexistence, but there is alpha-niche construction (extension) and contraction. Below-ground, plants construct the niche for plant root parasites and indirectly for saprophytes (Leake 1994). Their roots leak mineral nutrients such as nitrogen and phosphorus (Section 2.2), potentially creating a niche for other species (notably N_2 fixation by legumes). Above-ground, alpha-niche construction is better known: canopy trees create a shade/temperature/humidity niche for understorey plants, and the fact that many are restricted to forest understories suggests they are unable to survive without those modifications. Trees also create the niche for climbers and epiphytes via their support, and for many epiphytes by water and nutrient stemflow. Aerial parasitic plants obviously occupy a niche that would not exist without other plants; the term 'niche construction' is very appropriate here. Less dramatically, but perhaps quite importantly, facilitation in various physical factors (Section 2.3), and interactions such as associational defence (Section 2.7), can extend a species' niche, enabling it to grow, or to grow to greater biomass, in sites where it otherwise could not (He and Bertness 2014).

Not only does the habitat change, but the plant can change its morphology and its physiological functioning plastically. Even if the plant's phenotype remains constant, competition can shift it towards a different niche, for example being forced to take up nutrients from a different soil depth (O'Brien et al. 1967) or in a different chemical form (nitrate/ammonium/glycine: Miller et al. 2007).

3.2.4   Alpha-Niche Conclusions

Models of species coexistence based on differences in resource use are easy to form, and it is easy to see that species differ in characters, but it is more difficult to envisage real-world differences in resource use. Even then it is surprisingly hard to find documented examples. Yet this is negative evidence. Niche differentiation may still be present, especially since niches can be based on interactions between axes (Clark et al. 2007). The best evidence comes from experiments such as that of Dornbush and Wilsey (2010), removing the opportunity for one type of differentiation and testing for reduced benefits from niche differentiation: in stability, coexistence (species diversity), resilience, overyield, etc. Dornbush and Wilsey reduced the opportunity for rooting-depth differentiation, but other possibilities include lighting from below to reduce the light gradient, or allowing only one type of N source.

3.3   Environmental Fluctuation (Hourly to Decadal Change)

Differences in species' responses to environment fluctuations can permit coexistence, whether those fluctuations are predictable or unpredictable, annual or on other timescales: the 'gradual change' mechanism of Connell (1978). The fluctuation can be one that affects vegetative growth (for example, the vernal flora of forests and spring ephemerals of semiarid areas[4]), or flowering/fruiting niche.

Many authors have claimed that simple variation in the environment and therefore in demographic parameters would allow long-term coexistence (e.g. Connell 1978). For example, Gigon (1997) wrote: 'The fluctuations and their interferences mean that no species encounters optimal growth conditions for a prolonged period of time. Therefore no species can outcompete the others. Fluctuations are thus decisive for the coexistence of species.' This is incorrect. For coexistence, the long-term growth rate of each species has to be RGR = 0.0 (r = 0.0, λ = 1.0), and this is the arithmetic average of its RGR in each period (the geometric average of λ). Variation in growth rate will not make it more likely that long-term r is 0.0, in fact a value of exactly 0.0 due to such averaging is infinitely unlikely. There has to be an interaction between growth and resource supply for environmental fluctuation to cause coexistence, and there are two mechanisms for this: relative non-linearity and the storage effect. These two mechanisms can be seen as special cases of temporal alpha-niche differentiation (Chesson 2000a).

3.3.1 Relative Non-Linearity

The Relative Non-linearity mechanism operates when two species are competing for the same resource, **R**, they respond differently to levels of **R**, and they respond by differently shaped relations (Chesson 2008). For example, the three scenarios in Figure 3.1 count as different shapes. A test for relative non-linearity of shapes is to plot the values of RGR of one species at each level of resource **R** against the values of the other: if the result is anything but a straight line, the species are relatively non-linear. However, even then the shapes have to be such that the species with the lower interference ability has a more convex curve, or exclusion of that species, not coexistence, will result (Chesson 2000a).

To see the mechanism, take the Graph 3 in Figure 3.1. If [**R**], the level of resource **R**, is constant at the mean value (R_{mean}), species **E** has a higher RGR than **F**. However, with high environmental fluctuation around the mean, the mean growth rate of **F** would be higher due to the effects of non-linear averaging ('Jensen's inequality'). Thus, low fluctuation in [**R**] gives **E** an advantage, high fluctuation gives **F** an advantage. The reason this matters is that at low [**R**] species **E** grows faster than **F** and therefore depletes the resources; at high [**R**] it grows little more than at R_{mean}, and leaves much of the **R** unutilised. Both ways, when species **E** is in the majority, it exacerbates the fluctuations in [**R**], thus favouring **F**. Conversely when it is in the minority, fluctuation in [**R**] is lower, which favours itself: increase-when-rare is achieved.

In contrast, species **F** grows little at low [**R**], and will hardly deplete **R**. At high [**R**] it grows disproportionately fast, absorbing **R** and therefore reducing [**R**]. Both ways,

---

[4] Though there is often an accompanying variation in reproduction.

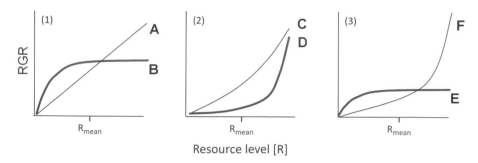

Figure 3.1 Pairs of two species showing relative non-linearity.

when species **F** is in the majority, it damps down the fluctuations in [**R**], thus favouring **E**. Conversely when it is in the minority, fluctuation in [**R**] is higher, which favours itself: again increase-when-rare is achieved. Overall, coexistence is achieved due to the effects of one species on resource variation tending to favour the other species, combined with appropriate differences in the curvature of the growth responses.

3.3.2   Storage Effect

The second way that environmental fluctuations can cause coexistence is the storage effect (Chesson 2008). There are five requirements for the storage effect to operate:

1. There must be overlapping generations. In annuals the seeds produced in year 1 must be able to germinate at least in both years 2 and 3. A perennial achieves storage by year-to-year persistence, overlapping for different plants. Storage ensures that population gains during favourable periods are protected from losses during unfavourable periods. The 'storage effect' is named thus for historical reasons, but storage is a rather trivial requirement for plants, found in almost all species.
2. The species must be competing for a particular resource.
3. They must be affected by an environmental (i.e. non-resource) factor, and respond differently to it in germination or growth ('annuation', Section 4.2).
4. There must be covariance between the environmental factor and the intensity of competition. This is almost inevitable, because when the plants are denser and/or larger, competition will be more intense. That is, in 'favourable' conditions competition will be greater.
5. There must be subadditivity (= buffering, = an interaction between the effects of environment and competition). That is to say, when environmental conditions are favourable to growth, the effect of competition on RGR is greater. So, whilst '3' refers to the intensity of competition, '4' refers to the effect of competition.

In years (or other periods) when the environment is favourable for a species that is in the majority, it cannot take much advantage of the conditions because it is competing mainly against itself: **X** in Figure 3.2.

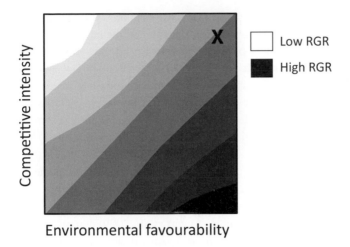

Figure 3.2 The effect of competitive intensity and environmental favourability for a species on RGR. For symbol 'X', see the text.

Calculations by Chesson (2008) and experimentally parameterised simulations by Holt and Chesson (2014) indicate that the storage effect is a considerably stronger force than relative non-linearity. The latter is likely to contribute only when interference abilities are almost equal (mechanism 9) or when it reinforces other stabilising mechanisms.

3.3.3   Conditions Required

Whilst it is true that environmental variation can cause coexistence, it can also promote exclusion by interference or have no effect at all on coexistence/exclusion, depending on the biological response of the species to the environment and to competition (Chesson 1990). Moreover, the environmental variation cannot be too subtle, or neither the relative non-linearity nor the storage effect can operate; but on the other hand it cannot be too severe or the whole of the population of one species may be killed, or the population so reduced that stochastic demographic extinction occurs.

For neither non-linearity nor the storage effect do the species need to differ in the resources they use, the requirement is that because of environmental fluctuations they use them at different times. The two mechanisms clarify that the timescale upon which such fluctuation can cause coexistence is that timescale on which depletion of the particular resource can occur. Light intensity can change instantaneously, it cannot be stored from one second to another and its effects in producing photosynthate are quite short term, so within-day fluctuation could be necessary and sufficient. Water depletion could occur over a few days, and nutrient depletion over a few months. Soil nutrients often become more available in the spring due to mineralisation over winter accompanied by minimal uptake, forming a reservoir which can be depleted during the period of active growth, so with nutrients as the resource, the storage effect can operate on within-season or between year variation.

Temporal environmental variation can also combine with spatial environmental variation to allow coexistence via, for example, a spatial storage effect (Chesson 1985), as seen in the models of Muko and Iwasa (2003), Roxburgh et al. (2004), Berkley et al. (2010) and Bode et al. (2011), and the empirical study of Sears and Chesson (2007). The role of spatial variability in promoting coexistence has been quantified mathematically (Chesson 2000b; Chesson et al. 2005) and includes, in addition to the spatial storage effect, the concept of 'fitness-density covariance' whereby total population growth is modified by the covariance between spatially varying growth and population density.

Since environmental fluctuation occurs on all scales, and since species must, and do, react differently to them, we suggest that this is an important cause of coexistence. Some of the components of both mechanisms are either common knowledge, or have often been demonstrated. For example, Angert et al. (2009) demonstrated the elements of the storage effect among annuals in an Arizona desert, i.e. increase-when-rare due to differential response to yearly variation in rainfall, for example high-SLA species responding more to heavy rainfall. However, full confirmation will come when models of the mechanism have been parameterised (Chesson 2008). The closest approaches so far have been for the storage effect: by Adler et al. (2006) in a Kansas prairie, by Verhulst et al. (2008) with shrubs in a Mexican desert, by Usinowicz et al. (2012) for tree species of tropical forest on Barro Colorado Island, by Holt and Chesson (2014) in the Chihuahuan Desert, and by Adler et al. (2009) in Idaho sagebrush, the latter finding the effect very weak. The need is for higher-quality parameterisation of the models, and application to more field situations.

3.4   Pest Pressure (Heterotroph Challenges)

Both pathogens and herbivores (from insects to large mammals) have the potential to give an increase-when-rare process, the 'frequency-dependent predation' mechanism of Chesson (2000a) (We use 'pest' to cover both pathogens and herbivores). Three conditions are required:

Impact: The pest must significantly reduce the fitness (growth and/or survival) of the plant species.

Specificity: The impact of the pest must be on a limited range of hosts, ideally limited to one plant species within the community. It could be sufficient for the species with lowest interference ability to have no specific pest/disease, but to benefit when the others are suppressed by the pest/disease.

Abundance-dependence: The pest must have less impact when the plant species is sparse than when it is abundant: an abundance-dependent effect. The requirement is for a lower *impact* on the RGR (growth and reproduction) of a species when sparse, but this will presumably be through reduced infection/infestation.

If these three conditions are met, when any one of the plant species in a mixture becomes more abundant, the host-specific pest (*Specificity*) will move more rapidly among its host population and the degree of infection/infestation will increase (*Abundance-dependence*). This will reduce the abundant plant species fitness (*Impact*). This will not directly impact other species, or will do so only to a lesser extent (*Specificity*). Conversely, when a species becomes sparse (low in abundance) and in danger of being eliminated, infestation by its specific pests will decrease and its fitness increase, giving the increase-when-rare effect.

Any pest must have a carbon requirement, which is almost bound to result in lower vegetative and/or reproductive growth for the plant, meeting condition *Impact*. Many pests show considerable *Specificity* to a species or group of plant species. *Abundance-dependence* can be caused by the positive plant–plant interactions via heterotrophs that we discussed in Chapter 2 (Section 2.5). We shall discuss pathogens (i.e. diseases) and herbivory separately since they have different patterns and processes.

There are sometimes Jansen–Connell patterns, i.e. maximum regeneration at intermediate distances from conspecific adults, that cannot be attributed to a particular pest. This is only circumstantial evidence for the pest pressure mechanism. The pattern could also result from any likely dispersal kernel plus interference (competition or autoallelopathy) from the mother plant and interference-based density-dependent mortality among seedlings. Stoll and Newbery (2005) were inclined to attribute such a pattern to adults draining nutrients from seedlings via CMNs.

On the other hand, sometimes greater seedling density has been found near a conspecific adult, presumably because there was insufficient density-dependent mortality to erase the seed dispersal shadow. Hyatt et al. (2003), in a thorough review of the literature, found no evidence for an effect of distance from conspecifics on seed survival in either temperate or tropical communities, though there was a tendency for seedlings to show higher survival at distance, with hints that this occurred especially in tropical forests. This matches the conclusions of Wright (2002) who, with a rather different review approach, found considerable evidence of low growth performance of saplings near conspecific adults in tropical forests. This would be an important increase-when-rare process in diverse, stable communities of trees, specifically tropical rain forest, but it seems to be far from universal.

3.4.1   Pathogens

The *Impact* of fungal pathogens – below-ground, in the plant's photosynthetic and reproductive systems, and in the plant systemically – can be considerable, though it is not often documented. Mihail et al. (1998) found in a greenhouse experiment with the annual legume *Kummerowia stipulacea* (Korean clover) that the fungus *Rhizoctonia solani* reduced plant density by 40 per cent whilst the fungus *Pythium irregulare* reduced it by 80 per cent. Mitchell (2003) found in an oldfield grassland at Cedar Creek that 8.9 per cent of the leaf area was infected by fungal pathogens that decreased leaf life, in turn decreasing root production by 25 per cent, whilst herbivores had no effect.

There is a remarkable paucity of knowledge on whether pathogens show *Specificity* to hosts. In the Cedar Creek oldfields, most fungal pathogens infect only one plant species

(Mitchell et al. 2002). However, Gilbert's (2002) review of polypore fungi, many of which cause tree mortality either directly or indirectly via windthrow, indicated that the degree of host specificity varies widely between ecosystems. He cited two tropical forests in Central America where few or none of the polypore fungi were host specific, but a mangrove forest where all the common fungi were. However, even where the same species of fungus infects several plant species, there may be ecotypes of the fungus virulent on particular species (e.g. Konno et al. 2011).

However, pathogen *Abundance-dependence* is often possible. Most ephemeral pathogens are transmitted aerially, including rusts and smuts, and their populations are often greater when their hosts are more dense. An example is *Ustilago violacea* smut on *Silene alba* ($\equiv$ *S. latifolia*; white campion), for which Thrall and Jarosz (1994) experimentally compared the behaviour of the host and pathogen populations to theoretical models, confirming abundance-dependence. An excellent study by Burdon et al. (1992) described the mortality of *Pinus sylvestris* (Scots pine) caused by the snow blight fungus *Phacidium infestans* as being mostly abundance-dependent, with greater mortality in subsites where the host had been denser the previous year. This abundance-dependence has sometimes been shown to lead to a lower pathogen load in mixtures, which must imply some host specificity. For example, Mitchell et al. (2002) examined 147 plots in an experiment at Cedar Creek established by Tilman and coworkers, sown and weeded to species richness from 1 to 24 species. The percentage of each leaf visibly infected was estimated, using calibrated cards as a guide. Infection dropped as species richness increased, the 24-species plots having only 37 per cent of the foliar fungal pathogen load of the mean of the monocultures (though more than the least-infected monoculture). Similarly, Mitchell et al. (2003) analysed another Cedar Creek experiment sown and weeded to 1–16 species, and found that diversity decreased the load of all 16 pathogens examined, the total load in the 16-species plots being only 34 per cent of that in the mean monoculture. In Section 4.8.4 we described the 'selection' artefact in overyield and invasion-resistance experiments. A similar artefact would be possible here if the species less susceptible to disease had thereby an interference advantage and increased its proportion in the mixture, so that the mixture had a lower mean pathogen susceptibility and thus a lower pathogen load. However, Mitchell et al. (2002, 2003) present evidence that this was not the cause of the effect they found. The effect may not be a distance/dispersal one; the associated species in a mixture can affect the transmission of a pathogen indirectly, via interference with the susceptible plant species or via alternation of the microclimate. However, the traits of associates could affect the susceptible species directly, e.g. a sticky or furry leaf surface might be more effective in trapping spores. Little seems to be known of this.

Below-ground pathogens could cause similar effects. Bever (2003) modelled this and Petermann et al. (2008) produced evidence using soil sterilisation, without being able to pinpoint the pathogens responsible. This is an area ripe for more research.

The conclusion so far must be that the opportunity for fungal-based pest pressure coexistence varies between ecological systems, but more information is needed on all three conditions: impact, specificity and abundance-dependence.

3.4.2   Herbivory

Herbivores come in all sizes, specialisations and guilds. On vegetative parts there are leaf eaters, sap suckers and other stem borers, and root eaters. On reproductive systems there are flower exploiters, frugivores and granivores. Plant species are variously adapted to herbivory, with chemical and physical defences, life histories, and growth patterns that have seemingly evolved to deal with the challenges. The potential mechanisms for plant–plant interactions via herbivory are similar to those for pathogens. However, whilst pathogens often reduce the functional efficiency of plant parts, many herbivores simply remove plant material so that the plant needs to regrow to replace tissue and thus its resource-capturing apparatus. In many systems, herbivores exploit plant populations in an abundance-dependent way. In any case, a selective herbivore can hold down a potentially dominating plant species and thus allows subordinate species to survive (Grover 1994). Carnivores can affect plant–plant interactions via herbivores (Section 2.7.3), and Calcagno et al. (2011) suggested from modelling that it is possible for them to permit plant coexistence, given the right behaviour by the herbivores and by the carnivores, and the right interference abilities of the plant species. In the following sections we deal with the abundance-dependent culling of disseminules (seeds, etc.) and seedlings (Section 3.4.3) and removal of adult vegetative plant material (Section 3.4.4).

3.4.2.1   Herbivory of Disseminules and Seedlings

Seeds, as well as seedlings and vegetative disseminules, are a rich nutritional resource and often heavily predated, having *Impact* on plant populations.

Disseminule herbivory often shows *Abundance-dependence*. *Cygnus bewickii* (swans) eat the turions (disseminules, fleshy buds) of *Potamogeton pectinatus* (pondweed) in the autumn. Jonzén et al. (2002) demonstrated clear density-dependent control of *P. pectinatus*: denser patches of turions were exploited, reducing their density, whilst areas of low turion densities were unexploited and their plant density increased. Edwards and Crawley (1999) examined four herbs of British meadows and found that granivory by rodents was abundance-dependent, but the consequences for adult abundance differed. Densities of species with larger seeds (*Arrhenatherum elatius*, oat grass; *Centaurea nigra*, knapweed) appeared to be reduced by predation, but in the smaller-seeded *Rumex acetosa* (sorrel) and *Festuca rubra*, seedling survival increased, compensating for predation. Again, Ehrlén (1996) found that in *Lathyrus vernus* (spring pea), although seed predation by a beetle (in the Bruchidae) was correlated with seed density in small plots and with inflorescence size, this had no consistent effect on plant population recruitment. Thus, even the occurrence of abundance-dependent seed predation is no guarantee that it will control the population and hence contribute to species coexistence. The abundance-dependence of seed predation can even be negative, perhaps due to predator satiation (Takeuchi and Nakashizuka 2007).

Another limitation of disseminule/seedling herbivory as a mechanism of coexistence is that most mammal granivores are not specific to one species, i.e. the condition of *Specificity* is not met. There is huge literature assuming that fruit/seed size restricts the beak-size range of possible bird granivores, but not to one plant species. However, invertebrate seed predators can be specialists, for example species of bruchid beetle to the seeds of one or a few species of Fabaceae (legumes) and of dipterans to various species of Asteraceae (e.g. *Centaurea nigra*). Indeed, Pocock et al. (2012) observed in surveying a Somerset farm, UK, that bird and rodent granivores were very non-specific, whereas insect granivores had very high specificity. Herbivore specificity is also quite high in some tropical forests (Novotny and Basset 2005).

Invertebrates can eat seedlings. Their palatability relative to adults appears to vary ontogenetically and between species, but the phase is too transitory for herbivores to specialise on.

3.4.2.2 Herbivory of Vegetative Parts

Vertebrate herbivores can have a considerable *Impact* of herbivory on vegetative parts (Jones 1933) and both above- and below-ground insects (Brown and Gange 1989b).

The *Specificity* of herbivores varies widely. Many large non-ruminant animals such as *Loxodonta africana* (African elephant), even though they have preferences, will readily eat a wide range of species. More importantly, some such as *Equus* spp. (horses) often graze finely patterned vegetation at a relatively coarse scale, necessarily taking in species with a range of palatability. Other ungulates such as *Ovis* spp. (sheep) and *Bos* spp. (cattle) are more selective. Many insect herbivores feed on only one plant species, or a few closely related species that may not be found together (Novotny and Basset 2005). This is true for many Lepidoptera larvae such as *Tyria jacobaeae* (cinnabar moth) on *Jacobaea vulgaris* ($\equiv$ *Senecio jacobaea*, ragwort), some tiny gall wasps (Cynipidae), most aphids and leafhoppers (both in the order Hemiptera) and most leaf miners (in Diptera) (Schoonhoven et al. 1998; Pocock et al. 2012), though Novotny and Basset (2005) reported wide variation in sap-sucker specificity between studies, and Pocock et al. (2012) found only moderate specificity in leaf miners. The specificity of root herbivores is little known, though Pokon et al. (2005) found that larvae of Chrysomelidae beetle species in a New Guinea rain forest were consuming the roots of 3–10 tree species.

There is little evidence for positive *Abundance-dependence* in vertebrate herbivory. Poli et al. (2006) showed that, on the contrary, heifers preferentially grazed on the minority species in an artificial pasture, as if to balance their diet. If this happened in natural communities, it would counteract the pest pressure mechanism. In fact, there is a basic problem with vegetative herbivory, that the herbivores are longer-lived than the plant modules that comprise their food. Therefore, rather than undergoing population decline when their preferred food becomes scarce as the pest pressure mechanism requires, herbivores may remain and actively seek it.

It seems that the pest pressure mechanism could operate with vegetative herbivory in some circumstances. That vertebrate grazing often increases species diversity is well known, as shown by the rabbit exclosures erected by Tansley and Adamson (1925). However, this is often misunderstood. The effect is surely that when the sward is higher, light competition is more important giving the opportunity for the feedback between the outcome of competition and competitive ability (cumulative competition;

Section 2.2.1.2), thus promoting competitive exclusion. With insects, Carson and Root (2000) found that periodic plagues of folivorous chrysomelid beetles checked populations of the dominant *Solidago altissima* (goldenrod) in an oldfield in New York state, USA, and thereby increased plant species diversity and facilitated succession (i.e. tree invasion), yet there is no indication that the beetle outbreaks were caused by the abundance of the host plant: a necessary condition for the pest pressure mechanism to operate. Thus, simple grazing of dominants operates by equalising the interference ability of the species, basically the equal chance mechanism (Section 3.10), allowing stabilising mechanisms of coexistence to operate, including pest pressure by other means.

3.4.3 Interactions among Processes

There can be complex interactions among pests. An example is seen on Dutch sand dunes, on which there are relatively uniform soils with spatial sequences very likely to mirror succession. The vigour of the pioneer *Ammophila arenaria* (marram grass) declines in older, more inland stands, due to pathogens and nematodes (van der Stoel et al. 2002). Individual fungal species have little effect, but synergism in mixtures of fungi reduces *A. arenaria* performance by 20 per cent.[5] Herbivorous nematodes on their own reduce growth only early in the season and early in succession, but their impact adds to that of the fungal pathogens. This allows *A. arenaria* to be suppressed by *Festuca rubra*, the next dominant in the succession, especially when the substrate is nutrient-poor (de Rooij-van der Goes 1995; van der Putten and Peters 1997). Later still in succession, there is often a mosaic of *F. rubra* and *Carex arenaria* (sand sedge), where Olff et al. (2000) discovered that each species had phases of increased and decreased vigour, replacing each other, and that this process was associated with pest phases, particularly of the nematodes. Each species seems to be affected by different groups of pests, leading to the changing mosaic aspect of the vegetation, which might look superficially like cyclic succession (Sections 3.9 and 4.4).

Two mechanisms operating simultaneously might be expected to have a more stabilising effect than either alone. However, Kuang and Chesson (2010) demonstrated that pest pressure weakens competition, weakens the covariance between response to environment and response to competition, and therefore weakens the stabilising effect that the storage effect has in permitting coexistence.

3.4.4 Pest Pressure Conclusions

The pest pressure mechanism seems most likely to operate via diseases. It may operate via herbivory of reproductive and vegetative material in some circumstances, though almost all the studies have been on seed and seedling survivorship, not on adults.

The *Impact* of all kinds of pests for many species in many communities cannot be doubted.

---

[5] Compared with growth in sterile soil.

For host *Specificity*, there is only sporadic information. Fungal pathogens are more likely to be species-specific, but it is not yet possible to make generalisations. In a study of seedlings of eight tree species in temperate forest in Japan by Yamazaki et al. (2009), the fungi causing seedling death were generalist ones, though they speculated on local genotypic adaptation to hosts. This left open the possibility in their system for mortality via bacterial or viral pathogens. With a few notable exceptions there may be little host specificity in seed predation, though seed size and availability at different times could produce some. There is very little information on the specificity of seedling herbivory by rodents, and that for invertebrates varies not only between groups but between reports. Many larger herbivores are selective.

Critical information relating to *Abundance-dependence* is how often survivorship is lower at higher species density and closer to conspecific adults; and whether this can be attributed to disease or herbivores. Yamazaki et al. (2009) examined this for eight seedlings of tree species in temperate forest in Japan. Deaths due to both disease and herbivory were higher in denser populations and among those closer to adults for the majority of species, but not all. Preferences by large herbivores are unlikely to be reinforced when the target species is more abundant; in some circumstances it may be relaxed.

Pest pressure is likely to be an important mechanism of species coexistence, but in many cases evidence is lacking.

3.5   Circular Interference Networks

An interference network between a set of species is said to be transitive if the species can be arranged in a 'pecking order', such that a species higher in the order can always exclude by interference one lower down. The opposite situation, non-transitivity, is the existence of circular relations somewhere in the network (Figure 3.3). If circular relations exist, they would contain an increase-when-rare mechanism (Laird and Schamp 2006): as species **A** starts to displace species **B**, species **C** increases because it has high interference ability against **A**, but then it in turn is replaced by **B**, completing the cycle. Revilla and Weissing (2008) modelled the specific case of a circular interference network based on nutrient-limited growth, though in their model coexistence involved oscillations and was critically dependent on the extent to which species took up which nutrients.

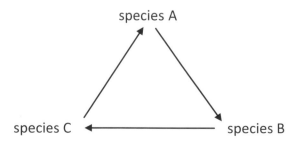

Figure 3.3 A circular interference network between three species.

Table 3.2 Which species has the higher interference ability? The starting biomass for both species was 1.00; cf. Connolly (1997)

Species	Biomass in monoculture		Biomass in mixture	RGR in mixture	Outcome
A	3.00	Decrease - - - - - - - - - - - - - ▶	2.77	1.02	Winner in mixture
B	2.64	Increase ━━━━━━━━━━━▶	2.71	1.00	

However, the real-world relevance of the concept is not obvious. First, we note that the question can be asked only in one environment, for interference abilities must change with the environment (Fynn et al. 2005); that is a major reason there is different vegetation in different places. Second, the species that dominates the mixture will be the one with the highest RGR, but as displacement proceeds, the proportions of the species will change and as a result the relative RGRs of the species (i.e. their interference abilities) may change. The critical criterion will be the RGR of each species when it is a very minor component of a mixture, i.e. its increase-when-rare ability. Therefore, the eventual result must be judged in terms of exclusion by interference. Yet this chapter presents 12 mechanisms by which species can coexist, not exclude each other. The question of transitivity cannot be asked when one of the other 11 is operating.

In order to determine whether interference is circular, it must be measured. Several studies have done so by comparing species' performances in mixture with those in monoculture. Connolly (1997) pointed out logical flaws in this. Correction can be made for the 'size-bias' that he discusses, but the basic error has been the monoculture comparison. It is tempting to conclude that if species **A** grows more slowly in mixture than in its monoculture whilst species **B** grows faster in mixture than in its monoculture, **B** has the higher interference ability. Yet Connolly's table (Table 3.2), over the undefined period of his artificial data and assuming a starting biomass of 1, gives an example where **A** does worse in mixture than in monoculture, and **B** does better in mixture than in monoculture, yet **A** has the faster RGR in mixture ($\log_e 2.77 - \log_e 1 = 1.02$) than Species **B** ($\log_e 2.71 - \log_e 1 = 1.00$) and will come to exclude its competitor from the mixture (subject to the conditions mentioned above). If **B** disappears from the mixture it can hardly be said to have had the higher interference ability.

It turns out that what is essential in designing such an experiment is not the monocultures, as many had thought, but sequential harvests so RGR can be calculated. This invalidates almost all the studies of transitivity done so far. Waiting, perhaps for close to infinite time, will reveal which species has the higher growth rate as the mixture approaches a monoculture. This is looking like one of those community ecology questions that is impossible to answer.

At the moment, it is interesting to look at the imperfect evidence available. Buss and Jackson (1979) claimed several competitive cycles for coral reef sedentary organisms, as seen in static evidence for overtopping. Likewise, Russ (1982) claimed non-transitive relations between species in the overgrowth of sedentary marine organisms observed

Table 3.3 Pecking order of interference ability from Mouquet et al. (2004), strong interference ability at the top

High density	Low density
Holcus lanatus	*Holcus lanatus*
Rumex acetosella	*Rumex acetosella*
Cerastium glomeratum	*Cerastium glomeratum*
Anthoxanthum odoratum	*Anthoxanthum odoratum*
Festuca rubra	*Festuca rubra*
Arabidopsis thaliana	*Lamium purpureum*
Lamium purpureum	*Arabidopsis thaliana*
Veronica arvensis	*Veronica arvensis*

colonising experimental plastic sheets in the sea in Australia, though no circular relation can be made out of his results.

Turning to higher-plant work, Mouquet et al. (2004) grew eight herbaceous meadow species in replacement mixture in all possible pairs. Using relative yield (RY_{AB} = biomass of species **A** when growing with species **B**/biomass of **A** in monoculture), if the species form a pecking order (i.e. a transitive hierarchy) it should be possible to arrange them so if species **A** is further up the pecking order than species **B**, $RY_{A,B} - RY_{B,A}$ is always positive. In his experiment, at both low and high density, it almost is, and with a very similar order (Table 3.3).

At each density, there is one negative $RY_{A,B} - RY_{B,A}$ indicating a conflict with the pecking order, it is between species not contiguous in the pecking order, but it is of size -0.05 and -0.06, respectively, which is clearly within the experimental error.

Freckleton and Watkinson (2001) reanalysed the results of a competition experiment by Goldberg and Landa (1991). Seven species, all from oldfields or pastures but not from any particular community, had been grown in shallow boxes in a greenhouse for five weeks. The 'per individual equivalences', i.e. the competition coefficient (α_{ij}) for a species growing in competition with another compared with that growing with itself (α_{ii}), enable us to try to fit the species into a pecking order. The best order (Figure 3.4) has only two discrepancies from perfect transitivity, and those small in magnitude and possibly attributable to experimental error.

A study that returned a clear answer to the question of transitivity is that of Roxburgh and Wilson (2000a). It relates to a real community, since the seven species used in the interference experiment were taken from the University of Otago Botany Lawn, New Zealand (Section 5.13). The use of 10 replicates allowed significance tests. The seven species could be arranged in a pecking order to which all significant competitive relations conformed (Figure 3.5). In fact, relations between *all* pairs of species, significant or not, conformed.

The experimental design of Keddy et al. (1998) comprised planting 18 'wetland' species into five receptor monoculture swards of similar species. The 18 species tended to respond similarly to different receptor swards, e.g. Kendal's coefficient of concordance took a rank of 0.7 (1.0 = complete agreement as to which target suffered more/

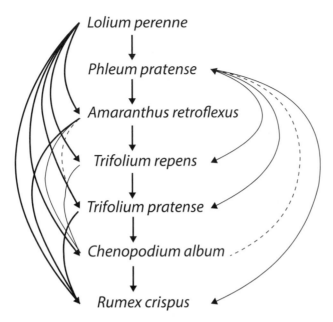

Figure 3.4 The pecking order of interference relations in seven oldfield/pasture species. From the coefficients of Freckleton and Watkinson (2001) calculated from the interference experiment of Goldberg and Landa (1991). ➙ = a species higher in the pecking order displaces one lower down with either a difference in equivalence sign or a ratio of equivalences >3.0, ➙ = displacement to a lesser degree, ⟵ = similar, but the lower species displacing the higher one.

less), which was highly significant. Some of the variation in invasion ability across receptor swards could be due to experimental error (there was no replication), but some results are impressive, e.g. the rank of *Carex crinita* (sedge) varied only from 14 to 17 across the five receptor swards (18=suppressed most), and *Lythrum salicaria* (purple loosestrife) varied from 4 to 7 (1= suppressed least).

The effects of interference can perhaps be seen more realistically in mutual invasion experiments. Silvertown et al. (1992) used data from an experiment where several species had been planted in adjacent hexagons, and invasion between hexagons recorded. Examining the difference between the invasion of Species **A** into Species **B** and that of Species **B** into Species **A**, replacement rates could be calculated. A pecking order for the conditions obtaining in the experiment can be formed from these results (Figure 3.6), with no qualitative discrepancies.[6] There are quantitative discrepancies. Since *Holcus lanatus* (Yorkshire fog) could invade *Poa trivialis* (**1**) and *P. trivialis* could strongly invade *Lolium perenne* (ryegrass) (**2**), the expectation would be that *H. lanatus* would be able to invade *L. perenne* even more strongly, but in fact their invasion rates were exactly balanced (**3**). Moreover, although *Agrostis stolonifera* (creeping bent) at the top of the order could invade *Cynosurus cristatus* (dog's tail) at the bottom, the rate of replacement was less than for other pairs (**4**).

---

[6] Though *L. perenne* (ryegrass) and *C. cristatus* (dog's tail) could equally well exchange positions.

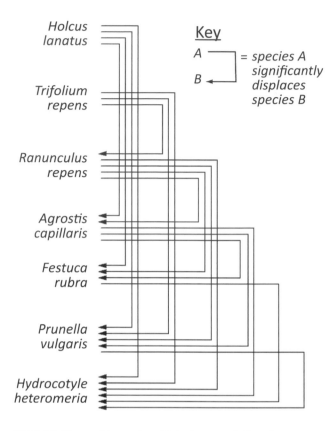

Figure 3.5 The pecking order of interference relations in seven species from the University of Otago Botany Lawn. There is an arrow from the species of a pair with the statistically higher interference ability to that with the lower.
Reprinted with permission from John Wiley & Sons Inc., from Roxburgh and Wilson (2000a).

In a similar experiment Silvertown et al. (1994) used only four species, so there was less opportunity for non-transitivity, but in any case there was none in any of the four grazing treatments (Table 3.4).

A complication with assigning interference abilities is that all plants have to go through a juvenile phase, which will differ in traits from the adult phase; relative interference abilities may therefore differ between the two phases. Zhang and Lamb (2012) reported an interference hierarchy among the seedlings of seven prairie perennials, but non-transitive relations among their adults, with a different situation for interference 'effect' and 'response'. Wang et al. (2010) found interference 'effect' and response' of two species against 22 others, some from one prairie type, others not. 'Effect' and 'response' were different, but correlated. However, the concern here, and elsewhere, is with the net competitive difference as defined above, not with distinguishing effect and response components.

Attempting to pinpoint the traits determining interference ability, Vermeulen et al. (2009) found that out of 10 genotypes of *Potentilla reptans* (cinquefoil) growing together

Table 3.4 The interference pecking order of four species in four treatments in Silvertown et al. (1994) (*Schedonorus arundinaceus* ≡ *Festuca arundinacea*)

Summer sward grazing height	Winter and spring	Invasion ability: greater → lesser
3 cm	Grazed	*Lolium perenne → Festuca rubra → Schedonorus arundinaceus → Poa pratensis*
3 cm	Ungrazed	*Festuca rubra → Lolium perenne → Poa pratensis → Schedonorus arundinaceus*
9 cm	Grazed	*Festuca rubra → Lolium perenne → Schedonorus arundinaceus → Poa pratensis*
9 cm	Ungrazed	*Lolium perenne → Festuca rubra → Poa pratensis → Schedonorus arundinaceus*

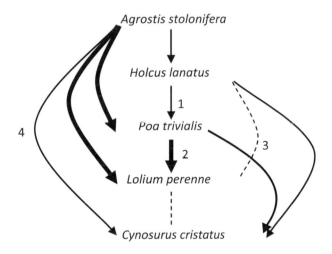

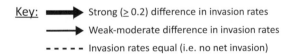

Key: ➡ Strong (≥ 0.2) difference in invasion rates

──▶ Weak-moderate difference in invasion rates

- - - - - Invasion rates equal (i.e. no net invasion)

Figure 3.6 The interference pecking order from invasion rates in data of Silvertown et al. (1992).

in experimental plots for five years the genotype dominating the mixture, Genotype I, had the lowest light-saturated photosynthesis (P_{max}) but the lowest dark respiration. According to their model it had lower than average productivity (carbon gain per above-ground mass) in the upper strata (layers 0–3), above-average productivity in the lower strata, and at the bottom of the canopy (layer 8) it had the highest productivity, indeed most other genotypes showed negative carbon gain. Its leaf area was split evenly between the three major strata, whereas two of the other four major genotypes had a smaller proportion of their leaf area at the base of the canopy (though height distribution could be a result of competitive ability as much as a cause). This could imply that the high interference ability of Genotype I was a combination of height with ability to survive at

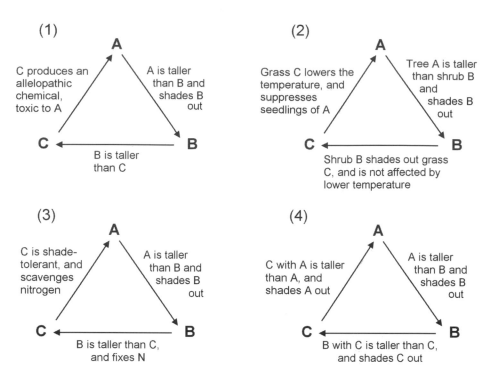

Figure 3.7 Four possible causes of non-transitivity between three species: **A**, **B** and **C**.

the bottom of the canopy, the latter a Tilman R* argument (Section 6.6.2). Competitive effect and response are complicated, especially because of the plastic response to competition. It is the net effect, the RGR in mixture, that we should concentrate on.

It is interesting to wonder what ecological processes could give rise to non-transitivity (Figure 3.7).

In scenario (1), we use an allelopathic chemical produced only by **C** and toxic only to **A**. Lankau and Strauss (2007) demonstrated such a cycle with two chemodemes of *Brassica nigra* (black mustard) and one out of several species able to complete the triangle. This does produce a cyclic network, but in general species-specific allelopathy is difficult to demonstrate and examples are rare.

Scenario (2) is similar, except that the third factor is lower temperature (Ball et al. 2002) rather than a toxin. In (3), the question is why **C** can suppress **A**; presumably the shade tolerance of **C** minimises the competition for light, so competition for N becomes important, and **C** has the lower Tilman R* (Section 6.6.2). Why cannot **C** suppress **B**? Perhaps because it is shorter and so cannot compete for light, and its low R* for N does not help because **B** can fix N. Does this work? Probably. In scenarios (1) to (3), not all pairs are interfering using the same resource/factor.

Could there be a 3-species solution using competition for light (Scenario [4])? How can there be heights of **A**>**B**>**C**>**A** ? Differential plasticity allows such magic: in this case probably by red:far-red effects (Section 2.2.4), but again, this introduces a second factor: light spectrum in addition to light intensity.

These thought experiments are rather convoluted, which suggests that circularity will be uncommon, and this is confirmed by the analyses above. Circular (i.e. non-transitive) interference networks have rarely, arguably never, been observed in plants. In retrospect they should not have been expected, because ecologists had not thought what mechanisms would cause them, and such mechanisms are difficult to envisage. This is almost certainly not an important mechanism of coexistence.

3.6   Allogenic Disturbance (Disrupting Growth, Mainly Mechanically)

Disturbance can have the same effect as climatic variation (Roxburgh et al. 2004), but the original Intermediate-timescale disturbance mechanism, as defined by Connell (1978), is a patch mechanism: within an area there is a mosaic of patches at different successional stages (i.e. times) since disturbance, therefore with different suites of species (Wilson 1994a), the 'regeneration niche' mechanism of Bengtsson et al. (1994), but often referred to simply as the 'intermediate disturbance hypothesis' (Shea et al. 2004).

3.6.1   Conditions Required

3.6.1.1   Spatial Focus

This mechanism of coexistence can apply only when considering a spatial focus that is larger than the size of a disturbance, and thus includes patches of differing time since disturbance. Large-scale disturbances (up to several square kilometres), such as those causing cyclones, fires or drought in the seasonal dry evergreen forest of western Thailand, are not relevant here, though gaps there caused by the death of a single tree are (Baker et al. 2005).

3.6.1.2   Species Specialisation

Allogenic disturbance can permit coexistence only if there are, within the spatial focus being considered, gaps of various age differing in species composition from each other, with the extremes represented by distinct pioneer and climax species, i.e. *r* versus *K* species, R versus C-S (Section 6.5). However, succession in gaps is secondary, with less stressed soil and climatic conditions, so this cannot be assumed. Brokaw and Busing (2000) suggested that in many forests few of the tree species are gap specialists, and that many species have some role in gap colonisation. Indeed, Cowell et al. (2010), examining old-growth deciduous forest (*Fraxinus americana*, green ash, *Acer saccharum*, sugar maple, etc.) in Indiana, USA, reported that tree species composition was not affected by gaps. Even when gap-dependent species are present, they can be sparse: Peterson and Pickett (1995) found that after windthrow disturbance in a North American conifer/deciduous forest some climax species regenerated from seed and some from already present seedlings, but pioneer shade-intolerant species were sparse, apparently due to a lack of propagule input. The extreme situation – autosuccession – in which the climax species immediately reestablish after a disturbance, is known from disturbances in mesic areas such as after windthrow in temperate *Nothofagus* rainforest in New Zealand (Cockayne 1926), but it is specially found under environmental stress, as predicted by C-S-R theory (Wilson and Lee 2000; Section 6.5).

In tropical rain forests the greater species richness might include a good number of gap specialists. Indeed, Hubbell (2005) demonstrated for Barro Colorado Island tropical rainforest a close negative correlation among species between survival rate in shade and growth rate in full light (in gaps), though admitting there were rather few specialist gap species and their abundance was low. Wright et al. (2003) confirmed this for the same forest, showing a continuous distribution between gap-colonising species and those that avoided gaps, but with the majority being rather indiscriminate in gap preference. Similarly, Lieberman et al. (1995) found that 87 per cent of the tree species in Costa Rican tropical forest had no significant canopy gap/matrix specialization. Poorter et al. (2005) found that only one of 47 species in a Liberian tropical rainforest was a shade species for its whole life, and only one a light species for its whole life. It is clear that most species in tropical rainforest are intermediate in this respect. This suggests that allogenic disturbance may not be an important mechanism of coexistence in the very biome where ecologists most expect it. For temperate evergreen forest in Chile, Lusk et al. (2006) found a continuum, except that three gap-specialist species were distinct. Yet in some temperate forests there may be greater opportunities for autogenic disturbance to increase species richness: Poulson and Platt (1996) demonstrated in Michigan that the size of the gap affected which species reestablished, such that single treefalls favoured *Fagus grandifolia* (American beech) but multiple fall gaps favoured *Acer saccharum* (sugar maple). This is not a question of gap versus non-gap, but also of differences between types of gap.

Much of the discussion has concerned gap-specialist tree species. However, many other life forms, such as herbs and lianas, can be gap-dependent (e.g. Goldblum 1997), and forest studies very rarely report the species composition for the whole gap community, i.e. for all strata. Schnitzer and Carson (2000) estimated that up to 65per cent of the flora (all life forms) of tropical forest on Barro Colorado Island might be gap specialists.

Gaps are by no means restricted to forests. Grubb (1982) suggested that in roadside communities around Cambridge, England, the climax dominant among the grasses was *Arrhenatherum elatius* (oat grass), but *Dactylis glomerata* (cocksfoot) and *Plantago lanceolata* (ribwort plantain) retained their place in the community by being the first to invade small gaps. King (1977) found that of the 58 major plant species in 13 grassland sites in southern England three were significantly 'much more abundant'[7] on ant-hills than in the surrounding grassland.

3.6.1.3   Frequency of Disturbance

The mechanism requires also a sufficient frequency/intensity of disturbance. Disturbances that are too frequent/intense will hold the whole area in an early-successional

[7] Unfortunately, cover was estimated visually, rather than measured objectively.

state, those that are too infrequent/mild will often allow a whole area to have recovered from disturbances, i.e. be in climax state (Connell 1978).

3.6.2   Mechanism

The increase-when-rare coexistence process with allogenic disturbance arises because, when rare, a species specialising in a particular patch type will increase when a patch of that type appears, as it will have more of its preferred resources initially available, with less intraspecific competition.

Although allogenic disturbance acts primarily as a between-patch mechanism, i.e. small-scale beta niche, disturbance also generates environmental fluctuation, notably in light in forest gaps, but also in mineral nutrients and water (e.g. Beckage et al. 2008). This can form the basis of relative non-linearity and temporal storage effect mechanisms (Section 3.3), operating at an alpha scale, i.e. without requiring differentiation between patches (Roxburgh et al. 2004; Miller and Chesson 2009; Gravel et al. 2010).

In a sense, allogenic disturbance should not be counted as a mechanism of coexistence; it is small-scale beta-niche differentiation. We do so here because it is so frequently seen as one, because of the impossibility of defining the target scale for coexistence, because disturbances occur on all scales so that however small a scale is examined there will still be disturbances within it, and because when ecologists look at vegetation and wonder about the Paradox of the Plankton, it is one of the answers. The characteristics of disturbance as opposed to environmental fluctuation are: (a) plants not only fail to reproduce but are killed, at least above-ground; (b) most species are killed, not just those that cannot tolerate a particular stress; (c) the event is sudden and (d) the environmental effect is temporary, i.e. it is a pulse perturbation, so the original species can reestablish the composition of the patch. However, the real difference is that allogenic disturbance is a between-patch mechanism. Chesson and Huntly (1997) recognised this, suggesting the term 'successional mosaic hypothesis' (cf. 'intermediate disturbance hypothesis') to describe the process.

3.6.3   Importance

Disturbance is common, creating gaps over the landscape at a range of spatial grain from worm casts (about $0.0009 \, m^2$) up to meteor hits (many square kilometres). Fossorial rodents, ants and termites act at scales that can be important for single ramets. A good example is McGinley et al.'s (1994) description of nutrient-enriched harvester ant mounds in western Texas.

However, ecologists rarely have any idea how much of the small-scale patchiness that they see around them comprises different stages of recovery from allogenic disturbance. How much of a community comprises gap? In a forest, since old trees tend to fall, and since disturbance is endemic, the answer is 'all', but the question really is, in how much of the area is species composition still affected by gap formation, i.e. not at climax?

Since forests comprise patches with a range of postdisturbance ages, it is difficult to say what percentage of a forest's area is 'gap', but estimates have been made (Table 3.5).

Table 3.5 Records of the fraction of forest landscapes comprising gaps

Forest type	Location	Area of the forest covered by gaps	Gap area	Subcanopy spp. recorded?	Source
Fagus/Fraxinus/ Quercus/Ulmus	Denmark	2%	Mean 478 m^2	No	Emborg et al. (2000)
Mixed evergreen coniferous/ deciduous angiosperm	Northern Japan	28% (formed at 0.8% of the area per year)	Mean 326 m^2	No	Kubota (2000)
Picea abies	Northern Finland	43%	20% <100 m^2	No	Caron et al. (2009)
Deciduous, with *Fraxinus americana, Acer saccharum*, etc.	Continental-climate Indiana, USA	39% gap (formed at 1.5 of the area per year)	Mean 295 m^2	No	Cowell et al. (2010)
Mixed *Fagus sylvatica*	Slovakia	11.3% gap	46% 5–50 m^2, 22% 50–100 m^2	No	Kucbel et al. (2010)
Quercus petraea/ Fagus sylvatica	Romania	12.8% of the area comprised canopy gaps	60% <100 m^2, 34% 100–300 m^2	No	Petritan et al. (2013)
A range of forest types	Czech Republic	25–50% 'growth stage', defined by the size distribution of live & dead trees	Mean between types 570–800 m^2	No	Král et al. (2014)

There may be a tendency for gaps to comprise a larger percentage of the forest at higher latitudes, but different ways of interpreting 'gap' make this tentative. How much these gaps contribute to alpha richness, i.e. to solving the Paradox of the Plankton, depends on whether the associated subcanopy flora is different in gaps, which none of these studies recorded. Emborg et al. (2000) did define the 'innovation' phase (2 per cent of the area) as containing herbs and shrubs, which later phases did not. The other question is whether the gaps are of sufficiently small grain to promote coexistence. Taking 20×20 m as an arbitrary coexistence focus for forests, disturbances in several of the forests in Table 3.5 are sufficiently local. Nevertheless, the information is frustratingly incomplete.

Where gap formation has been followed over time, the length of the gap cycle can be estimated: 66 years (Indiana; Cowell et al. 2010), 123 years (Japan; Kubota 2000), 284 years (Denmark; Emborg et al. 2000). These calculations assume that gaps occur in older areas rather than at random, but this may not be far from the truth.

The work of Šamonil and colleagues in *Fagus sylvatica* (beech)/*Picea abies* (Norway spruce)/*Abies alba* (silver fir) forest of the Žofínský Prales and Razula reserves, Czech

Republic, gives some of the best evidence (Šamonil et al. 2013a). Demographic censuses were made in 1975 and at intervals since. In general, the major canopy species reinvade any single- or few-tree gaps, though other sparse trees, such as *Sorbus aucuparia* (rowan) and *Salix caprea* (goat willow), are present after major storm blowdowns (which are not relevant to our question of local-scale coexistence), and to some extent in smaller gaps (Šebková et al. 2012). Šamonil et al. (2013b) estimated mean time between disturbances at the two sites of c. 870 years and 1250 years, respectively. Trees also die standing, about 1.5 times as often as they die due to a disturbance. This indicates that the trees live for c. 400 years before they die one way or the other, though of course small, suppressed trees die younger. Indeed, the oldest trees present of the three major canopy species were 450–500 years old. Unfortunately there is no record of whether different understorey species are present in gaps.

In pastures too, it is possible that much of the variation in species composition is due to past disturbances where vegetation cover had been regained and obvious pioneers have been eliminated, but differences in species composition remained. Perhaps ecologists do not realise this because they often fail to recognise mid-succession species as such (Veblen and Stewart 1982). Perhaps King (1977) was able to recognise only 3 pioneer species out of 58 in grassland because patches at an intermediate stage of recovery from ant disturbance were not recognisable. Research to investigate the importance of disturbance in coexistence would not be too hard with permanent plots in herbaceous communities and using dendrochronology as well as permanent plots in forests.

In their review of the topic Shea et al. (2004) documented many studies that had described patterns of peaked diversity under intermediate disturbance regimes, but that efforts to identify the mechanisms underlying those patterns were sorely lacking. Autogenic disturbance is probably important for coexistence, and the theory is well-founded, but there are few data.

3.7   Interference/Dispersal Tradeoff

This concept originated simultaneously with Skellam (1951) and Hutchinson (1951). It has been known under a variety of names (Wilson 1990), including 'Life History Differences', 'fugitive coexistence' and the endearing if not entirely accurate 'Musical Chairs' (Crawley 1986). Consider a model in which two annual species occupy single-plant safe sites. Species **C** has the greater interference ability, and eliminates the weaker **D** if it reaches a site, but it has less efficient reproduction/dispersal than species **D** and therefore fails to reach some sites. Species **D** has better dispersal and is therefore available to colonise most of the sites that **C** has not reached. This gives increase-when-rare because if **C** becomes sparse, there are many empty sites for its offspring to occupy and its population growth rate increases; similarly if **D** becomes sparse, there are many sites left over by **C** for it to occupy. The mechanism can be distinguished from ('1') Niche differentiation in that no differences between species in resource use are required. It can be distinguished from ('5') allogenic disturbance in that: (a) the gaps are caused by monocarpic or seasonal death, not necessarily by external disturbance, though that is possible[8] and (b) species **C** is limited only by dispersal, not by its ability to tolerate the environment of the gap. It can be distinguished from ('9') equal chance in that, though there is a random element, the dispersal and interference abilities of the two species are very different.

There have been many mathematical models of the interference/dispersal tradeoff mechanism, e.g. Levins and Culver (1971), Nee and May (1992) and Tilman (1994). There is an assumption of a negative correlation, due to a tradeoff, between interference ability and dispersal ability, but Ehrlén and van Groenendael (1998) surveyed the literature and found that this was common. Turnbull et al. (1999) demonstrated the mechanism experimentally by sowing seven species from a limestone grassland, ranging from a seed mass of 0.013–0.16 g, back into that grassland. When the seeds were sown at a high density, 83 per cent of the resulting plants were from the three species with the largest seeds, but when a low density was sown, this percentage was reduced to 49 per cent. This is entirely compatible with the interference/dispersal tradeoff mechanism: when there were enough seeds to reach almost all microsites, the three big-seeded, strongly interfering species occupied them, but when fewer seeds were sown, there were microsites not occupied by the big three, which the light-seeded, probably well-dispersed species could occupy. The unlikely interference/dispersal tradeoff mechanism was demonstrated! However, seed size is not necessarily correlated with early interference ability (Eriksson 2005).

The importance of the interference/dispersal tradeoff is uncertain, and more research would be valuable. Negative correlations between dispersability and interference ability, such as that found by Ehrlén and van Groenendael (1998), have been found using proxies for both variates, such as reproductive effort for dispersal and seed size for interference ability. Direct measures would be excellent, though hard to obtain. Further experiments of the Turnbull et al. (1999) type would be useful. Observation of natural gap colonisation in the field would be difficult, but also very useful.

3.8   Initial Patch Composition

The coexistence model of Levin (1974) is that two species occupy small, transient patches. Some patches will by chance have more ramets of one species than the other. The species in the majority will suppress the other in that patch if intraspecific interference is less than interspecific interference, the 'Heteromyopia' mechanism of Murrell and Law (2003). The latter condition is beloved of ecological modellers, but it seems unlikely in the real world. It could theoretically arise with mutual species-specific allelopathy, but overall we do not believe this model is applicable to plants (or, perhaps, at all).

---

[8] Indeed, interference/dispersal trade-offs form the basis of many theoretical studies of coexistence under allogenic disturbance (Roxburgh et al. 2004; Shea et al. 2004).

3.9   Cyclic Succession: Movement of Community Phases

This well-known topic will be covered in more detail in Section 4.4. The increase-when-rare mechanism is similar to that of ('4') circular interference networks, but differs in three ways:

(a)   Circular interference networks comprise transitions between individual species whereas cyclic succession involves transitions between whole communities, though in many of Watt's (1947) examples of mosaics that are caused by cyclic succession there are communities that comprise one species.

(b)   Both involve reaction, but in circular interference networks the modification may be temporary, for example shading or aerial autoallelopathy, whereas the reaction in cyclic succession will generally modify the soil, chemically or in the soil microflora.

(c)   There could be cyclic succession between just two phases, whereas there cannot logically be a circular interference network with fewer than three species.

A mosaic arises because a phase of the cycle that is replaced at one point appears elsewhere; we therefore count it as a mechanism that uses movement to escape exclusion by interference. This also means that the mechanism is dependent on the spatial focus: the scale examined has to be one that includes patches of the mosaic in different phases. However, as we discuss in Chapter 4, evidence for cyclic succession has been remarkably elusive.

3.10   Equal Chance: Neutrality

It is a longstanding idea that there is an element of chance in which a species occurs at a spot (Lippmaa 1939). Combined with near-identical interference abilities this gives an equalising mechanism: any one of a number of species is equally likely to occupy a particular microsite, due to chance dispersal (Schulz 1960). This has been invoked especially for tropical rain forests (e.g. Schulz 1960; Hubbell and Foster 1986). Sale (1977) described it as a 'lottery', Connell (1978) proposed it as the 'equal chance' mechanism of coexistence and Barot and Gignoux (2004) as 'small fitness differences' with 'demographic stochasticity' and 'ecological drift': all helpful terms.

It is impossible to prove the operation of chance, but some have implicated it. Ultimately every plant has established itself by a process that can be explained by its dispersal, tolerances (beta niche) and the environmental conditions prevailing during its lifetime, but the reasons for a species' presence in a particular spot are usually obscure. In that sense, chance does not exist, but refers to processes that are so complex as to be unpredictable in practice. Dispersal by eddy diffusion is one example. Propagule germination and young-plant establishment can depend on equally unpredictable climatic and disturbance events, and in a longer-lived plant these historical causes will no longer be observable. Chance is then a tempting explanation. Equal chance implies equal interference abilities – the 'equivalence of competitors' concept – that whilst interference is often intense, many species are similar in their interference ability (Goldberg and Werner 1983).

In New Zealand, Veblen and Stewart (1980) used equal chance as a speculative explanation for the colonisation of canopy gaps by any of the trees *Dacrydium cupressinum* (rimu), *Weinmannia racemosa* (kamahi) or *Metrosideros umbellata* (southern rata), depending on seed/seedling availability, mast seeding and the ability of many New Zealand tree species to remain as suppressed seedlings.

Equal chance would result in variation in the species composition of communities that was impossible to correlate with any environmental factor, present or past. Some have used this kind of negative result as evidence of chance. McCune and Allen (1985) in forests in Montana, USA; Allen and Peet (1990) in forests in Colorado, USA; and Kazmierczak et al. (1995) in kettle-holes in Poland all found only weak correlation between species composition and the environment and invoked chance. In such work, the weak correlation could be because: (a) there were important environmental factors that had not been measured, (b) some factors measured gave a non-linear response not allowed for by the analysis, or (c) factors were operating at a different spatial scale than that measured or (d) there were historical factors not now measurable. Using equal chance as an excuse for failing to find vegetation/environment correlations is the last resort of the scoundrel.

The most well-known invocation of chance is the Island Biogeography model of MacArthur and Wilson (1963), based on probabilistic immigration and extinction. However, Kelly et al. (1989) and Tangney et al. (1990) could find little evidence for its operation in Lake Manapouri islands, New Zealand. Nesting and chequerboarding have been seen as evidence of community structure and therefore their absence as evidence of chance (Wilson 1988d; Sanders et al. 2003). There are too many assumptions in this argument. Wilson et al. (1992a) sampled the algal flora of virtual islands: intertidal rock pools within a limited area selected for habitat uniformity. The distribution of species agreed closely with that expected at random in nesting, chequerboarding, incidence functions and the distribution of associations. The simplest explanation is that differences in specific composition between the pools are caused by chance, but that is no proof; it is a minimalist default. The best example of chance – no difference between species if one can ever have an example of no difference – is from Munday (2004), who investigated two small congeneric coral reef fish species. Neither in field removal experiments, nor in lab colonisation, nor in field distributions, was there any evidence of niche differentiation or difference in interference ability beyond a priority effect.

The models of terHorst et al. (2010) suggested that coexisting species should either evolve towards divergence in alpha niche, or else converge in resource use, giving interference equivalence. This convergence could happen to whole species, but with the usual problem that most species occur with different associates. It could also happen locally, the opposite of character displacement. Whether local evolutionary convergence could happen faster than exclusion by interference is very doubtful.

Even equal chance's strongest advocates have been equivocal. Hubbell (2005), having emphasised differences in niche between Barro Colorado Island tropical rain forest species, eventually attributed coexistence to dispersal and recruitment limitation, i.e. equal chance. However, he immediately discussed negative density dependence, which the equalising equal chance mechanism cannot give. The equal chance mechanism is simply a statement that all other mechanisms operate more readily if the species differ little in interference ability.

3.11   Spatial Mass Effect

The spatial mass effect refers to the maintenance of a population of a species by constant immigration into a patch where, because of one of filters '**B**' to '**D**' of Section 1.5, either it cannot reproduce, or at a rate insufficient to maintain the population without repeated immigration (Zonneveld 1995). This has also been called the 'sink effect' and 'vicinism'. It allows coexistence because a population of a species with negative growth rate at the site is not excluded, but subsidised by more vigorous populations nearby.

Seed immigration is the most likely type of subsidy, but it is difficult enough to monitor occasional seeds blowing in, and even more difficult to demonstrate that the population into which they are blowing would have RGR < 0.0 without that subsidy. The immigration could also be by rhizomes or stolons. *Populus tremuloides* and related species (aspen) produce root suckers (Barnes 1966), and these can appear beyond the canopy of the tree where there is no chance that they will survive to be self-supporting, let alone sexually reproductive, for example in a lawn.

The spatial mass effect has rarely been quantified. Kunin (1998) examined boundaries between plots with different fertiliser treatments in the 150-year-old Park Grass Experiment. There was a very sharp pH change, within 50 cm of the boundary. Although there were many exceptions, the majority of plots examined (34 out of 51, two-tailed $p = 0.024$) showed higher species richness towards the boundary. The effect was seen especially where the two adjacent plots differed more in species composition. The spatial mass effect is a very likely explanation. It can be seen clearly in extreme cases where the recipient (sink) population does not reproduce at all, like the 13 species of angiosperm that grow in the Lost World Cavern, northern North Island, New Zealand, without any of them ever setting seed (de Lange and Stockley 1987). When the situation is not so clear-cut, careful demographic analysis is required. For example, Watkinson (1985) demonstrated that in a *Cakile edentula* population at the landward end of a sand dune system in Nova Scotia mortality was greater than reproduction; clearly the population survived only because of subsidy of wind- and wave-borne fruits from the more vigorous population near the sea.

We believe the spatial mass effect is widespread. More data and demographic models like that of Watkinson (1985) are needed. The mechanism is difficult to categorise: like an equalising mechanism there is no increase-when-rare process, but like a stabilising mechanism it can maintain coexistence indefinitely – although we note the immigrant propagules must come from another stable community, almost always a multispecies one where one of the stabilising mechanisms of coexistence must operate.

3.12 Inertia

Spatial and temporal inertia are other equalising mechanisms, slowing exclusion by interference and possibly allowing stabilising mechanisms to operate.

3.12.1 Temporal Inertia

Temporal inertia can be caused by perennial persistence, local dispersal, or a switch.

Perennial persistence: Temporal inertia (Cowles 1901) can be caused by an individual adult plant persisting for some time when conditions are temporarily or permanently unsuitable for it. Inertia caused by the persistent life of trees must be common. Several examples can be found in the literature. Abrams and Scott (1989) describe a situation where, until disturbance occurs, early-successional trees dominate the canopy suppressing young plants of later successional stages beneath them, and their model diagram shows the high species richness resulting at this stage. This is one meaning of the Initial Floristic Composition model of Egler (1954). The decade-long dominance phases, with smooth increases and decreases, that Watt (1981) found in the Breckland may be partly due to such inertia in successive 'regimes' of *Festuca ovina* tussocks, *Pilosella officinarum* ($\equiv$ *Hieracium pilosella*) clones and then *Thymus polytrichus* ($\equiv$ *Thymus drucei*) subshrubs.

In dryland vegetation, where the rainfall is erratic and temperatures high, temporal inertia is obvious. Extremely long-lived ramets of slow growth might establish only during a rare event, perhaps a flood or a 1/100 year wet season, occurring stochastically yet within the lifespan of the plants. Annuals might grow after rain events, but persist between rains via the seedbank.[9] The establishment of many succulents is a very rare event, as exemplified by *Agave macroacantha* in Mexico (Arizaga and Ezcurra 2002). Sufficient rainfall for the development of surface root hairs in established succulents is also rare, yet by means of massive storage, CAM photosynthesis and often a high albedo, they can withstand years of unavailable moisture. Deep-rooted shrubs and trees survive by using water from deep in the soil. Clarke (2002) described woody dryland vegetation in southwestern Australia where no natural recruitment of shrubs was observed over five years. However, the rare event that allows establishment might be a different grazing regime, as Prins and van der Jeugd (1993) found in Tanzania. There, two pandemics in the herbivores in the 1880s (rinderpest) and 1961 (anthrax) temporarily reduced browsing and allowed even-aged stands of *Acacia tortilis* (umbrella thorn) to establish. These are now a conspicuous and apparently integral part of the vegetation of national parks in the area, yet are present through inertia, without recruitment to maintain the populations.

Local Dispersal: Inertia by perennation may not apply to all the species in a community, since many contain species that differ markedly in establishment and survival probabilities, perhaps with perennials of lifespan differing by one or two orders

---

[9] Noting seedbanks can also contribute to stabilising mechanisms, Section 3.3.

of magnitude. When a tree falls there is a natural tendency for its propagules or suppressed seedlings ('oskars') to be the most numerous in the canopy gap created. Dalling et al. (1998) found tree seedlings on Barro Colorado Island to be denser near to a conspecific adult. The species there differed little in the effect of light intensity on growth and they declared that differential responses to soil and topography were rare. Moreover, the correlation between parent and juvenile positions was weaker for species with small disseminules, probably more widely dispersed. These effects comprise a dispersal switch (Section 4.5.4.7), causing inertia at the population level.

Switch: A plant may construct a niche that favours its own juveniles and thus prolongs the life of the population: an environmental switch (Chapter 4), but with the environmental modification insufficient to retain the species in the site long-term. This is the delaying outcome of a switch.

The ingress of species with superior interference ability is slowed by temporal inertia. In a sense, it is a temporal mass effect: the persistence of modules from previous times. Inertia due to the genet/ramet persistence of a single tree, a dispersal switch with seeds, and the persistence of a clone through vegetative reproduction (basically a vegetative dispersal switch), are all basically the same, since trees are populations of modules and the concept of 'individual' has no consistent meaning (Chapter 1). Inertia may not have a sudden end; there may be a reproducing but gradually declining population (Harper 1967). Priority effects (Section 5.4.2) can be seen as a particular example of temporal inertia.

Temporal inertia must almost always occur. It could be observed through experiments that changed the environment and observed how long plants/populations persist.

3.12.2 Spatial Inertia

Many of the 12 mechanisms here give aggregation, perhaps caused by: (a) establishment of the population by dispersal, when several propagules arrive together, say on an animal (an 'ecological founder effect'), (b) phalanx invasion, or (c) a changing environment that disfavours the species, with some patches remaining marginally more suitable. This spatial aggregation is basically the same as temporal inertia, delaying exclusion by interference since it occurs only at patch boundaries.

Stoll and Prati (2001) demonstrated beautifully the slowing of exclusion by interference caused by experimental aggregation. Among four annuals they found that the species with least interference ability (*Cardamine hirsuta*, bitter-cress) decreased its biomass over the experiment to 6 per cent of its monoculture value when planted in a random arrangement with other species, but only to 26 per cent in an aggregated arrangement. The species with highest interference ability (*Stellaria media*, chickweed) increased its biomass to 324 per cent of its monoculture value in a random arrangement but to only 239 per cent when aggregated.[10] This would be a most potent mechanism for delaying exclusion by interference of a subservient species, as for *C. hirsuta* above.

---

[10] All this is in the high-density treatment.

Rebele (2000) found a similar, but very slight, effect in an outdoor mesocosm experiment using mixtures of *Calamagrostis epigejos* (reed) and *Solidago canadensis* (goldenrod). Naturally, such effects can be seen in the field too, e.g. Raventos et al.'s (2010) study of *Ulex parviflorus*-dominated Mediterranean-climate shrubland in Spain. More experiments of this type are needed.

Spatial inertia is seen also in the experiment of Thórhallsdóttir (1990), planting a hexagonal grid of adjacent outdoor plots. Each plot contained one of six meadow species: *Agrostis stolonifera* (creeping bent), *Holcus lanatus* (Yorkshire fog), *Cynosurus cristatus* (crested dog's tail), *Poa trivialis* (meadow grass), *Lolium perenne* (ryegrass) and *Trifolium repens* (white clover). Silvertown et al. (1992) ran simulations to see in retrospect what effect aggregation would have, given the invasion rates that Thórhallsdóttir found between the pairs of grass species. When the species were intermixed in a random pattern, the species with the weakest interference ability, *Lolium perenne*, had almost disappeared after 50 time periods (reduced from 20 per cent to 1 per cent), but with the species 'planted' in bands, depending on the order of the species in the bands, it decreased only to 9 per cent, stayed at 20 per cent or even increased slightly to 21 per cent.

Aggregation might also delay exclusion by interference via effects on herbivory (Parmesan 2000), fire spread (Hochberg et al. 1994) and other environmental factors.

3.12.3   Conclusions

For either type of inertia, we have to ask: what was the original coexistence due to? If there is no coexistence, Inertia cannot prolong it. In a sense, inertia is not an answer, but a sign that the question is wrong, the timescale being used is too short for the life history of the plants, or too short for the effects of aggregation to disappear.

3.13   Coevolution of Interference Ability

Aarssen (1983) suggested that in a mixture of two species there would be stronger selection pressure on the species with lower competitive ability,[11] which would cause it to become the stronger competitor of the two: 'Superiority in competition therefore alternates between…members of the two populations'. It is difficult to believe in these continual increases in competitive ability, as Aarssen (1985) has since concluded. Vasseur et al. (2011) modelled a more likely mechanism of coexistence by coevolution: given a genetic tradeoff between traits conferring interference ability against plants of the same species and those conferring interference ability against neighbouring species, then when a species is sparse in the community there will be selection pressure on it to evolve increased interference ability against other species, and thus to increase when rare, at the expense of its self-interference ability. However, in Vasseur et al.'s model

---

[11] Presumably this operates with any type of interference.

there was only one neighbouring species, and one could envisage traits that increased interference ability against it. In a multispecies community, could a sparse species increase its interference ability against all of them simultaneously? terHorst et al.'s (2010) model (see Section 3.10) is also believable, with explicit changes in resource use rather than just an arms race for the same resources, though any change in interference ability must represent to some extent a change in resource use.

The major problems with such mechanisms are: (a) whether genetic change could happen rapidly enough, presupposing there is sufficient and relevant genetic variation within each population on which selection can act (Becks et al. 2010) and (b) whether it does in practice, especially since the plastic response to interference can give a buffering effect.

Rapid selection was seen in the experiment of Walley et al. (1974), selecting within six months the few copper-tolerant genotypes of *Agrostis capillaris* ($\equiv$ *A. tenuis*; bent) from commercial seed. In the field, Snaydon and Davies (1982) found genetic change in *Anthoxanthum odoratum* (sweet vernal grass) within six (probably 3–4) years of a change in liming treatment in plots of the Park Grass experiment. This was surely, like Walley et al.'s study, selection from existing genetic variation, not of mutations. But is the selective differential among associated species anywhere near the strength of that from a copper-laden soil, or even an increase in pH? The selection pressure in these cases was probably abiotic, though Snaydon and Davies (1982) suggested that in the Park Grass case it might be change in interference from other species: coevolution at least in the *A. odoratum*. If so, this is coevolution at the speed required by the Aarssen and Vasseur models.

If such selection operates in practice, small-scale and population-level genetic differences within species that corresponded to their associates would be expected, though such biotic selection may be thwarted by fluctuation in species associations. In any case, evidence is very sparse. Lüscher et al. (1992) reported adaptation of *Trifolium repens* (white clover) to different ecotypes of *Lolium perenne* (ryegrass) in French cattle-grazed pastures, though the complications found subsequent to a comparable report by Turkington and Harper (1979b) of adaptation of *T. repens* to different grass species serves as a warning that there might be alternative explanations. However, neither the Lüscher study, nor that of McNeilly and Roose (1996) in 40+-year-old English pasture, could find evidence of coadaptation between neighbouring ecotypes of associated *L. perenne*. Martin and Harding (1981), collecting seed from populations of two species of *Erodium* in California and growing the species in mixture in a greenhouse, found a suggestion of eventual exclusion by interference in one mixture of allopatric populations, of stability in one mixture of sympatric populations, and no evidence of either in two other mixtures. This hints at coevolutionary adaptation, but in view of the small number of populations the conclusion must be tentative. The best evidence is experimental, and Aarssen (1989) indeed found that the interference ability of *Senecio vulgaris* (groundsel) increased over two generations relative to a standard genotype of *Phleum pratense* (Timothy grass) with which it was growing, though evidence for a reciprocal change in *P. pratense* was not sought. Experiments on genetic changes when a species is confronted with different associates are difficult because the changes are probably long-term, because the changes may be subtle and because there may be high error variation.

Thus, evidence is extant, but sparse, for the component processes required for coexistence by coevolution: rapid evolution and coevolution to associated plant species at the same trophic level. We cannot credit Aarssen's mechanism of a continuously escalating arms race, but Vasseur's mechanism is possible, as is that of terHorst. It seems likely that such coevolutionary forces are too weak to be a stabilising mechanism, but could be an equalising one.

3.14   Conclusions

We believe our review covers all the mechanisms by which species might coexist in stable mixtures. Chesson's classification into stabilising versus equalising mechanisms has focussed attention on the fact that some proposed mechanisms of 'coexistence' cannot, in fact, enable long-term coexistence. It has also highlighted what few had recognised, that even though the equalising mechanisms cannot on their own enable stable coexistence between two species, they can reduce the difference in interference ability between species to the extent that a stabilising mechanism can operate.

We must speculate on the importance of each mechanism in order to build-up in our minds a vision of the plant community. The lack of empirical support for initial patch composition (Box 3.1, Mechanism **7**) and coevolution of interference ability (**12**) reflects their lack of realism, though experiments to test them would be possible and interesting. Strict equal chance (**9**) is very unlikely and chance is impossible to prove, but evidence for plants as good as Munday (2004) obtained for fish would be fascinating. Circular interference networks (**4**) are an attractive idea, but essentially undiscovered. They are unlikely to be important though better-quality information is needed, using the species from a natural community and examining how RGR changes as exclusion by interference is approached. Cyclic succession (**8**) is more believable and from time to time fashionable, but it is seldom observed and has never been proven to have taken place. Since Egler's (1977) diatribe against space-for-time substitution there have been many more permanent plots set up and these may eventually give evidence for or against it. A cyclic mechanism via autoallelopathy may be widespread but the soil is an intractable and infinitely complex medium where clear chemical pathways and effects are difficult to prove. There is accumulating evidence of changes in the soil microflora, and possibly the soil fauna, that could operate this way (Section 2.2.2). Interference/dispersal tradeoffs (**6**) and temporal and spatial inertia (**11**) are probably widespread, but there is miniscule evidence for either. Temporal inertia is hard to measure, but evidence may come when a change of environment, natural or experimentally imposed, affects a permanent plot. Stoll and Prati (2001) gave neat evidence of spatial inertia from a very artificial system, and more evidence from more realistic systems would be useful. Inertia is an equalising mechanism. Spatial mass effects (**10**) are also likely widespread, but the main measurement difficulty is that the tail of the typically leptokurtic dispersal function is hard to quantify. The answer may lie in monitoring field experimental mesocosms as receptors, or mesocosms transplanted into the field as propagule sources.[12] We can put together a story on pest pressure (**3**) from separate pieces of work (Section 2.7), at least for pathogens, but more experiments applying pesticides to natural communities would be valuable, with examination of all the processes in a single system to aid interpretation through the nexus of interactions involved. Allogenic changes imply that environmental fluctuation (**2**) has huge impacts on plant communities, but it will not always meet the strict criteria for permitting coexistence. Further work to parameterise the models of Chesson (2008) would be valuable. How much of the alpha-niche differentiation (**1**) that can be seen in plant communities is actually causing coexistence is hard to know. Experiments with treatments preventing niche differentiation would be useful, e.g. using shallow boxes to prevent differences in rooting depth or long-term growth-cabinet work with and without seasonal differences. Ecologists rarely have any idea how much of the small-scale patchiness that they see around them is due to allogenic disturbance (**5**), but the research would not be too hard, with permanent plots in herbaceous communities and using dendrochronology as well as permanent plots in forests.

Based on the evidence derived from the present literature, and very tentatively, we list below the mechanisms (numbered as in Box 3.1) in decreasing order of importance:

Major importance
      Alpha-niche ifferentiation (Mechanism 1), stabilising
      Environmental fluctuation (Mechanism 2), stabilising
      Pest pressure (Mechanism 3), stabilising
      Spatial mass effect (Mechanism 10), equalising/stabilising?
      Allogenic disturbance (Mechanism 5), stabilising
Moderate importance
      Interference/dispersal tradeoffs (Mechanism 6), stabilising
      Inertia (Mechanism 11), equalising
Possibly non-existent
      Cyclic succession (Mechanism 8), stabilising
      Circular interference networks (Mechanism 4), stabilising
      Equal chance (Mechanism 9), equalising
      Coevolution of similar interference ability (Mechanism 12), probably
         equalising
      Initial patch composition (Mechanism 7), stabilising

In a changing, disturbed world it will be increasingly difficult to determine where stabilising mechanisms are operating and where equalising ones are, and therefore the search is for stabilising processes rather than achieved stability. The temporal turnover of species in communities depends on some species invading, others disappearing, so that many species in a community may be present by courtesy of one of the equalising mechanisms and will ultimately be doomed. The multiplicity of mechanisms should

---

[12] This is a type of phytometer, an underused technique.

allow the coexistence of a very large number of plant species. The question then becomes: 'Why are there so few species in most habitats?'. A plant, however, is sedentary and extends over a spatial volume, necessarily exposed to a wide range of environmental conditions. It therefore cannot be confined to a precisely defined niche. It is the interplay between the potential for high plant diversity in restricted niches and the necessity for plants to tolerate a wide range of environments that encourages us to look for patterns in plant communities. If adaptations to available niches were most of the reason for every species' occurrence, our enquiry in this book would be less interesting.

Having addressed here the mechanisms that allow species to coexist, we next discuss, in Chapter 4, the processes involved in the accumulation of species into communities.

4 Community-Level Processes

4.1 Introduction

We started our hierarchical description of the makeup of plant communities with the individual plant (Chapter 1), through their interactions in Chapter 2, with a focus on the mechanisms of their coexistence in Chapter 3. In this chapter we address the processes involved in their assemblage. We deal with the interplay of species rather than individuals, for we find species a more recognisable entity and the assembly of species into communities is, after all, the subject of this book. Processes acting on communities, whether leading to change or stasis, can be due to forces external to the species mix (allogenic) or within it (autogenic). We consider allogenic forces, because the environment and its variation are the basis of species filtering and sorting (Section 1.5). However, we concentrate on autogenic forces, those operated by the plants' reaction on the environment, as the cause of the great majority of interactions between plants, and therefore of community-level processes (Clements 1916; Section 1.3).

Processes within the community can have only two outcomes: change or stability. Change should be the expectation, because the plants alter their own environment by reaction. However, stabilising processes do exist, even if strict stability has never been proved. The major contrast (Box 4.1) is between a species/community changing the environment to its disadvantage (i.e. altruistic facilitation leading to succession-towards-climax, the concept though not the term of Clements (1916), or changing the environment to its advantage (i.e. a switch: Odum 1971; Wilson and Agnew 1992). In either case, there can eventually be negative feedback, stabilising an endpoint. We also consider the cyclic succession of Watt (1947), the third type of autogenic community change, with no endpoint (Figure 4.1).

4.2 Allogenic Change

The external force behind allogenic change is normally climatic change. Such change on the timescale of thousands of years, as seen via the pollen record of lakes and mires, has lessons for us in terms of community structure (Section 5.9). Although the climatic changes have generally been assumed to be driven by Milankovitch cycles it has been suggested that peat accumulation in bogs is the primary driver of climatic change, perhaps making cyclic succession that is vast in both spatial and temporal scale (Klinger 1996).

Box 4.1 Types of community change

Allogenic change: the external environment changes, causing species/guild **X** to increase or decrease

Autogenic reaction by species/guild **X** gives a relative disadvantage to itself if:
  • the effect is density-independent: facilitation and/or autointerference
  • the chain of change is linear = succession-towards-climax (Figure 4.1a), then the effect disappears at low density of **X** (negative feedback) = stability
  • the chain of change is circular = cyclic succession (Figure 4.1b)

Autogenic reaction by species/guild **X** gives an absolute or relative advantage to itself:
  • = switch (Figure 4.1c), then the effect disappears at low density of **X** (negative feedback) = stability

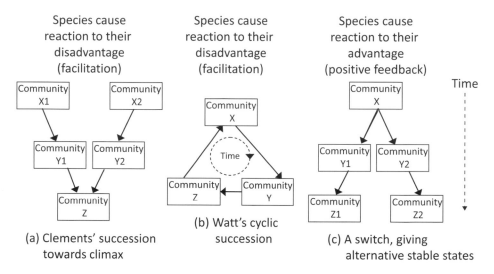

Species cause reaction to their disadvantage (facilitation)

Species cause reaction to their disadvantage (facilitation)

Species cause reaction to their advantage (positive feedback)

(a) Clements' succession towards climax

(b) Watt's cyclic succession

(c) A switch, giving alternative stable states

Figure 4.1 (a–c) Three types of autogenic vegetation change through time. In 'Watt's cyclic succession' some of the 'communities' can be bare substrate.

Fluctuations in the species composition of plant communities from year to year have been named 'annuation' (Clements 1934, pp. 41–42), the 'pulse phenomenon' (Penfould 1964) and 'pulse-phase' (Collins et al. 1987). In some systems annuation is well known, e.g. the response of desert annuals to wet and dry years, and responses to years of exceptional frosts, but investigation of subtler annuation has been hindered by the paucity of datasets with continuous and long records of approximately equilibrium vegetation. Although ecologists expect such fluctuations to be caused by climatic changes, correlations are often hard to see. Allen et al. (1995) found correlation with the rainfall at one of the New Zealand dry grassland sites they examined, but not at four other, nearby sites; and Collins et al. (1987) failed to find correlations between individual species and the climate in dry Oklahoma grassland. However, Peco (1989) was able to show that low rainfall in a single year in Mediterranean savannah produced retreat in whole-community successional change.

There are three notable datasets in which annuation can be examined: Bibury, Breckland and Park Grass. At Bibury, southern England, Dunnett et al. (1998) correlated climatic fluctuations with vegetation changes over 38 years in roadside vegetation. Wet growing-seasons favoured more productive species (C of C-S-R theory: Section 6.5), whereas warm, dry summers increased species typical of environmentally stressed or disturbed sites (S or R). Experimentally determined physiological responses of species to temperature were related to these annuation patterns, but the physiological differences were amplified in the field community (Dunnett and Grime 1999). If we compare geographic distributions across Britain (Preston et al. 2002) with the annuation that Dunnett and Grime observed, *Achillea millefolium* (yarrow) was notably more abundant in unusually warm summers (Grime et al. 1994), and it is distributed almost throughout the British Isles, absent only in a very few cold/wet parts. However, *Ranunculus repens* (buttercup) and *Trifolium repens* (white clover), with almost identical geographical distributions to *A. millefolium* implying similar environmental responses, were less abundant in those years. Similarly, *Brachypodium pinnatum* (tor grass), a species very frequent in chalk/limestone areas with warm summers (July mean $>15°C$), was more abundant in warm summers, but another grass with a very similar climatic and edaphic distribution, *Bromopsis erecta*, was less abundant. This is puzzling.

One of the most fascinating datasets in ecology is Watt's (1981) 38-year record from calcareous 'Grassland A' in Breckland, eastern England. There was a sequence of dominants, from 1936 to 1950 the small-tussock grass *Festuca ovina* (sheep's fescue), from 1950 to 1958 the creeping herb *Pilosella officinarum* (= *Hieracium pilosella*, hawkweed), from 1959 to 1964 the latter two codominant with the procumbent shrub *Thymus polytrichus* (thyme), and from 1965 to the end of recordings in 1973 *T. polytrichus* on its own. These three species all showed a steady increase to a peak and then a steady reduction from it, though when *H. pilosella* decreased it held on to some microsites rather tenaciously. Watt explains these dramatic changes by very subtle changes in the weather, but as at Bibury the correlations are not compelling and in truth these fascinating changes have never been explained. Like Dunnett and Grime (1999), Watt saw competition as important, which it probably is, but build-up of herbivores and disease of species may have played a role in the decreasing phases, and spatial and temporal inertia in slowing these decreases (Section 3.12) must have smoothed the changes.

Usher (1987) analysed Watt's (1960) rather similar 22 years of observations in Breckland acid grassland with Markovian models. However, the predictions of the models are difficult to believe, e.g. that the 'Festuca' cover type, which had been increasing rather steadily over the 22 years up to 37 per cent, would somehow decrease to 20 or 28 per cent in a steady state. Markovian models can make poor long-term predictions, often conservative ones due to spurious reverse transitions caused by minor ecological fluctuations or by recording errors (but of course not in Watt's case!) (Mark and Wilson 2005). Indeed, Usher's matrix suggested that change could occur in either

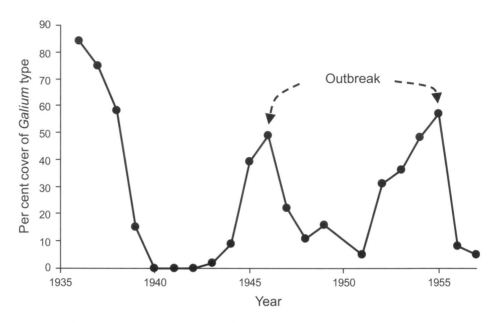

Figure 4.2 Changes in the percentage of the 'Galium' cover type in Watt's acid Breckland grassland.
Data from Usher (1987).

direction between almost any pair of cover states. Further, if transitions or even states come into play that are unrepresented in the history, the predictions will be nonsense. The third problem is that such models assume constant transition probabilities, which is not only intrinsically unlikely but for Watt's record clearly untrue (e.g. the 'Galium' cover type: Figure 4.2). Neither Watt nor Usher had any ecological explanation for these changes. They are also puzzling.

The Park Grass experiment at the Rothamsted Experimental Station, southern England, was established in 1856 and offers an even longer record. Dodd et al. (1995) took 60 years from that record and found that of 43 species recorded on non-acidified plots, 10 showed outbreaks – rapid increases in abundance from a low level followed shortly by an equally rapid decrease to a similarly low level, all within a few years. These species tended to be self-fertilised and suppressed by interference and they interpreted them as ruderals. Outbreaks can be seen at Bibury too (Dunnett and Willis 2004) in species such as *Urtica dioica* (stinging nettle) and *Chamerion angustifolium* (rosebay willowherb), and on shorter timescales in species such as *Galium aparine* (goosegrass) and *Rumex obtusifolius* (dock). Dunnett and Willis say many in the latter group have ruderal traits, confirming the conclusion of Dodd et al., though *U. dioica* and *C. angustifolium* do not. Outbreaks can be seen in Watt's (1960) acid grassland (Figure 4.2). The decade-long changes in Watt's (1981) calcareous site could be compared to such outbreaks, but happening on a longer timescale because of the inertia of the shrubs *Thymus polytrichus*, tussocks (*Festuca ovina*) and clones (*Pilosella officinarum*). Outbreaks of a few years can be caused by single-year climatic variations via community memory in the seed bank or in the persistence of perennials, but explaining decade-long outbreaks is difficult.

The elusive Carousel Theory (van der Maarel and Sykes 1997) addresses change at a fine scale. It appears to say that even if a community remains in equilibrium on a large scale and most species are capable of occupying most microsites, species' presence on a small scale changes from year to year, i.e. in any year each species will leave some potential sites unoccupied, at random. This begs the question of why a species sometimes chooses to be absent and why overall frequencies do not change. The logical implication is that there is niche limitation to the number of species that can coexist locally, and there is evidence for this (Wilson et al. 1995a). van der Maarel and Sykes' observations of turnover were made on a fine scale, e.g. 3×3 cm, which is less than the size of most of the ramets and certainly less than their sphere of influence. The apparent turnover may be a consequence of plant geometry, for example that one year a leaf of a plant overhangs the quadrat, and next year the same plant is present in the same place but its leaves lie in a different direction. The first priority in understanding communities at this scale is to solve the Paradox of the Plankton, and then to find how these mechanisms of coexistence and those of exclusion by interference interact to produce fine-scale patterns. Self-allelopathy by *Hypochaeris radicata* (cat's ear; Section 2.2.2) is one example of the thousands of interactions that must be involved. The Carousel Theory is descriptive, whereas the need is to understand the mechanisms involved.

Allogenic, non-directional changes in plant communities ('annuation', 'pulse-phase') are difficult to deal with. It is clear that they are surprisingly common in vegetation that ecologists tend to assume is constant. However, the changes are driven by the many environmental factors, and additionally there is interference and facilitation between species, and interactions with other trophic levels: pathogens, insect herbivores, pollinators and many others. There will also be inertia and switches operating. Probably for these reasons, all attempts to understand the cause of allogenic change at any particular site have, basically, failed. They will continue to fail until there is a complete model of all the species – plant, animal and microbe – with their response to the environment and interactions with each other and their reactions on the environment. It is a huge task just to find why species composition varies on, e.g. Bibury roadsides, but it is part of the movement away from considering communities as static entities to accepting that change is normal.

4.3   Succession-towards-Climax

4.3.1   Concepts

The term 'succession' applies only to community change caused by the reaction of species on the environment (especially the soil) that results in their demise and/or facilitates the establishment of others. It can be cyclic (Section 4.4), or it can be directional (Section 4.3.3). A monotonic, unbranched gradient in species composition from pioneer stage to climax through associes ('seral stages' we would say today), the

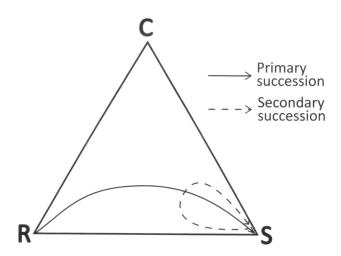

Figure 4.3 The pathways of succession in a stress habitat under Grime's C-S-R theory.

norm for Clements (1916), was termed by Egler (1954) 'relay floristics' and by others 'facilitation succession' or 'succession-to-climax'. Since disturbance will often happen before climax is reached, and since climatic change will make climax a moving target, 'succession-towards-climax' is a better term. As always, Clements saw much, and synthesised his observations into a general pattern. Like all generalisations, there are exceptions, which Clements saw as well as anyone and reported. However, the general pattern that he emphasised is a true one.

4.3.2   Processes

Typical pioneers of secondary succession are well known as short-lived ruderals with low interference ability but high seed output: 'r' (by analogy with the Lotka–Volterra logistic equations: Krebs 1978) and 'R' (ruderal in C-S-R theory where C stands for competitive and S stands for stress tolerant: Grime 2001; Section 6.5 and Figure 4.3). They are replaced by longer-lived species producing fewer but larger seeds and often with vegetative reproduction. But under Grime's theory whilst secondary succession starts in the R corner, primary succession starts in the S corner and, after a trip towards the C-S-R centre, ends up also back in the S corner (Grime 2001; Figure 4.3). Grubb (1987) agrees, pointing out that in primary successions with limited water and nutrients the pioneers are often S species.

Before Egler (1977) pointed out the fraud of space-for-time substitution – 'chronosequences' that have no chrono – there were few attempts to obtain real evidence on succession. One genuine record of primary succession is that on Mount St Helens, Washington State, USA, where ecologists have been studying primary succession on tephra over lava since the volcanic eruption in 1980. Del Moral (2007) concluded that convergence, predicted by the succession-to-climax theory, was seen there only locally. In Buell-Small oldfields in New Jersey, USA, the genuine temporal record of secondary succession shows that superimposed on a steady change in species guilds – annual and perennial herbs, then biennials, then all herbs decreasing as shrubs, lianas and a little later trees increase (Meiners et al. 2007) – there were more dramatic increases and decreases of particular species that were apparently triggered by droughts (Bartha et al. 2003). Change was not continuous.

Although the reaction that every plant has on its environment operates both above- and below-ground, the principal reactions in mesic successions are below-ground. Nutrient pools and soil structure change as roots exploit and die in the soil, whilst the soil fauna uses these resources and incidentally forces nutrient cycling. There is a trend towards soil stratification imposed by surface litter deposition and water movement. Soil systems contain a complex of biotic elements, all of which are leaky, so all components can affect all others through diffusion and by mass flow of solute. Moreover, organisms are typically more closely intermixed below-ground than above-ground. Fungi play a vital role in the nutrition and performance of most higher plants. They also have the potential to provide intimate connections between roots allowing the transfer of mineral nutrients, carbon, possibly water and possibly defence signals (Burke 2012; Babikova et al. 2013).

However, the process behind succession-towards-climax – reaction causing altruistic facilitation (Section 2.3) – has rarely been documented. Walker and Chapin (1986), interpreting zonation up the banks of an Alaskan river as a chronosequence, found that although *Alnus tenuifolia* (alder) raised soil nitrogen by N_2 fixation this facilitation was overwhelmed by shading and root competition for N, giving no net facilitation. *Picea glauca* (white spruce) eventually dominated by its ability to establish in the shade, recalling the tolerance model of succession of Connell and Slatyer (1977). In contrast, Walker et al. (2003a) gave experimental evidence for succession after a volcanic eruption in North Island, New Zealand; the net effect of the N_2-fixing shrub *Coriaria arborea* (tutu) on the succeeding dominant tree *Griselinia littoralis* (broadleaf) was positive, with considerable soil enrichment outweighing the effect of shade.

In spite of all the discussion of succession, remarkably little is known about the processes involved. Realisation of this problem has led to a search for long-term records that can be used to answer the questions, and to the setting-up of permanent plots. This is highly commendable, though it will need to be combined with field and greenhouse experiments to give a full picture.

4.3.3   Ecosystem Trends and Pathways

The new species entering later in succession are usually more functionally diverse, larger and longer-lived than their predecessors, and therefore biodiversity and biomass increase with time (Clements 1916). In Chapter 1 we listed the necessary parts of ecosystems: an energy input apparatus, the capital and transfer of energy, the cycling of elements and rate regulation (Reichle et al. 1975). In succession, available niches are filled, and in particular the canopy closes and leaf area index increases, increasing the energy input apparatus. Biomass increases, usually through storage of carbon in wood. The biodiversity of heterotrophs facilitates nutrient cycling (the resource economy).

Nutrient capital in the system is part of the rate regulation, other rates being set by the physiology (autecology) of the species.

4.3.3.1   Retrogressive Succession

Monotonic increase in ecosystem status is not inevitable. Allogenic disturbances can trigger decreases in all the trends. Autogenic retrogressive succession is also possible. Wardle et al. (2004) compared sequences from six 'chronosequences', up to 4 million years of substitute-time, in different parts of the world and on different substrata; all showed a decrease in biomass. The timescale on which this process would occur makes it difficult to support with evidence, but they say that 'For all the sequences, the decline in basal area became noticeable within thousands to tens of thousands of years after the start.' A paludification switch can cause reduction in biomass and lead to blanket bog and perhaps a raised bog (Section 2.4.3.1; Section 4.5.4.5); Clements (1916, pp. 161–162) cheated, describing this as an effect of disturbance, i.e. flooding. Podsolisation is a rather similar process in which the litter produced by some species, especially gymnosperms, is low in bases and acidifies the soil, resulting in the loss by leaching of the nutrients. Such litter is often slow to break down, sequestering the few nutrients remaining and probably further reducing biomass production.

The patterns are not as clear for soil nutrients as they are for biomass. The sequences described by Wardle et al. (2004) differ in the trends of litter N and P, litter decomposition, etc., though most conform to podsolisation. A change from N-limitation to P-limitation later in succession is generally assumed, and a decrease in soil P is the clearest prediction of Peltzer et al. (2010) for the 'retrogressive' phase of succession, but the reality is unclear. It was verified for a sand dune 'chronosequence' in South West Australia using *Brassica napus* (rape) phytometers (Laliberte et al. 2012). One of Wardle et al.'s (2004) examples is that of Walker et al. (1981), a 'chronosequence' of sand dunes in eastern Australia, and indeed Walker et al. comment that P migrates down the profile during their sand dune succession, which, if the sequence is really a time one as they claim, is true. They also claim that there is a reduction in total P, but extracting the data from their figures the amount of total P through the profile shows a non-significant tendency to increase (Figure 4.4).

Another example given by Wardle et al. is that of Coomes et al. (2005) from the Waitutu terraces in southern New Zealand, contrasting forest sites with shrubland on older terraces. The shrub sites are all in one part of the 'chronosequence' and the forest in another, the terraces manifestly differ in altitude and therefore in temperature, rainfall and thus in leaching, and the terraces were formed from different glaciers depositing potentially different material and having it reworked at different times (Ward 1988). In any case, although in the Waitutu terraces there is a decrease in total P between the forest and shrubland, it is small (shrubland being 73 per cent of forest) and non-significant, the difference in total N being even smaller (84 per cent).

Overall, depletion would be expected for nutrients such as phosphorus that cannot be fixed from the atmosphere except via dust and precipitation, and that can probably be leached faster than they can be weathered from rocks. Tilman's (1982) interpretation for Cedar Creek oldfield grassland, that the resource under competitive demand should

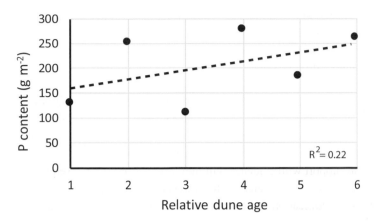

Figure 4.4 Soil P content in sand dunes of Australia; data extracted from Walker et al. (1981).

change from light to nutrients, supports this. Succession towards forest should have a special dynamic since the system is storing nutrients as biomass, potentially depleting soil resources.

Another biomass-decreasing process that could occur late in a succession is salinisation, caused by deep-rooted plants drawing on a subsaline groundwater. Climatic change and soil salinisation were probably the cause of loss of Casuarinaceae forest and an accompanying increase in salt-tolerant chenopods in southwestern New South Wales, Australia, after 4500 years BP (Cupper et al. 2000). The salinisation could have been caused by the transpiration of the forest, in which case the process seems to be autogenic. However, biomass loss caused by autogenic succession seems to be rare. Clements was correct that biomass generally increases in succession.

4.3.3.2   Alternative Pathways

Most earlier workers followed Clements (1916) in envisaging a single pathway to a single endpoint of a 'climax community' in successions. This happened in the Buell–Small oldfields, even if progress was intermittent. However, elsewhere there may be many potential pathways through vegetation types. For the 'chronosequence' in Glacier Bay, Alaska, Fastie (1995) suggested alternative successional pathways in tree species, but admitted that the initial conditions were 'dramatically different' at sites starting their succession at different times, suggested the pathways were dependent on nearby seed sources, and saw the alternative pathways as differences in the rate of species replacements rather than a different species sequence. In spite of the title of Fastie's paper it is doubtful whether these are alternative pathways.

Landform is not the only determinant of successional sequences: Allen (1988) suggested that the causes of differing trajectories of secondary succession in Wyoming's sagebrush rangeland included anthropogenic topsoil alteration, mycorrhizae and alien invaders. Even if all these are uniform, there may also be alternative stable states (ASS; Section 4.6). Still, for all the talk of alternative pathways there is little good evidence, a general lack of alternative pathways must be accepted as the null hypothesis.

Convergence towards climax was part of Clements' (1916) concept. There is little evidence for or against. del Moral (2009) found an increase in correlation of species composition with environmental factors from 1993 to 2008 in potholes on Mount St Helens (using RDA constrained ordination), though this confounds a possible decrease in random influences with a possible increase in control by the local environment. Belleau et al. (2011) used surveys over a span of c. 22 years to extrapolate to changes over 500 years, using a Markov model. The results suggested convergence from four original forest types to open/partially-open *Picea mariana* (black spruce) stands, though the predictions of Markov models are often unrealistic (Section 4.2). Perhaps the best evidence comes from clear-cuts in Amazonian tropical forest, where sites became less similar in tree species composition over two decades of succession (though this was a combination of observations over 11 years and comparisons between different sites: Longworth et al. 2014). Perhaps this just means that pioneers (R species) tend to be consistent over habitats, as C-S-R theory could predict (Section 6.5).

4.3.4   Conclusion

Too many ecologists have reacted to the difficulties of studying vegetation change by rejecting a caricature of Clements' concepts, especially rejecting the word 'climax', and using this as an excuse for lack of investigation. In fact there is very little evidence on the processes of succession. There is essentially none for the existence or absence of alternative pathways. The evidence against convergence to climax would be the existence of alternative stable states, and in spite of much discussion there is practically no evidence of them from higher-plant communities (Section 4.6). Scientific investigation is required, not statements that Clements' concepts were a 'totally incorrect view of life', 'Clements was wrong', with 'the word "climax"...so seeped in...Freudian imagery that its use is best avoided' (Attiwill and Wilson 2003).

4.4   Cyclic Succession: Fact or Romantic Fiction?

4.4.1   Concept

A quite different concept is Watt's (1947) assertion that vegetation/biomass/diversity cycles are widespread, and the usual cause of vegetation mosaics. His British Ecological Society Presidential Address encouraged ecologists to think in dynamic terms and to consider autogenetic community heterogeneity (Section 7.3.2). It fired field ecologists worldwide to look for examples of cyclic succession. The process is that a patch of a major species (maybe the only one) becomes unable to sustain its own presence or abundance and dies, 'moving' laterally, its place being taken often by members of a different guild (e.g. shrubs replaced by bare ground with lichens), the cycle ending at an environment suitable for recolonisation by the first species (Figure 4.1).

Cyclic succession, like succession-towards-climax and a switch, is based on reaction, but unlike succession-towards-climax the state transitions are circular, and unlike a switch the reaction favours another species/community, not the one causing it. This reaction can be altruistic facilitation of the next species, though most of Watt's (1947) examples seem to be of a species unable to maintain its own habitat, rather than directly facilitating the establishment of the next species in the cycle. For example in his Dwarf Callunetum, *Calluna vulgaris* dies out from the windward side (due to desiccation?, or because its size disrupts the airflow?); in Grassland A (Breckland) the *Festuca ovina* tussocks suffer 'death and decay' (the reason is unclear; possibly soil P is exhausted). This parallels the blowdown of *Abies* spp. (fir) trees in wave regeneration: blown over when they become too tall and their sheltering trees are removed (Sato 1994).

The reaction can be on any environmental factor, such as pH, P, wind, via litter or autoallelopathy. Similar effects could be caused by plant pathogen increase at high density of a species; this would probably not be strictly classed as succession because no reaction on the physical environment is involved. It can be difficult to distinguish between these effects in the field.

4.4.2 Patchiness

Patchiness in vegetation comprising shrubby and herbaceous phases has often been interpreted as cyclic. However, solid evidence is hard to find. There are many 'examples' of cyclic succession evidenced by differences in species composition between a forest canopy and the seedlings/samplings below it, but these assume constant climate and ignore that species differ in their lifetime survivorship curves. Seven suggestions have been made for cyclic successions in New Zealand forests, but there is very little supporting evidence for any (Wilson 1990). For example, Mirams (1957) and others suggested that *Agathis australis* (kauri, Araucariaceae), once a valuable timber tree, is unable to regenerate under its own canopy due to the amount, pH and 'harshness' of its litter, and is replaced by angiosperm trees under which the soil changes are reversed and *A. australis* can reinvade. There is only a little evidence for this (e.g. Wyse and Burns 2013) and Ogden (1985) suggests, with evidence, a quite different explanation: that in spite of its huge size *A. australis* is essentially a mid-successional tree, cohorts declining after a few 500-year-long generations.

In arid lands, environmental extremes lead easily to facilitation and hence perhaps to altruistic facilitation, driving a cycle. Perhaps the closest to a demonstration of cyclic succession was provided by Yeaton (1978) in the Northern Chihuahuan Desert, Texas (Figure 4.5). He examined plant densities, the remains of dead plants below others, the vigour of associated plants, the incidence of rodent burrows and soil erosion. The evidence suggests that *Larrea tridentata* (creosote bush), a prolific seeder, can establish in bare ground. Birds perch in its branches and drop seeds of that species and of a cactus, *Opuntia leptocaulis*. The *O. leptocaulis* establishes and grows well under *L. tridentata*, perhaps because of the fine wind-blown soil that accumulates there. The *L. tridentata* bush later dies partly because of competition for soil water by the cactus with its mat of shallow roots. *Opuntia leptocaulis* then dies, partly because rodents burrow below it, and also because soil erosion is liable to occur below its shallow root system without the stabilisation that the deeper roots of *L. tridentata* had provided.

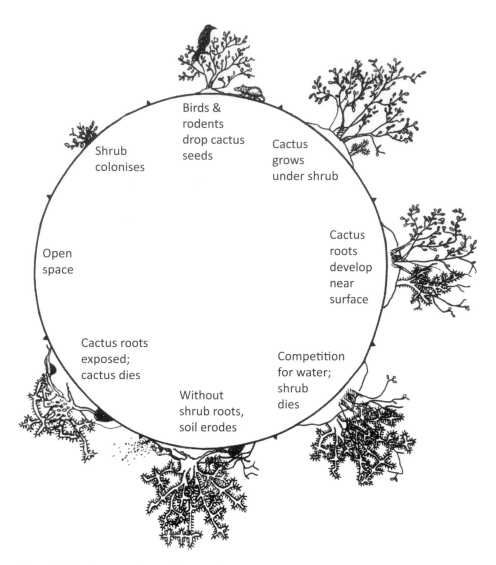

Figure 4.5 The *Larrea tridentata* (creosote bush)/*Opuntia leptocaulis* cyclic succession proposed by Yeaton.
Reprinted with permission from John Wiley and Sons Inc., from Yeaton (1978).

This returns the patch to open ground. The data are observations at one time – space-for-time substitution – rather than a temporally observed cycle, but they build a convincing story. The mechanisms are speculation, but believable.

4.4.3   Small-Scale Negative Feedback

Reactions of species on the soil might occur quite locally, and if negative, lead to either succession-towards-climax or cyclic succession. Proof would be that: (a) species reacted on the environment (probably the soil) and (b) that this reaction is the direction that disfavours the species making it. Demonstrating that the soil is different beneath different species does not prove reaction, but Binkley and Valentine (1991), Alriksson and Eriksson (1998), Pelletier et al. (1999), Ehrenfeld et al. (2001) and Fujinuma et al. (2005) present convincing evidence for it (Section 1.3.2). The second requirement, negative feedback, has certainly been shown, though the exact mechanism is usually unclear. Petermann et al. (2008), in an unfortunately unreplicated experiment, found that 23 out of 24 European grassland species (but not from any particular community) performed more poorly in soils in which that species had previously grown, the 24th showing higher growth in soil conditioned by itself. These negative feedbacks were absent when the soil was sterilised, implying an effect via pests, though they could not pinpoint the organisms involved, for example bacteria, fungus-like organisms or nematodes. Harrison and Bardgett (2010), in northern England, sowed nine species of herbs in 42-litre pots outdoors, in a randomised, replicated experiment. After 16 months' growth, the same species sown into soils conditioned under the same nine species all tended to do worse in their own soil, significantly so for seven of the species. The workers presumed the effects to be due to pathogens, and indeed total fungal and bacterial presence was greater under some plant species than others. Grayston et al. (1998) showed that the soils in which four herbaceous species had grown for four weeks differed considerably in the functional types of microbes present. However, negative feedbacks are not necessarily due to species-specific pathogens: McCarthy-Neumann and Kobe (2010), working with six tree species from tropical rainforest in Costa Rica, found that extracts from soils taken beneath a species and conditioned by seedlings of the same species predominantly had the most negative effect on seedlings of the same species (in 16 cases more negative on the same species, in 2 more negative on another species, in 12 no significant difference). Yet these effects remained after extract sterilisation, indicating that the effects were due to nutrient depletion or allelopathy, not to species-specific pathogens.

Such negative feedback could result in succession-towards-climax, but most of the species in the experiments above seem to be near-climax communities, so it is more likely that they would cause cyclic succession, be it on the scale of one or a few plants, rather as Newman and Rovira (1975) suggested for the demonstrated autoallelopathy of *Hypochaeris radicata*. There has been speculation of cycling at this spatial grain. Turkington and Harper (1979a) suggested that there was cyclic succession between patches of grasses and clover in a pasture, but they provided only very sketchy evidence. Herben et al. (1990), in very careful work in a mountain grassland in Czechoslovakia, found no such cycles.

4.4.4   Conclusions

Cyclic succession is difficult to detect. Many apparent examples of cyclic succession vanish on close examination. Almost all the evidence has been obtained by space-for-time substitution; no cycle has ever been confirmed by observation over time. Indeed, none of Watt's (1947) examples were based on solid evidence, and most have been proved wrong on inspection over long time scales. The cycle involving a hummock–hollow alternation in bogs had been proposed long before Watt's paper, and it has appeared in textbooks, but the weight of evidence is now for the opposite process: a switch, stasis rather than alternation (Wilson and Agnew 1992; Belyea and Clymo 1998). Watt (1947) described cycling driven by the growth phases of *Pteridium aquilinum* (bracken) where a vigorous front of *P. aquilinum* can invade *Calluna vulgaris* (heather) but the *C. vulgaris* could invade an older *P. aquilinum* stand (Watt 1955). However, Marrs and Hicks (1986) found that after several decades the changes that Watt predicted had not happened. In the control plot of acidic grass-heath in Breckland, eastern England, Watt (1947) had proposed that patches of *A. capillaris* ($\equiv$ *A. tenuis*) and *A. canina* (bent) spread, die out in the middle to form a ring, and the patch is eventually recolonised making a cycle. In Watt's (1960) records, those species do indeed appear and then disappear within 10 years in two places in the plot, but they persisted for 20 years elsewhere in the plot. Perhaps these are outbreaks rather than cycles.

We also have to ask why the systems with proposed cycles do not settle down to an equilibrium. In Watt's (1947) *Pteridium aquilinum/Calluna vulgaris* example, why is the equilibrium not scattered *C. vulgaris* with *P. aquilinum* of moderate vigour? In the bog hummock–hollow alternation, the hollow species are supposed to draw level with the hummock ones, and then forge on up until they become hummocks. Why do they not stop when they draw level and the surface is no longer moist enough for fast growth or for hollow species to persist? Cyclic succession is an appealing idea, but some of the proposed mechanisms have been illogical. The most likely line of research is into local cycles driven by species-specific pathogens or other soil changes, but our science is a long way from accumulating for any one system the reaction on the biotic or abiotic factors of the soil, the negative feedback on the species effecting the reaction, and the temporal and spatial cycles in species composition. Cyclic succession may exist, but there is no evidence of it yet.

4.5   Switches: The Positive Feedback Processes

4.5.1   The Concept of the Switch

The term 'switch' was coined by Howard Odum (1971) for the positive feedback process between biota and environment. We see the vegetation–environment switch as a key to plant community ecology. We have emphasised reaction, the effects that plants have on the environment and hence on each other (Clements 1916; Section 1.3). At the community level, there are two opposite results of reaction: (1) The community makes the conditions relatively *less* favourable for itself. This is the basis of Clements' theory of succession-towards-climax and of cyclic succession, both based on altruistic facilitation. (2) The community or a key species makes the conditions *more* favourable for itself, relative to other species. Under '2', reaction reinforces the community's hold on the site by positive environmental feedback (Wilson and Agnew 1992). This switch process has been known by a variety of terms such as 'constructiveness', 'positive feedback', 'self-intensifying effect', 'self-reinforcing trend' and 'bootstrapping'. Switches have been understudied and underemphasised, probably because the opposite process, Clements' theory of succession-towards-climax, still has a strong hold in the ecological psyche even as some affect to reject it. The two elements of a switch are:

Element i. Community **X** changes the environment (reaction).

Element ii. This change is relatively more favourable for community **X** than for community **Y**.

It is very difficult to find hard evidence for both elements 'i' and 'ii' for any single situation, though convincing evidence is emerging for a few cases. The environmental factor changed is usually a physical one, but some switches operate via the biotic environment, i.e. using the 'plant-to-plant interactions mediated by heterotrophs' of Section 2.7 (and Section 4.5.4.7). It would be theoretically possible for a switch to operate via autogenic disturbance (Section 2.5), for example via strangler vines, though we know of no example.

4.5.2   Types of Switch

Wilson and Agnew (1992) identified four types of switch:

1.   One-sided switch: Community **X** changes the environment (e.g. increases the nutrient status or lowers the soil pH) in patches where it is present.

2.   Zero-sum switch: Community **X** changes the environment in its patches and thereby also changes the same environmental factor in the opposite direction in the patches where it is not present (community **Y**), for example by channelling water or nutrients into those patches. We call this 'zero-sum' here because it is caused by there being a limited amount of water or nutrients. (Wilson and Agnew 1992 used 'reaction' switch.)

3.   Symmetric switch: Community **X** changes an environmental factor in its patches and Community **Y** simultaneously changes the same environmental factor in its patches, but in the opposite direction, e.g. one community raises the soil pH and the other lowers it.

4.   Two-factor switch: Community **X** changes an environmental factor in its patches and Community **Y** changes a different environmental factor in its patches, e.g. one community promotes fire and the other casts shade.

4.5.3   Outcomes of Switches

A sharp vegetational boundary without an obvious environmental cause hints that a switch may be operating. However, there are four possible ecological outcomes of the operation of switches (Figure 4.6):

1.   Stable mosaic outcome: In a previously uniform environment, a switch of types 2–4 can create a stable mosaic of communities, separated by sharp boundaries. On a newly formed or just-denuded surface, Community **X** establishes at some

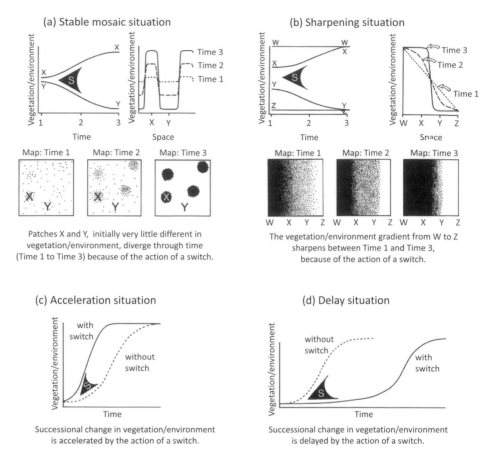

Figure 4.6 Four possible outcomes of vegetation switches. (a) stable mosaic situation; (b) sharpening situation; (c) acceleration situation; (d) delay situation. In 'a' and 'b', the figures start when a small initial difference has already appeared.
Reprinted from Wilson and Agnew (1992), with permission from Elsevier.

sites by chance dispersal or by temporary environmental differences such as disturbance, and its reaction ensures it holds the sites. The environmental differences can also be intrinsic but subtle, magnified by reaction, effectively small-grain boundary sharpening ('2') below. Elsewhere, Community **Y** establishes, and a different reaction holds its sites. These are alternative stable states (Section 4.6). A one-sided switch cannot give a stable mosaic because community **X** can further invade the unmodified environment.

2.   Sharpening situation: A switch (of any of the four types) can convert a gradient in environment and vegetation into a sharp vegetation boundary.

3.   Acceleration situation: If a switch (of any of the four types) is operated by an invading species/community, the invasion can be accelerated. The invading species/community **Y** changes the environment to make it more suitable for community **Y**, and less suitable for the previous community, **X**.

4. Delay situation: Existing species could delay vegetation change via a switch (of any of the four types). For example, a switch could delay succession, prolong the effect of initial patch composition, or delay the response to climatic change. This operates by community **X** changing the environment to keep it more suitable for community (e.g. seral stage) **X** and less suitable for the succeeding community, **Y**. The delaying effect of switches is one mechanism of temporal inertia (Von Holle et al. 2003; Section 3.12.1).

If a switch results in ASSs, they are technically stable in the Lyapunov (=Liapunov) sense, i.e. the state will be restored after a small perturbation. However, a large perturbation may cause a change to another of the stable states, the process of 'state change' (cf. Brooks et al. 2004).

4.5.4   Mediating Agents

We may generalize that there are seven types of agent (mechanisms) that can cause a switch. We outline examples briefly here, especially those reported in more recent literature, to complement the fuller review in Wilson and Agnew (1992).

4.5.4.1   Mechanisms Involving Regional Climate

Suggestions of switches on a grand scale are relatively recent. For example, Miller et al. (2005) proposed that human firing of the vegetation in the centre of Australia in the early Holocene changed the largely forest/shrub cover to more open desert-scrub. This was both more fire-promoting and more fire-tolerant, keeping the fire frequency high. A general circulation model suggested that resulting increased surface albedo and decrease in surface roughness would reduce the extent to which monsoonal precipitation penetrated inland, and this climatic change would further increase the fire frequency. These processes would hold the vegetation cover in its changed state.

Models indicate that forest vegetation, with different albedo, wind resistance (roughness), energy flow and evapotranspiration (O'Brien 1996), results in a region-ally warmer climate. In boreal areas at least, this will favour the growth of forest and give a regional- or global-scale switch (Bonan et al. 1992; Otto-Bliesner and Upchurch 1997; Bergengren et al. 2001). According to Gaia theory, this is the reason earth is habitable at all, unlike Venus or Mars (Lovelock 1979). In the tropics, the rise of angiosperms with their faster transpiration may have resulted in a cooler and moister climate, enabling their spread (Boyce et al. 2010). Other climatic, regional-scale feedbacks have been proposed, e.g. involving mires, monsoons and desertification.

Even though regional-level climatic switches involve the same agents as those in other categories, we list them separately for their spatial extents, orders of magnitude greater than our other examples. They are necessarily more speculative, being based on word or simulation models.

4.5.4.2   Mechanisms Involving Water

Water is an obvious mediating agent in switches. Trees, and to a lesser extent stiff, erect tussock grasses and even soft flag-form tussocks and short grassland, can trap 'occult precipitation' from fog/clouds. The increased water input allows the vegetation to grow faster, to be less affected by drought and therefore to expand (Figures 4.7 and 4.8). Naturally this is a feature of dry/arid climates. A similar switch is theoretically possible through P input via fog or dust (DeLonge et al. 2008), but there is little evidence.

Scott and Hansell (2002) describe *Picea glauca* (white spruce) islands in Manitoba (Canada) tundra where (isolated) young trees trap snow, which ameliorates low moisture and temperature. They grow well and peripheral branches root to form woody islands. However, as they age more, branches droop, needles abrade and the canopy thins, allowing lichen heath to establish and permafrost to enter the island. So the young trees operate the switch, but old trees do not. It is unclear how the young trees establish. This is a one-sided switch, but unusually for a one-sided switch it leads to a mosaic because the processes driving the switch no longer operate as the trees age.

The obviously patchy nature of much vegetation under low and erratic rainfall regimes points to the possibility of switches, most notably 'tiger bush': lines of dense shrub/tree plant growth at right angles to a gently sloping aspect with the areas between sparsely covered by grasses and their associates, or bare (Tongway et al. 2001). Similar patchiness can be seen in deserts, savannahs, shrublands and grassland, especially on gravelly soil. We (Wilson and Agnew 1992) interpreted such patterns as ASSs of sparse and dense vegetation caused by a switch. During storms water infiltration is impeded in sparse areas, especially if there is a static bare soil surface, perhaps with a stone armour or a cryptogamic crust (Verrecchia et al. 1995). Water runs off and onto the dense zones. There, litter reduces the impact of raindrops on the soil surface and roots provide soil channels for infiltration, which can be increased by rodent burrows, soil caches and ant nests, giving much higher soil infiltration rates (Berg and Dunkerley 2004). McDonald et al. (2009) validated this model for a Texas desert: up to 24 per cent of heavy rainfall ran off sparse bands onto dense bands (under showers the excess water can be evaporated from the bare areas instead), giving them double the water input in some years. Nutrients in the runoff water and in aeolian sediment that also moves from sparse to dense bands increase their nutrient capital (Field et al. 2012). Water and nutrient input both reinforce the density of the band vegetation, completing the positive feedback to give an autogenic pattern. Van de Koppel et al. (2002) and Shnerb et al. (2003) modelled the process, showing that the formation of tiger bush occurs in a specific range of precipitation. Wilson and Agnew (1992) interpreted this as a zero-sum switch, because the increase in water and nutrients by the dense phase reduces that available to the other phase.

Other water-mediated agents included the nutrient/pH/waterlogging switch that forms ombrotrophic bogs (Section 2.4.3.1) and water/salt switches that maintain saltmarsh pans.

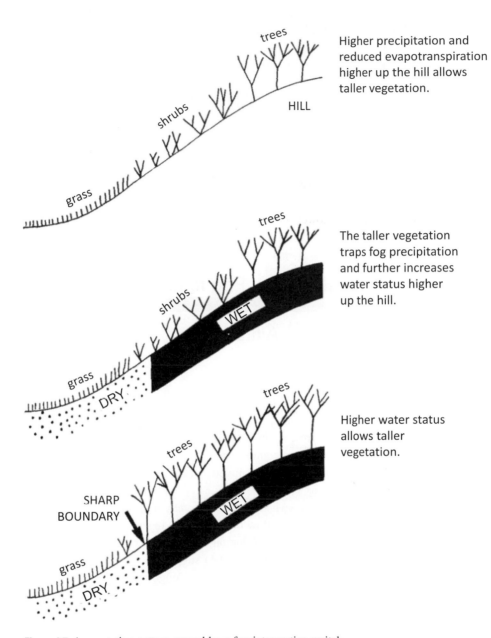

Higher precipitation and reduced evapotranspiration higher up the hill allows taller vegetation.

The taller vegetation traps fog precipitation and further increases water status higher up the hill.

Higher water status allows taller vegetation.

Figure 4.7 A vegetation pattern caused by a fog-interception switch. Reprinted from Wilson and Agnew (1992), with permission from Elsevier.

4.5.4.3   Mechanisms Involving Other Aspects of Microclimate

Montane closed-canopy forest usually ends at an abruptly upper treeline. A tree cover has a lower albedo, which tends to give a lower mean temperature in the plant cover and in the soil (Körner 2003a), and whilst radiation is more readily absorbed it is also more

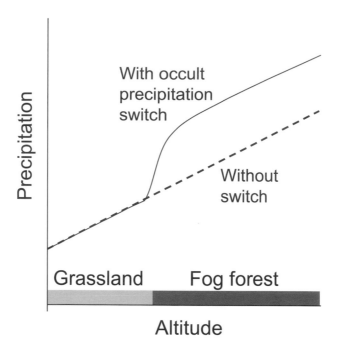

Figure 4.8 Change in precipitation caused by a fog-interception switch.

readily lost (Grace 1983). However, the crucial factor is frost. Most tree species are less tolerant of frost as juveniles than as adults. They can therefore regenerate under their own canopy where frost is ameliorated but only with difficulty beyond: a temperature switch. Such switches are common among shrubs and trees in alpine and arctic areas, for example the maintenance of inverted treelines in Australian alpine environments (Moore and Williams 1976).

Light can mediate a switch, mainly whereby a species, by its canopy, causes low subcanopy light levels that allow that species to regenerate but prevent regeneration by others. This could be the explanation for the rare cases of monodominant tropical forests; for example, Torti et al. (2001) suggest monodominance of the legume tree *Gilbertiodendron dewevrei* (limbali) in the Congo is due partly to low light below it, caused by its dense canopy and deep litter layer. Similarly, *Pteridium aquilinum* (bracken) often produces very dense and impenetrable litter, suppressing other species and forming in monospecific stands. Such light switches seem to be very common.

4.5.4.4   Mechanisms Involving Physical Substrata

Plants can have long-term effects on soil particle size and porosity, and hence soil stability (Angers and Caron 1998). Hellström and Lubke (1993) describe a moving dune system in South Africa that has been colonized by the Australian shrub/small-tree *Acacia cyclops* ('rooikrans' in South Africa). It has stabilised the sand and increased soil nitrogen to the extent that removal of *A. cyclops* will not return the dune to its original state: a physical/chemical substratum switch. Another sediment-entrapment

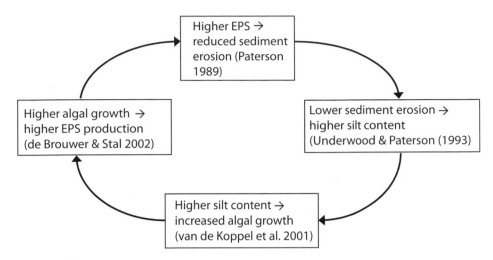

Figure 4.9 A sediment-mediated switch. EPS: extracellular polymeric substances.

switch occurs in estuarine salt marshes (Baustian et al. 2012), often resulting in single-species stands, e.g. of *Spartina anglica* (cordgrass) or *Puccinellia maritima* (a saltmarsh grass). If the entrapment continues this can start a succession, but sometimes a stable vegetation type results. At the other extreme of particle size, roots (perhaps especially ecto-mycorrhizae) and litter can weather rock, improving both physical and chemical aspects of the substratum and thus allowing more plant growth: a switch (Ehrenfeld et al. 2005).

Probably the best-documented case of a sediment switch is for algal growth on estuarine sediments (Figure 4.9). Diatoms (Baccillariophyceae) and cyanobacteria (= Cyanophyta = bluegreen algae) produce extracellular polymeric substances (EPS), largely carbohydrates. These bind the sediment together, leading to lower erosion of the silt component (though the latter seems to be the weakest link in the chain of evidence). This encourages diatom growth. If there were a limited amount of silt in the estuary, the switch would be zero-sum, resulting in a permanent, if shifting, mosaic. That would give bimodality in the frequency of two alternative states, one of the signs of a switch, and this has been observed (van de Koppel et al. 2001).

4.5.4.5 Mechanisms Involving Substratum Chemistry

Patches of more mesic conditions in a nutritionally poor arid environment have been called 'resource islands' and 'islands of fertility', and the plants involved 'nurse plants'. A tree or shrub can facilitate plants and animals beneath it by providing shelter from high temperatures and photoinhibition, and through trapping wind- or water-borne material around its base and acting as a focus for animal activity, thus increasing the nutrient status and water-holding capacity. The plants and animals it facilitates can themselves benefit the nurse plant, enhancing nutrient cycles and even providing favourable sites for its seedlings. This is a composite switch: soil nutrients, physical soil composition, water, wind, temperature, light, grazing and animal dispersal. Such processes have often been described for isolated trees and shrubs in dry grassland. In a series of papers, Belsky (e.g. 1994) has shown how the shade zone of *Acacia tortilis* trees in Kenya is an enhanced soil environment with lower evapotranspiration, dampened temperature fluctuations and a more adequate population of soil microorganisms than the intertree open grassland. This in turn attracts large mammals who defecate, adding to the effect. When seedling establishment takes place around an isolated tree to form a copse, a switch is occurring (Wilson and Agnew 1992). This is a classical cause of vegetation patchiness in a landscape. At the other extreme of rainfall, Agnew et al. (1993a) described a switch mechanism of virgin gymnospermous rain forest in western New Zealand, where the nutritionally rich seedling sites were on fallen tree trunks. Trees were prevented from falling into an adjacent heath-mire by a fringe of tall myrtaceous shrubs. In another example, Lee et al. (2012) suggested that *Microstegium vimineum* (stiltgrass), exotic in Indiana, USA, somehow promoted nitrification in the soil, resulting in the soil N being mainly in the form of nitrate, which was to its benefit compared to native herb species.

Thus, there are several routes to nutrient enhancement that are capable of mediating switches.

In other switches, nutrients are depleted. In pygmy forest in northern California, Northup et al. (1998) found that the leaves and therefore the litter are high in polyphenols and condensed tannins. The decomposition of this litter is slow and the soil is therefore low in mineral nutrients, especially N. These conditions are tolerated by the species that grow there but not by possible invaders. A nutrient-based switch was also suggested by Wurzburger and Hendrick (2009): litter of *Rhododendron maximum* produces tannin–N complexes that are taken up more effectively by the ectomycorrhizal roots of *R. maximum* than by the surrounding species. And in *Picea sitchensis* (sitka spruce) forests in northern Sweden, the low shrub *Empetrum hermaphroditum* (crowberry), already present as understorey, is abundant after cooler forest fires (Mallik 2003). The phenolics in its litter lower the pH (sequestering available nitrogen and leaching out metallic ions) and inhibit the development of conifer mycorrhizae – conditions that it can tolerate – constituting a switch.

Switches in which a species reduces the pH and tolerates the resulting pH environment seem common, e.g. many Ericaceae and *Pinus* species. The acidification can be effected by either their living or dead parts. In Britain, *Calluna vulgaris* (heather) litter is acid, peat-forming and maintains the acidic, oligotrophic nature of the *C. vulgaris* habitat. Many switches combine a lowering of the pH with a lowering of nutrient availability, as in the *Empetrum hermaphroditum* example. Low pH is part of a similar switch in paludification and bog growth (Wilson and Agnew 1992; this volume, Section 2.4.3.1).

Species that interfere with their neighbours by allelopathy, benefitting themselves, and thereby increasing their allelopathic influence, also operate a switch (Odling-Smee et al. 2003). Heavy-metal hyperaccumulators occurring on metaliferous soils accumulate metals (Zn, Cd, Cr, Ni, etc.) to very high concentrations and are tolerant of such levels. Concentrations in their shoots have been recorded in the order of 10,000 times that in the soil solution for Zn and 600 times for Cd (Knight et al. 1997). The litter of these plants might raise the available concentrations of such elements in the surface soil to levels that they can tolerate, but other species cannot (Boyd and Jaffré 2001), reinforcing their abundance in the site – a switch. However, the only rigorous investigation of this, by El Mehdawi et al. (2012), found that the non-hyperaccumulator *Astragalus drummondii* grew better with selenium-hyperaccumulating *A. bisulcatus* than with itself, especially on seleniferous soil: the opposite effect for the proposed switch. An analogous salt-mediated switch operates for a species that absorbs saline water from the lower layers of the soil, increases the salinity of the surface soil by its litter and is able to tolerate this better than its associates. The opposite switch would be if a species reduced the ambient concentration of a toxin (by uptake or detoxification), thus a population at low abundance would be inhibited but at high abundance the population could maintain itself (van der Heide et al. 2010). There is some evidence for this in microbiology, and van der Heide et al. (2008) found evidence for it in ammonium toxicity of *Zostera marina* (eelgrass).

A much studied, though complicated, switch occurs in some shallow lakes that can have two ASSs: clear and turbid (Scheffer et al. 1993). In the clear state, submerged aquatic vegetation is abundant. Nutrients are low, especially P but also N, limiting phytoplankton growth, and supporting only few fish (Figure 4.10). Macrophytes such as members of the Nymphaeaceae (water lilies) can shelter algal herbivores such as *Cladocera* species (water fleas), allelochemicals produced by *Chara* spp. might suppress the algae, and these factors keep the algal populations low (Stansfield et al. 1997; van Donk and van de Bund 2002). The macrophytes and the sparsity of fish suppress water movement, allowing sediment to settle on the bottom. Both features keep the water clear. In the turbid state, fish movement causes water movement, which stirs up sediment giving eutrophy and therefore phytoplankton growth, but the sediment reduces light penetration and therefore suppresses submerged macrophyte growth. Food chains are critical. Jones and Sayer (2003) suggest that, in eastern England, fish control the epiphytic grazers, which control the periphyton (algae attached to macrophytes), which controls the macrophytes. That is to say, with too many fish, the periphyton kills the macrophytes, and with no fish the macrophytes are dominant. The switch-mediating agents are basically nutrients, light and fish. Although there are variants, this is an exceptionally well-documented switch.

van Nes et al. (2002) showed by modelling that this clear/turbid mechanism can produce ASSs (Section 4.6). The most critical parameter in their model was the degree to which macrophytes lowered the local turbidity. ASSs developed most readily in shallow lakes with uniform depth, or where there was little mixing, or to a lesser extent where there was little mixing of water between areas of different depth. Evidence that ASSs exist alongside each other in the real world can be seen in similar lakes, some in one state and some in the other, in natural state changes, and in state changes effected by experimental manipulations (e.g. Mitchell 1989; Scheffer et al. 1993). Lakes naturally accumulate nutrients from inwash, and trap them through sedimentation and in macrophytic vegetation. The critical point at which a change from one ASS to another occurs appears to depend on the initial state and the sedimentation rate (Janse 1997), which must have to do with nutrients. Reversal may or may not occur readily (Körner 2001).

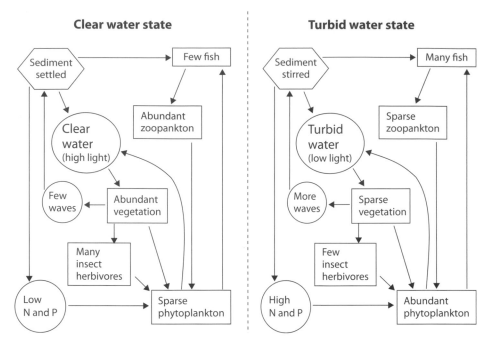

Figure 4.10 A switch mediated by nutrients, light and fish in lakes.

For example, Blindow et al. (1993) described two moderately eutrophic, shallow lakes in southern Sweden that have changed between the two states several times during the past few decades. In both lakes, water level fluctuations were the most common factor causing shifts. Blindow et al. (2002) examined one of these lakes in detail and found that the chlorophyll:total-phosphorus ratio (dissolved, suspended + plankton) declined from April to September, strongly suggesting that phosphorus was not limiting and that the phytoplankton population must be controlled by some other factor, such as carbon limitation and/or allelopathy by the submerged charophytes. Hargeby et al. (2004) recorded a change to the turbid state after there had been 30 years of the clear-water state. It occurred over two years with mild winters that increased fish populations, thus increasing bioturbation and nutrient cycling, combined with cool springs, which hindered the establishment of macrophytes.

Experiments are the key to understanding the workings of communities. Schrage and Downing (2004) removed 75 per cent of the benthivorous fish from a lake. One immediate effect was a decrease in suspended sediment (probably due to reduced fish movement) and reduced phosphate and ammonia (probably due to both the reduced sediment load and reduced fish excretion). There was an increase in zooplankton, which seemed to lead to a reduction in phytoplankton. This led to an increase in water clarity, which may have been the cause of an increase in macrophyte frequency. Thus all the elements of a turbidity switch are present. However the lake reverted after two months to its previous state, so an ASS was not induced.

4.5.4.6   Mechanisms Involving Disturbance

Autogenic disturbance can mediate switches. For example, lianas can facilitate the formation of forest gaps (Section 2.5.3) in which their own life form is favoured, possibly constituting a switch. Seedling establishment in intertidal marshes is generally restricted by tidal disturbance, but Bertness and Yeh (1994) describe how the erect stems of *Iva frutescens* (marsh elder) can trap so much water-borne litter that the underlying marsh vegetation dies, allowing its own seedlings to establish. Reduction in evapotranspiration and thus surface salinity caused by the *I. frutescens* canopy then allows them to survive. This litter-initiated switch maintains *I. frutescens* thickets. Young et al. (2014) describe how on a Pacific atoll *Cocos nucifera* (coconut palm) produces copious, heavy litter that *C. nucifera* seedlings can tolerate (7 per cent mortality) but seedlings of dicot trees cannot (96 per cent mortality), another autogenic disturbance switch producing the monodominance observed in *C. nucifera* patches

Fire is a major disturbance creating vegetation edges because certain vegetation types are fire promoting but fire resistant and indeed dependent on fire for their persistence – the South African fynbos shrubland is an example – whilst other types – such as the adjacent closed forest – do not readily carry fire but are susceptible to it (Manders and Richardson 1992). The species of both vegetation types benefit from increased water availability (Manders and Smith 1992), but only during wet years can tree seedlings establish in the shrubland. The fire regime maintains the balance between the two types. Savannah/closed-forest ecotones, perhaps on an underlying soil water/nutrient gradient, are sharpened by a two-factor fire/shade switch. Fire is infrequent in the forest, but the forest species cannot expand into the savannah because fire is too frequent for them to reach reproductive age (Hoffmann et al. 2009). The critical savannah composition may be 40 per cent cover of trees (Abades et al. 2014). Neither can the closed-forest species invade closed forest because they are shade-intolerant. The operation of a switch is likely to lead to a bimodal distribution of vegetation types, and this is true for savannah/ closed forest (Hirota et al. 2011).

Humans also disturb. For example, they follow existing paths in vegetation. The effect on vegetation is to reduce plant height and productivity, which in turn invite foot passage, giving a switch between vegetation types (Wilson and King 1995).

4.5.4.7   Mechanisms Involving Heterotrophs

Mycorrhizae of different types, e.g. arbuscular and ectotrophic, could mediate a switch if one plant species were dependent on a type and encouraged it. For example, Mangan et al. (2010) grew seedlings of two tropical rainforest trees in a forest soil mixture for eight months, then a second batch of seedlings with soil+roots from the first experiment. The tree species *Apeiba aspera* grew larger with its own inoculum than with inoculum from the other species, and controls in the experiment make it likely this was due to the mycorrhizal flora. Callaway et al. (2004) found that *Centaurea maculosa* (a knapweed) in pots containing soil originally collected from Montana, USA, but in which that species had previously been growing for 3.5 months, grew faster than in soil that had borne *Festuca idahoensis* (blue bunchgrass) during those 3.5 months. Moreover this effect was removed by soil sterilisation. This is clearly a switch mediated by the soil microflora. However, most reported soil feedbacks that seem to be mediated by the microflora have been negative (Section 2.2.2), presumably due to pathogens. The difference in the Callaway et al. experiment may have been that *C. maculosa* is exotic in Montana, where the soil was collected, perhaps releasing it from its specific pathogens and allowing the underlying positive feedback to be seen. Indeed, no positive effects were seen using soil from its native range in France.

Animal invasion can alter plant communities, but this cannot comprise a switch unless the community type both attracts and benefits from the animal exploitation. Much of the material is speculative. We shall consider two major mechanisms: the effect of browsing and grazing, and the effect of frugivores on seed dispersal.

Grazers can be switch-mediating agents if they are encouraged by a particular type of vegetation cover, and promote it. For example, in grass openings in forest, large mammalian grazers crowd in the forest/grassland ecotone to avoid predators (Lamprey 1964). They eat or trample any outlying tree seedlings, preventing the forest from spreading and thus sharpening the ecotone. This has been speculated to be the cause of the apparent ASS of grassland and *Acacia* spp. savannah in East Africa (van de Koppel and Prins 1998). On abandoned agricultural land and rangeland in western Europe, horses and asses (*Equus caballus* and *E. assinus*) defecate in or around preferred sites that they fastidiously avoid for grazing, creating within the meadow thorny patches of *Prunus spinosus* (blackthorn) and *Rubus* spp. (brambles) that protect regenerating trees, leading to a mosaic of forest with grassland openings (Bakker et al. 2004; Kuiters and Slim 2003). This seems to be a zero-sum switch, because the thorny patches divert the equines onto the grassy areas. However, the grass area limits the animal population numbers, and they also defecate at the grass/ shrub ecotone expanding the woody patches: elements of a one-sided switch. Again, Augustine et al. (1998) describe how ASSs can arise when high numbers of white-tailed deer (*Odocoileus virginianus*) trample *Laportea canadensis* (Urticaceae) populations. At high *L. canadensis* plant densities, the plant population survives because *O. virginianus* avoid the area, whilst at initially low densities, *L. canadensis* is trampled as *O. virginianus* pass through and it disappears. This also seems to be a zero-sum switch.

Granivorous insects such as ants (Fomicidae) can mediate switches (Berg-Binder and Suarez 2012), as can herbivorous invertebrates. Futuyma and Wasserman (1980) studied the effect of the geometrid moth *Alsophila pometaria* (fall cankerworm) on *Quercus coccinea* and *Q. alba* (scarlet and white oak). Lower-altitude stands are dominated by *Q. coccinea*, and are separated by a narrow zone from higher altitude stands dominated by *Q. alba*. Most eggs are laid on *Q. coccinea* and hatch at the time of bud break. The larvae need young oak leaves to complete their life cycle, but they can be dispersed by wind to *Q. alba*, which breaks bud two weeks later. Thus in *Q. coccinea* stands, leaves on the few *Q. alba* are more severely damaged because they appear at a time of high demand from a large population of cankerworm larvae. Similarly, in *Q. alba* stands, the scattered *Q. coccinea* trees are more severely attacked because their leaves are available before those of *Q. alba*. Thus, the same herbivore limits the performance of the rarer of the two species in each type of stand. This process could result in ASSs, but this has not been demonstrated.

Plant species also have differential effects on the nematode community (Viketoft 2008), though the corresponding evidence that plant species favour the nematode species that favour them seems to be missing.

We described resource islands above with regards to nutrients, but patches of shrub/ tree seedlings can also be produced by perching frugivores. If the plant species is a preferred food of the frugivore, germinable seeds can be dropped during eating or defecated around the tree. This is a dispersal switch: the tree benefits the frugivore population and the frugivores facilitate tree establishment. It is a one-sided switch, for seeds can be carried by/in birds to establish an isolated tree, which can develop into a tree patch (Archer et al. 1988; Herrera et al. 1994). Animals such as squirrels (*Sciurus* spp.) could operate a switch, benefitting from nuts as food, but benefitting the plant by burying the nuts in sites that are safer from other predators and probably favourable for establishment, perhaps in new areas (Vanderwall 2001). A suite of Fabaceae (legumes) in African savannahs have carbohydrate-loaded pods that are eaten by browsers, the seeds within still attached after any dehiscence. The seeds within are refractory and they germinate faster after scarification, which can occur during their passage through the gut. Deposition in dung is clearly helpful in seedling establishment. Many browsers concentrate their dry season feeding on such legumes. Many of these are trees (*Baikiaea* spp., *Cassia* spp., *Pterocarpus* spp.) in yearly-burnt deciduous forest/savannah called miombo (Stokke 1999), and the process maintains the vegetation type. By this process, isolated trees of *Acacia tortilis* (umbrella thorn) and *A. erioloba* (camel thorn) in rangeland can develop into thickets. Miller (1995) and Dudley (1999) have shown that the pod resource is indeed heavily utilized by browsers and ungulates, and that there is an enhanced germination effect, but other suites of rodents, birds and beetles are also involved in a complex system.

Occasionally the mediator could be the absence of effective dispersal of other species: Murillo et al. (2007) found higher rates of vertebrate seed predation within *Paspalum* spp. tall-tussock patches than in surrounding short-grassland matrix, which could inhibit the invasion of other species, perhaps because the tussocks provided refuge from carnivorous predators.

A pollination switch could operate in which increase in the population of a pollinator specific to a host (perhaps an orchid) increased the seed set of a species, leading to greater abundance or a greater spread of that species, in turn providing more nectar/ pollen for the pollinator, increasing the population of the pollinator: positive feedback.

4.5.5   Switch Evolution

Wilson and Agnew (1992) suggested that switches would be more common than Clementsian altruistic facilitation, because there would be selection for genotypes with reactions that favoured the species producing them over those with reactions that disfavoured them. This is a Gaia-type idea (Lenton and van Oijen 2002), be it a rather mild one. At first sight, the argument seems logical. All plants, like other organisms, effect reaction. For example, a plant is bound to shade another plant below it. Random mutation and characters that were the result of selection for other reasons could happen to give the plant a reaction favourable to itself and hence a switch. Natural selection must operate on that genetic variation. There are hints of the evolution of switches in the endemic varieties that Northup et al. (1998) found on old soils in northern California, where it is possible that the ecotypes evolve towards soil modification, which leads to further ecotype differentiation, etc.

Using similar arguments, Odling-Smee et al. (2003) concluded that 'niche construction' [i.e. reaction] 'is a fact of life'. They seem less concerned with a population's affecting its own fitness, than with the causal change: current population → environment modification → next-generation population. Applying this to switch evolution, they say: 'We define *ecological inheritance* as any case in which organisms encounter a modified feature-factor relationship between themselves and their environment where the change in the selective pressures is a consequence of the prior niche construction by parents or other ancestral organisms'. They use the term '*positive* niche construction' for a switch, giving the examples of allelopathy and fire, and conclude that selection will occur for adaptations that allow organisms to carry out their (presumably positive) niche construction with greater efficiency. Laland et al. (1996, 1999) also analysed the effect of reaction on the selection process and concluded there could be strong selection for switches, perhaps leading to the fixation of otherwise deleterious genes.

We have to be careful here, almost to the point of disagreeing with ourselves (Wilson and Agnew 1992). Darwinian selection operates at the level of the individual genet. A mutant that produces a reaction favourable to that genet could lead to the evolution of a switch. However, this supposes an unlikely degree of independence between coexisting genotypes in their reaction and response – that the mutation benefits only that one ramet or patch, not its neighbours. We must be careful if we apply this logic to a mutant that produces a reaction that benefits a whole population of the species, because that mutation will have no selective advantage over the 'wild type'. Selection cannot operate if everyone benefits. Moreover, many of the plant features causing a switch will have a cost. For example, fire, allelopathy and herbivory defence switches can involve the production of secondary compounds, which are expensive energetically. If a character has a cost and benefits a plant's neighbours, it must be characterised as altruistic and can evolve only by a group selection mechanism (Wade 1978; Williams and Lenton 2008). The most likely group selection mechanism for plants is local kin selection, in which a plant benefits its neighbours who are likely to be somewhat related to it. However, such selection can operate only within a limited range of parameters, notably when interactions are very local (cf. the evolution of litter reaction. Section 2.4.4), and then it is a weak force (Wilson 1987; Silver and Di Paolo 2006).

We conclude that Darwinian selection would be possible were the reaction so local that the plant benefits only itself (e.g. produces litter just under itself), or at least with little benefit to neighbours of the same species. If the effect were a little more widespread, but local enough for most of the plants benefitting to be related to the mutant, the trait could possibly evolve by kin selection. The situation is not as clear-cut as Wilson and Agnew (1992) supposed. Perhaps most switches arise by accident, not by adaptation.

4.5.6   Conclusion

The urgent need is for documentation of all the steps of a switch from a single system. We commented (Wilson and Agnew 1992) that there was then no case where this had been done. Since then, further work has been done on the switch involving microscopic algae/sediment stability (Section 4.5.4.4), and so many of the processes involved in the lake turbidity switch have been established (Section 4.5.4.5), with Liapunovian stability but state change after a major perturbation, that this switch can be taken as established.

4.6   Alternative Stable States

4.6.1   Concept

In regions of apparently uniform terrain and climate, such as grassland on alluvial plains, there may be vegetation states that are different and persistent, but impossible to relate to variations in the underlying substratum or microtopography: alternative stable states (ASSs). The different states could coexist side by side in a mosaic, or they could exist in the same site at different times, each being Lyapunov stable (Section 4.8.2). Thus, a large pulse perturbation is required before a vegetation-environmental barrier can be crossed and a state change occur from one to another. There are three criteria for ASSs (Connell and Sousa 1983):

1.   The states, differing in species composition, occur in the same environment.
2.   The perturbation causing the original change in species composition in space or time appeared as a pulse, or by 'chance'.
3.   The states are stable.

These criteria are difficult. For criterion '1', each state must be held by interactions caused by its particular species composition, but almost all species' interactions are via reaction: i.e. a change in the environment. Thus, criterion '1' is impossible to meet literally. A realistic criterion would be that the differences in environment are small and of a type probably caused by the plants themselves.

For criterion '2', it is rarely possible to see the origin of the states in nature. A perturbation is normally taken to be an environmental or biotic condition outside the normal range, appearing stochastically in time. We have to distinguish between: (a) press perturbations, where the difference between treatments is continued, e.g. where there is continued grazing versus none, versus (b) pulse perturbations (shocks, instantaneous perturbations), where the change in environment is temporary. Some pulse perturbations really are instantaneous, e.g. the removal of plants. Other types of 'pulse' perturbation can be applied instantaneously but the environmental change persists for some time, such as a single fertiliser application. However, in the context of ASS the origin of the different states should be historical factors such as management or transient environmental differences that disappeared long ago, random dispersal, the season of establishment by the community founders, or environmental patchiness subtle enough to have been overwhelmed without the establishment of ASSs. The critical item is that no external environmental pressure is being applied that is different between the states.

For criterion '3' it is theoretically necessary to apply an infinitely small perturbation and show that the system regains its original state, at least after infinite time (Section 4.8.2). Persistence in the face of natural perturbations is often used as a substitute.

Alternative stable states are produced by positive feedback. Petraitis and Hoffman (2010) suggest that a switch is not necessary for the existence of ASSs, but it is difficult to see any ecological mechanism behind their formulae other than a switch. How can the system be pushed towards one state or the other without the positive feedback that comprises a switch? We thus conclude that the maintenance of ASSs in plant communities is inevitably due to a switch, one of types 2–4 (Section 4.5.2): zero-sum, symmetric or two-factor. Although ASSs are locally stable, a sufficiently large pulse perturbation can overcome the resistance and cause a state change (Section 4.6.5), as in the ball-and-cup analogy. The ASS concept is close to the theoretical concepts of 'Several stable points' (Lewontin 1969) or 'Multiple stable points from one initial condition' (Sutherland 1974), though in such papers the crucial involvement of reaction has not generally been elaborated. Ecologists should probably examine which type of switch is operating and through which mediating agent before they draw curves of hysteresis or do matrix algebra on 'multiple domains of attraction' (Gilpin and Case 1976; Schröder et al. 2005; Hansen et al. 2013).

4.6.2   Theory

Contrary to our wish to examine ecological mechanisms before generating theory, much of the consideration of ASSs has been by mathematics or simulation. Luh and Pimm (1993) and Samuels and Drake (1997) proposed the Humpty-Dumpty principle: that it might not be possible for communities to reassemble from their constituent species after a perturbation. A possible cause of a Humpty-Dumpty effect is that succession fails because intermediate species in the assembly process are missing (cf. Samuels and Drake 1997), though this seems a very rigorous interpretation of succession-to-climax. Alternatively, a Humpty-Dumpty effect might be because the original species composition was based on a switch, perhaps originated by the elusive process of 'chance'. Platt and Connell (2003) argued that secondary successions might be more liable to the initiation of ASSs, since where there is stochastic establishment by survivors, alternative paths of vegetation development may exist. Law and Morton (1996) modelled the formation of a mosaic of ASSs by this process. van Nes and Scheffer (2004) used a relatively simple Lotka–Volterra model and a community matrix (Section 4.8.2.1) to show that ASSs can arise in rich mixtures where intra- is less than interspecific competition, a process seen also in the model of Petraitis and Hoffman (2010), but one that seems unlikely in the real world (Section 3.8). Chase (2003) suggested that ASSs were more likely to exist when the regional species pool (if it is ever possible to define that) is large. He argued that with more species there will be more redundancy (i.e. species similar in alpha niche). The trouble with this logic is that if the species are so similar as to be redundant, the states will not be alternative in reaction. If they are somewhat different, one species will replace another on a path that approaches equilibrium convergence, so the states will not be ASSs. Similar reasoning can be applied to the suggestion that less dispersal will increase the likelihood of ASSs: a state is not stable just because species find it hard to disperse to the site.

It may be easy to model ASSs, but it is much harder to find evidence that ASSs actually exist.

4.6.3   Evidence: Aquatic

Most examples of ASSs cited in reviews are from freshwater laboratory microcosms, a few from mesocosms (Schröder et al. 2005). Even with those studies, a major problem has been lack of time for it to be clear that the communities are not converging. For example, an experiment by Robinson and Dickerson (1987) introduced species into glass beakers in two orders and at two slightly different rates, and found differences in final species frequencies, but the results were analysed on weeks 13–23 of the experiment when first introductions of a species occurred up to week 8, and reintroductions occurred right up to week 13. Drake (1991) performed similar experiments, again finding that the order of introduction of the autotrophs significantly affected their relative abundance, but the last introduction of an autotroph was on day 180 and the final census only 15 days later. *Chlamydomonas reinhardtii* was the species most likely to end up as a minor component, but this happened most notably in the two treatments where it was the first autotroph to be introduced, i.e. where there had been most time for it to have been excluded by interference. This gives the impression of an interference hierarchy more than of ASSs. Zhang and Zhang (2007) allowed 78 days between the last introduction and the last census and found 2 out of 15 green algal mixtures where the first-introduced species maintained dominance, in both cases *Selenastrum capricornutum*. This might be an alternative stable state. Tucker and Fukami (2014) added two species of yeasts simultaneously and two bacterial species simultaneously to micro test tubes containing a sugar/amino acid solution. When the two yeasts were added two days before the bacteria, one of them came to dominate the microcosm community and this dominance continued unabated for the next 30 days. When the two bacteria were added first, similarly one of them dominated for the next 30 days, with the two yeasts extinct from the mixture. The results were dependent on the temperature regime, but not much. Since colonies could establish in four days, 30 days represents many generations, so these systems, be they artificial, may be ASSs, noting there was no formal test for Lyapunov stability. Tucker and Fukami had no clear explanation for the mechanism. A switch must be operating, and that the dominating yeast species lowered the pH from 5.5 to 2.5 within 1.5 days strongly suggests one in pH. A nutrient switch might have been involved in the yeast-dominated stable state.

The most documented case of ASS is certainly the clear/turbid shallow lake switch described in Section 4.5.4.5.

4.6.4   Evidence: Terrestrial

What hard evidence is there for terrestrial ASSs? Schröder et al. (2005), in their review, could find only two field examples of convincingly demonstrated ASSs. One is the situation examined in several studies on Canadian high-latitude saltmarsh by R. L. Jefferies and coworkers, and summarised by Wilson and Agnew (1992) as a switch in 'Grazing and nitrogen cycling'. Schröder et al. cite a more recent paper (Handa et al. 2002) where the bare patches at the top of the marsh were hardly being recolonised, and vegetation could not be reestablished since the soil there had become hyper-saline due to greater evaporation rates (Srivastava and Jefferies 1995). This is akin to a switch of the saltmarsh pan type (Wilson and Agnew 1992), except that Srivastava and Jefferies suggest the salt is coming from buried sediments. Walker et al. (2003b) showed by modelling that nitrogen dynamics, with input from N_2-fixing cyanobacteria and N export in goose emigration, could also explain the ASSs in this system. The mosaic on the ground is a clear hint that a switch is operating, though state changes between the two ASSs do not seem to have been reported and it is not certain through which factor the switch is primarily mediated. Schröder et al.'s other example was an experiment by Schmitz (2004) in a Connecticut, USA, oldfield, removing a predator spider *Pisaurina mira*, thereby allowing an increase in population of the herbivorous *Melanoplus femurrubrum* (grasshopper), which hid in and ate *Poa pratensis* (meadow grass), releasing *Solidago rugosa* (goldenrod), which suppressed other forbs. Complicated? When predators were allowed back after two or three years, the cover of *S. rugosa* remained high for three or two years, respectively. Schmitz offers no explanation beyond 'loss of top predator control', so maybe the effect is temporary, with no switch present (i.e. no positive feedback) and therefore no ASSs.

In a possible example of a terrestrial ASS, Vandermeer et al. (2004) tracked regeneration for nine years after hurricane disturbance in Nicaraguan tropical forest and found floristic divergence, especially initially. This would be an illustration of 'multiple basins of attraction': ASS. Yet it is questionable whether the time span was long enough because in the last two years of observation species composition converged in four out of the six interstand comparisons.

Patterns in arid vegetation such as 'tiger bush' are now widely accepted to be ASSs due to a water-infiltration switch (Section 4.5.4.2 above), though naturally it is difficult to confirm that these are truly ASSs due to the long timescales involved. It has been suggested that there is upslope migration of the states due to water/nutrient input to the upper edge of the dense bands: does this contradict their status as ASS? Many switches have to do with water stress and there have been claims that most ASSs are in stressed, especially arid, areas; but Mason et al. (2007) concluded that the theory and available evidence suggested that ASSs were as likely to exist in habitats with low or moderate stress, perhaps more likely than under high-stress conditions. Perhaps patterns that are interpretable as ASSs might be easier to see in arid habitats.

Possibly the terrestrial situation where the switch mechanism behind ASSs has been most clearly demonstrated is in dune slacks on Texel, the Netherlands. Adema et al. (2002) found two seral stages side by side: a relict pioneer, low-productivity stage with *Samolus valerandii* (brookweed) and *Littorella uniflora* (shoreweed) and a late-succession, high-productivity stage with *Phragmites australis* (reed) and *Carex riparia* (sedge). The boundary between them had been static through 62 years of observation. Having investigated and dismissed all possible environmental reasons for this stasis they surmise that it must be due to some 'positive feedback mechanism', i.e. a switch, involving nutrient cycling or sulphide toxicity. Two early-successional species *L. uniflora* and *Schoenus nigricans* (bog rush), release oxygen into the soil allowing greater oxidation of ammonium to nitrate (cf. Adema and Grootjans 2003), and then denitrification to gaseous nitrogen (Adema et al. 2005). This could hold up succession. To complete the story it would be necessary to know that early-successional species are more tolerant of low N, and there is some experimental evidence that both *L. uniflora* and *S. nigricans* grow better at low N availability than *Carex nigra* (a sedge) and *Calamagrostis epigejos* (bushgrass), and more slowly at high N availability (E.B. Adema, pers. comm.).

4.6.5   State Changes

If ASSs exist in an area, a large enough perturbation should cause a state change between them. This is rather clear for clear/turbid lakes, but hardly ever for terrestrial supposed ASSs. There are many state-and-change flow diagrams in the literature, for example in papers comprising a whole issue of *Tropical Grasslands*, indicating the conditions under which each state change occurs. Unfortunately, all were speculation. Valone et al. (2002), investigating a grassland/shrubland mosaic in Arizona, USA, found that after 20 years without grazing there was little change in the shrubland. They suggested that there may be considerable lag times before any changes could be seen, but perhaps some larger perturbation is required. The experience of land managers is that intermediate states are much shorter-lived than final 'stable states', and state changes between them can be predicted from management changes. We are not doubting that state changes occur, we are just calling for attempts to document them in terrestrial systems.

4.6.6   Conclusions

Chase (2003) used examples to conclude that '… evidence suggests that community assembly can lead to multiple stable equilibria', and then continues with an assumption: 'Sometimes history matters, creating multiple stable equilibria…'. It may be that ASSs are the explanation for much of the patchiness that ecologists see in vegetation, due to alternative, divergent trajectories of change in vegetation in response to the same disturbance, but at different times with different biotic and/or physical environments (Walker 1993; Whisenant and Wagstaff 1995; Lockwood and Lockwood 1993), or different vegetation conditions at the time (Watson et al. 1996). This possibility has rarely been investigated. Walker and Wilson (2002) did so in dry grassland in southern New Zealand, and found no evidence for it. We believe switches are common and important in vegetation, and switches should often lead to ASSs, but there is very little evidence yet to support the latter's existence. To adapt Frank Egler's phrase, assuming that ASS exists is 'one of the more nauseating but delightful idiocies that stem from the heart, not the mind, of otherwise respectable scientists'.

4.7   Negative Feedback

Positive feedback means that a process **P** produces a condition/product **C** and that change in **C** *increases* the rate of process **P**: a switch. Negative feedback means that a process **P** produces a condition/product **C** and that change in **C** *decreases* the rate of process **P**. Negative feedback at the whole-community level occurs at the end of a succession (i.e. at the climax) or in one of several ASSs. The result is stability, and the speed at which the negative feedback occurs is resilience. We discuss these concepts below. However, we note that the negative feedback occurs primarily at the level of individual species, be they interacting ones. A negative feedback process in vegetation can be the same as a positive one: in an ASS situation, the positive feedback process that magnifies small initial differences, or that reinforces a state change, will be the same process that as a negative feedback enables the endpoint/new state to be stable.

4.8   Stability

4.8.1   Concept

4.8.1.1   The Nature of Vegetation

Our repeated question in this book is whether the plant communities seen in nature are arbitrary assemblages of species, such that another assemblage might well have existed at that time and place, or whether the assemblage is determined by the physical and biotic conditions of the site. The critical test between these possibilities is recovery from a pulse perturbation, i.e. stability. As always in ecology, this question can be asked, validly, at any spatial grain, quite possibly with different answers. There are three possible answers:

1.   Global stability, with one stable community state: There is only one possible endpoint of succession, to which the community always returns after a pulse perturbation. This is deterministic community structure.
2.   ASSs, with a few community states: There is a limited number of possible endpoints. This is the concept discussed above of ASSs caused by switches. The system is stable to small perturbations, but large ones induce a state change.
3.   A continuous range of community states: There is a large number of possible species assemblages, in terms of species presence and abundance, with all intermediates. There is no tendency to recover to the same state after a pulse

perturbation. The present assemblage exists for historical or even stochastic reasons, and after a perturbation or even as a result of drift another state might take its place.

We can doubt '3' from our common observation. If it were true, ecologists would have no predictive view of landscape, but in fact they generally know what species, or even species mixtures, to expect in a set of environmental conditions. They may sometimes be surprised, which is a prima facie case for '2', ASSs, and therefore for a positive feedback switch, which might with difficulty be demonstrated.

Global stability, '1', may often be present, but it is impossible to demonstrate. In order to do so, perturbations would have to be made from every possible starting assemblage of species and environmental conditions, to guard against local optima. We therefore have to restrict this question to Lyapunov stability, i.e. local stability, recovery from small perturbations. Even so, it is difficult to ask the question for communities subject to annuation (Section 4.2) or for seral communities of a succession-towards-climax (Section 4.3) unless the annuation/successional change is considerably slower than the rate of recovery from perturbation. Phases of cyclic succession (Section 4.4) could be treated in the same way at a small spatial grain, and stability could be defined for such communities at a grain that encompasses the range of separate phases, or the cycle can be seen as the stable state (Section 4.8.2.1).

4.8.1.2 What is 'Stability'?

Various features of a community have sometimes been described as 'stability'.

1. Stability *sensu stricto*, i.e. Lyapunov stability or neighbourhood stability: whether the community ever recovers from a small pulse perturbation.
2. Resistance: Lack of change upon a pulse perturbation. We separate it here into resistance to abiotic perturbation and resistance to invasion.
3. Resilience: Speed of recovery from a pulse perturbation.
4. Reliability: Constancy, 'temporal stability', lack of change, probably in spite of minor perturbation.

Seeing long-term continuance of a range of plant communities, e.g. different types of deciduous forest, fired grassland, closed evergreen forest, ecologists are tempted to see continuance, i.e. reliability from one time to another, as stability. However, the difficulty of defining this kind of 'stability' – how reliable is reliable enough? – plus a little mathematics envy, has led ecology into accepting Lyapunov stability (Section 4.8.2) as the true stability. Resistance is almost impossible to define (Section 4.8.3). Resilience (Section 4.8.5) is dependent on Lyapunov stability, for only if the community is recovering from perturbation to the previous state can we ask how quickly it does so.

Since all communities are at all times in a state of change, the assumption of equilibrium in these concepts is unrealistic. Nevertheless, the processes are vital to our knowledge of plant communities, for if there are no stabilising processes, communities hardly exist. We just have to consider stabilising processes rather than stability itself.

4.8.1.3   Stability in What?

All this leaves open the questions: change *in what*?, recovery *in what*? Here at least ecologists are free from the mathematicians to ask whatever ecologically meaningful question they feel pressing (Barros et al. 2016). The change/recovery can be in:

a. Species composition: Change/recovery in species presence and/or abundance.
b. Guild composition: The proportions of certain guilds.
c. Function: This can be productivity, biomass (usually in fact standing crop), the rate of nutrient cycling, water output, or any other result of whole-community processes.

Most general discussion of stability refers to species-composition change/recovery. The categories are hierarchically related. Constancy in species composition ensures (though it is not essential for) guild constancy and functional constancy. Constancy in guild composition ensures functional constancy, so long as the guild definitions are relevant to that function.

A recurring question in considering these concepts will be the effect of species richness: are communities with more species more 'stable' (*sensu lato*)?

4.8.2   *Stability* Sensu Stricto *(Lyapunov Stability)*

4.8.2.1   Defining Lyapunov Stability

May (1972) initiated the modern approach to stability with his community matrix analysis, based on the mathematical procedure of local stability analysis (= 'Lyapunov stability analysis' = 'neighbourhood stability analysis'). Such stability is defined as the ability to recover completely from an infinitely small perturbation after an infinitely long time. Though we have phrased this to make it sound a little ecologically unrealistic, some approach to the concept is essential if we are to discuss stability. Unless communities have no tendency to retain their state, they are hardly communities. Defined thus, stability is like pregnancy: either one is or one isn't. The 'infinitely small' is necessary to make the definition one of local stability, perhaps within one of several ASSs. The phrase 'after an infinitely long time' is necessary to avoid arbitrary cut-offs, as well as to distinguish it from resilience. Stability can be predicted mathematically from the community matrix: a summary of all possible pairwise species interactions in an equilibrium community, expressed as the effect that a small change in the equilibrium abundance of one species has on the equilibrium abundance of another, whilst holding the abundances of all the other species present at their respective equilibrium values. Stability is measured by the eigenvalues of the community matrix: if the real part of the largest eigenvalue, $R(\lambda)_{max}$, is less than zero, then the community is stable (Figure 4.11a). The community matrix introduces three important aspects of community structure. The number of species involved, S, is simple. The connectance is what proportion of the $S \times (S-1)$ off-diagonal matrix elements are non-zero (i.e. the species in that column does affect the species in that row). Thirdly, the interaction strength is the mean absolute value of these non-zero elements.

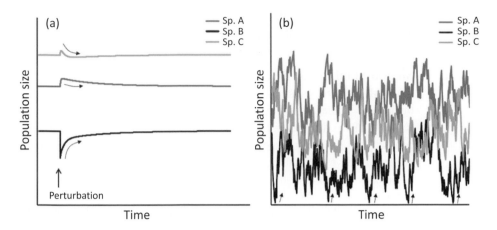

Figure 4.11 (a) Analysis of the maintenance of biodiversity by Lyapunov stability. In this case it is assumed that all species abundances have reached their equilibrium values (i.e. they are constant through time), and then an arbitrarily small perturbation is applied to the equilibrium. If all of the population densities recover to their preperturbation values (as in the figure), then the system is stable and biodiversity is maintained. (b) Analysis of the maintenance of biodiversity by Invasibility analysis. Under this paradigm there is no restrictive requirement for equilibrium, and the maintenance of biodiversity occurs when the long-term low-density growth rate of all species (the arrows forcing species B away from local extinction) is positive – with low-density growth rates reflecting the presence of stabilising mechanisms.

It is possible that the equilibrium around which a community is stable is not a fixed species composition but is a limit cycle (May 1973). This concept has been applied to cycles between predator and prey, such as the supposed *Lepus americanus* (hare)/*Lynx canadensis* cycle in Canada. Cyclic successions (Section 4.4) are basically the same process, since limit cycles can also be caused by stabilising negative feedback with a time lag (May 1973). When autoallelopathy is the supposed cause, it is assumed to operate until the self-poisoning plant dies, but effects of its litter could persist. In Yeaton's (1978) *Larrea tridentata/Opuntia leptocaulis* cycle (Section 4.4.2), the *L. tridentata* dies only when the cactus has become large enough to compete for water. In general, plants take time to grow and time to die, and soils take time to change ('temporal inertia'; Section 3.12.1). This gives ample opportunity for time lags to cause limit cycles.

The concept of Lyapunov stability can be compared to the more general criteria of biodiversity maintenance defined by 'invasibility analysis' (Figure 4.11). Under Lyapunov stability (Figure 4.11a) the relative abundances of the species are maintained following minor perturbation, and whilst the strict concept of 'equilibrium' is problematic in most situations, there is an expectation of consistency in the relative abundances of the species over time. Under invasibility analysis the presence of 'increase-when-rare' coexistence mechanisms (Chapter 3) simultaneously prevent species loss and maintain diversity, but allow the relative abundances of the species to vary over time, perhaps dramatically (Figure 4.11b). These are complementary tools for analysing community structure, with the stricter criteria of Lyapunov stability a subset of the more general criteria defined by the invasibility approach.

4.8.2.2   Should Stability Occur? (Simulations)

Since May's (1972) seminal paper, the community matrix has been used extensively in theoretical studies to predict what type of community would be stable (Hall and Raffaelli 1993). It turns out that, given a matrix comprising random values, modelled communities with many species and/or with many or strong interspecific interactions are less likely to be stable. In fact, it is very unlikely that a community of randomly assembled species will be stable at all. This represents an obvious conflict with our observation that multispecies communities commonly persist in nature. One explanation is that the species in natural communities have coevolved to be able to coexist, though with the usual problem that species occurring in many different species mixtures cannot coevolve to match each possible outcome. However, the introduction into the model of some basic ecological processes, especially simulating the assembly of a community by the immigration and/or extinction of species, allows complex multispecies communities to develop relatively easily with no coevolution (Tregonning and Roberts 1979; Taylor 1988). Coevolution does enable more species to coexist, though. In the simplest assembly models of Tokita and Yasutomi (2003) the community stabilised at about five species. When the introduced species were formed as minor mutants of a species already in the community, the number of species reached about 10. However, when the characters of new species arriving were based on one species already present, except for its interactions with one other (abundant) species (their 'local' model), the size of the stable community approached 50 and was still rising after 10,000 periods. Moreover, such communities were resistant to invasion by species not related to those already present. They interpreted this as the temporary persistence of species only marginally inferior in interference ability, that were able to exist stably once mutualistic partners appeared. This is stability by very simple coevolution.

4.8.2.3   Does Stability Occur? (The Real World)

Despite all the theoretical work on the community matrix, tests of stability in natural systems have been remarkably and surprisingly sparse. Volkov et al. (2009) attempted to estimate the values of the community matrix from static (i.e. spatial snapshot) information from the Barro Colorado Island plot. Other studies have searched for patterns among the pairwise species interactions in published food webs that are consistent with stability, as predicted by community matrix theory (Hall and Raffaelli 1993; Borrelli 2015). These analyses are sophisticated, but it is not possible to truly estimate the community matrix from observational data. The only valid method for estimating the community matrix for a given community is by the experimental manipulation of the organisms, quantifying the effect that each species has upon the growth rate of each other species. We are aware of only a few such studies (Thomas and Pomerantz 1981, with data on four ciliate protozoans from Vandermeer 1969; Seifert and Seifert 1976 with the insects of *Heliconia* spp. bracts; and those below).

Schmitz (1997) used data from field and laboratory experiments to quantify the interaction strengths between grasshoppers, an oldfield grass, three oldfield forbs and nitrogen supply. He used these values to parameterize the community matrix and thence to predict the response of each species to nitrogen and herbivore press perturbation. The predictions carried a high degree of uncertainty, but the observed results of the field perturbations were close to, or within, that range in six out of eight cases. They did not calculate stability themselves, but calculation of $R(\lambda)_{max}$ from their matrix indicates stability.

The most direct test of stability in a community has been on the University of Otago Botany Lawn (Roxburgh and Wilson 2000a). A prerequisite for the concept of stability is that the community should be at equilibrium, and indeed year-to-year fluctuation was small. Seven flowering-plant species were collected from the lawn and grown in deep boxes only metres from the actual community. The soil for the boxes was also collected from the lawn and kept in two strata. The seven species comprised 86 per cent of the total abundance of the lawn, and 92 per cent of the perennial abundance, and were grown in monocultures and in two-species and other mixtures. A community matrix calculated from interference coefficients predicted the mixture of species to be unstable, but very close to the stability/instability boundary. It was certainly closer to the stability/ instability boundary than matrices constructed under a range of null models. To test for the actual stability of the community, Roxburgh and Wilson (2000b) applied three types of pulse perturbation. After shading for six weeks and after mechanical perturbation (removal of vegetation), species composition recovered towards the original equilibrium state, although full recovery was not seen. Recovery from application of a herbicide specific to grasses was still quite incomplete after 2.5 years, apparently because the compensatory growth of dicotyledons made it difficult for the grasses to reinvade. This pattern of recovery is consistent with the prediction of marginal stability/ instability given by the community matrix analysis. Simulations (Roxburgh and Wilson 2000a) showed that near-stability of the community matrix could be attributed to the transitive interference abilities found (Section 3.5).

4.8.2.4   Prediction of Species Abundance

Multispecies communities are complicated. Simplistic extrapolation from interference abilities would be that species with a greater interference ability in two-species experiments would be more abundant in the community, and several authors have found such a correlation (e.g. Mitchley and Grubb 1986; Miller and Werner 1987). However, the strict hierarchy in interference ability found in the Botany Lawn pairwise box experiments, from *Holcus lanatus* (Yorkshire fog) with the highest interference ability to the herb *Hydrocotyle heteromeria* with the lowest (Section 3.5; Figure 3.5), had no clear relation to abundances in the lawn (Roxburgh and Wilson 2000b). Such a lack of correlation has been found elsewhere (Aarssen 1988). Roxburgh and Wilson suggested this was due to apparent mutualism in a multispecies mixture: the 'My enemy's enemy is my friend' effect (Stone and Roberts 1991). This effect will be more prominent with more species in the community. Apparent mutualism is implicitly included in the community matrix model; therefore, the model should have been able to predict abundances in the actual Botany Lawn community (subject to the restriction that only seven of the species were included in the model). For example, the model predicted that *Trifolium repens* (white clover), *Holcus lanatus* and *Agrostis capillaris* (bent) would be the three dominants, and these three are indeed among the four most abundant species in the lawn. From the pairwise experiments, *H. lanatus* headed the interference hierarchy. However, it was only the second most abundant species in an experimental seven-species experimental mixture and in the lawn (Roxburgh and Wilson 2000a), an outcome correctly predicted by the community matrix model.

4.8.3   Resistance to Abiotic Perturbation

Resistance is a difficult concept. It is the response of a system to a natural or anthropogenic pulse perturbation, or to be more exact the lack of response. But what is a perturbation? Suppose a light wind blows over grassland, with no effect on community composition. Obviously the change was too small to be called a perturbation. The trouble is that if a perturbation is applied and the community does not react, it is impossible to say whether that was because it was a very resistant community, or because the treatment applied was not a real perturbation in the first place. Even communities cannot be compared. If salt is applied to most communities, it will be a perturbation and plants may die. However, with a saltmarsh community, any obligate halophytes there will grow faster: it was no perturbation at all. This is analogous to the problems that Körner (2003b) saw in defining 'stress'. Nevertheless, the basic idea is important for our study of communities, that a community is hardly real if the least change in the environment will disturb it.

The concept of resistance can again apply in terms of function, guild or species composition, and even for individual species. We could expect some theory at the community level. There is little, but in Loreau and Behera's (1999) model with plants limited by nutrients the biomass resistance generally decreased with increased ecological difference between the species.

For evidence, we take two examples of perturbations: drought and trampling. Pfisterer and Schmid (2002), in mesocosms of grassland species, found that species-rich plots were less resistant to experimental drought in biomass, i.e. in function, and Rodríguez and Gómezal-Sal (1994) found a similar trend among 20 meadow sites in the Cantabrian mountains in Spain; in their case resistance in species composition. Other comparisons of field sites and those with experimental plots that differ in fertiliser application have yielded the opposite result from that of Rodríguez and Gómezal-Sal, though such comparisons confound species diversity with composition. Vegetation is complicated, so is its relation to soil, and the behaviour of each species depends on its evolutionary history, so it is little wonder that it is difficult to find generalisations.

There is an extensive literature on the resistance and resilience in response to human trampling. For example, Page et al. (1985) studied the above-ground productivity resistance of many species in Welsh coastal-dune vegetation over two years after press perturbations of quantified trampling, and compared the results with controls. They found that species could be assigned to three behavioural classes, partly related to their morphological guild/functional type. In the closed-canopy vegetation of the fixed dunes, graminoid productivity was the most resistant, and forb productivity the least. Forbs and bryophytes of unstable habitats, counter intuitively, recovered only slowly, but they have specialised reactions to particular environmental events. For example, trampling and the open ground that it produces triggered the corm-like base of *Bellis perennis* (lawn daisy) to produce stolons (A.D.Q. Agnew, pers. obs.). Other authors using greenhouse and field experiments (e.g. Andersen 1995; Cole 1995) have found that resistance to human foot and wheel pressure differs between plant guilds in similar ways, the geophytes and tussocky graminoids being consistently more resistant, upright forbs less resistant. Resistance seems to be individualistic to the species, as well as involving the impossible-to-answer question: 'Was it a perturbation in the first place'?

4.8.4   Resistance to Invasion

Another type of resistance is that to invasion. The effects of invasive non-native species on community composition are well documented. However, few studies have determined the mechanisms by which invaders drive these changes. The issues that affect invasibility are:

1. Diversity and species identity
    i.   The species diversity of the target community.
    ii. Whether it is species-saturated. An implication of low diversity is that there might be empty niches or whole missing guilds, with resources available to the invader, or there might be a niche that is not being exploited as aggressively as it could be. For example, *Spartina anglica* (cord grass) occupies the lowest zone on British saltmarshes, previously bare, though it has partly constructed its niche by reaction. It is difficult to find a demonstrated example of niche splitting, but this may occur often.
    iii. Interference ability: For example, introduced *Phragmites australis* (reed) replacing a mixed coastal marsh community in North America (Minchinton et al. 2006).
2. Disturbance
    i.   Whether disturbance is required before invasion can occur.
    ii. Whether continual disturbance is required for the invader to persist.
3. Reaction: A switch operated by the existing species gives resistance to invasion. On the other hand, a switch in any mediating agent (Section 4.5.4) operated by an invading species could create a new niche by reaction and maintain it. For example, species could increase or decrease the frequency/intensity of fires, directly or by suppressing other species, and to their benefit (Brooks et al. 2004).
4. Species interactions: Whether communities with low interaction strength and/or low connectance are more invasible, or possibly less so. There seems to be little work on this.
5. Coevolution.
6. Invaders as drivers versus passengers.

Often, invasion is taken to mean by species exotic to the region or continent, which confuses the ability of incomers to establish in a closed community with whether the incomers have intrinsic differences from native ones. We shall discuss only topics 1 and 2 above in this chapter. All the topics (1–6) will arise in consideration of exotic species in Chapter 5.

4.8.4.1   Diversity and Species Identity

Mathematical theory on this issue is not entirely settled. Case (1991) found that in simulated, randomly constructed Lotka–Volterra communities, resistance to an invader was higher when there were strong interspecific interactions and species richness was higher, the latter considered established theory by Shea and Chesson (2002). Intuitively, the explanation is that with high richness there is niche complementarity and thus limited resources left for an invader, with implied appeals to MacArthur and Levins' species packing. But then why do the residents have priority in getting the resources, except via inertia (Section 3.12) or via the cumulative competitive ability common with competition for light (Section 2.2.1.2)? In partial refutation of Shea and Chesson (2002), the careful plant community simulations of Moore et al. (2001) indicated that there is no simple relation between resistance and richness: it depends on the model and what causes the richness to vary.

Turning to reality, there have been many studies correlating species richness and invasion across sites, and also studies experimentally adding invaders to a range of natural systems. We can draw no conclusions from these because there are always environmental/historical differences between sites that confound differences in species richness. We like the idea that real communities can tell us more, but in this case artificial communities are needed.

For example, Dukes (2002) added *Centaurea solstitialis* (golden starthistle) seeds as invaders into microcosms in a Californian grassland, with a substratum of sand plus soil from the site, comprising grass and forb species from the site. Mixtures of two or more species reduced the invader's biomass more than did the average monoculture, but no mixture of up to 16 species inhibited invasion more than the most resistant single species. The latter species was *Hemizonia congesta* (hayfield tarweed), and it is, like *C. solstitialis*, a summer-active annual forb, as if that niche were full. No increase in resistance was seen beyond four species, though Dukes (2001) attributed this to an artefact of the experimental design. However, we are seeking a niche-filling effect, in which with more species more niches will be filled,[1] leaving few resources for an invader. As with overyield (Section 4.8.6), there is a possible artefactual 'selection effect' (Wardle 2001): when more species are sown, there is more chance that a species with high interference ability in the resident mixture is included and will take over the mixture. This mimics a niche-filling effect *if* that species also has high interference ability against the invader, but that is quite likely, for example a species with high cover or LAI reducing light intensity below itself. A counter to this artefact would be finding

---

[1] In this case perhaps the annual forb niche.

that a mixture is more resistant to invasion, not just compared to an average mixture with fewer species, but than *any* mixture with fewer species or any monoculture. The experiment of Dukes (2001, 2002) did not demonstrate this.

Sound evidence for an effect of richness on resistance to invasion comes from David Tilman's experimental site at Cedar Creek. Fargione and Tilman (2005a) planted and weeded 169 plots to richnesses from 1, through 2 to 16 species for 6 years, then opened them to colonisation for 2 years. In higher-richness plots there was significantly lower volunteer invader richness ($R^2 = 0.31$) and biomass ($R^2 = 0.18$). As usual, most of this effect on invader richness occurred at low resident richness, between one and two species (compared to four species in the experiment of Petermann et al. 2010). The effect seemed to be largely due to suppression of invaders by C_4 grasses, and the authors suggest their effect was via reduction in soil nitrate. They admit this is most easily seen as a selection effect. However, the invasion resistance of some mixtures was greater than that of the most resistant monoculture (*Schizachyrium scoparium*, bluestem, a C_4 grass), and resistance increased up to an eight-species mixture. Specifically, the percentage of plots that had lower invader biomass than did the best resident monoculture increased across the species richness gradient. This is all quite impressive evidence of the complementarity effect.

Wilsey and Polley (2002) attempted to outwit the sampling effect by experimentally varying the evenness of four grassland species (three grasses and one forb), keeping the richness constant. Reducing evenness led to a greater number of dicotyledonous invaders, but had less effect on monocotyledonous invaders. Guild-removal experiments could also give evidence on the issue with the species of a guild tending to replace others of the same guild, but the evidence from such experiments is often weak (Section 5.7.2).

Oversimplification to numbers can mislead, for species identity matters. For example, van Ruijven et al. (2003) sowed plots with one–eight species. They found an effect of species richness on invasibility, but it could be attributed to just two species, and mainly *Leucanthemum vulgare* (oxeye daisy). However this was not a selection effect, for *L. vulgare* was a low-yielder in monoculture. van Ruijven et al. suggested it may have suppressed invaders by harbouring root-feeding nematodes, and perhaps also transmitting viruses through them. Again, Crawley et al. (1999) found little overall effect of species richness in decreasing volunteer invasion, but there was consistently lower invader richness and biomass in plots containing *Alopecurus pratensis* (foxtail).

4.8.4.2   Disturbance

It is very frequently suggested that invasion is greater after disturbance (Davis et al. 2000). This is true, but misleading. Of course, there must be sufficient resources for an invader to establish, and autogenic and allogenic disturbances often release resources. For example Burke and Grime (1996) and Thompson et al. (2001) report on an experiment in a limestone grassland at Buxton, northern England, where they added seed of 54 species, all widespread native herbs that are commonly found in climates and soils like those at the experimental site but that were not originally present there. There was greater establishment of the 54 species when the soil was disturbed and fertiliser added, though the combination of disturbance and fertility levels in which invasion was maximal was unique for each invading species: species identity matters again.

Davis et al. (2000) generalised from experiments such as this to suggest that invasion requires resource enrichment or release, including that caused by disturbance. Care is needed! Why are invaders entering disturbed sites, and how permanent is the situation? We see four possible reasons:

a.   The environmental conditions have been changed by outside influences, and the invaders are better suited to the new conditions than were the residents. Possibly they evolved in similar conditions. In this case, disturbance is simply overcoming temporal inertia (Section 3.12) and the invasion will be permanent.
b.   The change in environmental conditions is autogenic, with the exotic invader operating a switch, changing the environment in a direction that gives it a relative advantage. In this case too, disturbance is overcoming inertia. The change is permanent, with the invaders acting as 'drivers' (Section 5.12).
c.   The conditions have not changed, but species that are better adapted happen to have evolved elsewhere; perhaps it is a worldwide super-species in those conditions. Disturbance just opens gaps to speed species replacement, again overcoming inertia.
d.   The invaders are not better suited to the preexisting environment, but are ruderals taking advantage of the temporary gaps. Many exotic species are ruderals, though it is not clear why (Section 5.12). Under this scenario the exotics will be present only whilst the disturbance is maintained, i.e. they are 'passengers'.

In cases 'a' to 'c' the invasion is essentially permanent.

4.8.4.3   Productivity

It might be expected that highly productive/high-biomass communities are more resistant to invasion, but this is not necessarily so, and invasion can even lead to decreased productivity (Petermann et al. 2010).

4.8.5   Resilience and Redundancy

Resilience is a simple concept: it is the speed of Lyapunov stability recovery after a pulse perturbation (if there is no Lyapunov stability, 'resilience' has no meaning). Since Lyapunov stability is the mark of a plant community, as opposed to a temporary collection of species, just possibly the speed with which Lyapunov recovery happens is a strength of community structure. Again, recovery can be measured in a system function such as productivity, or in guild/species composition. Again, these are hierarchically related: recovery in guild proportions should restore community function so long as the guilds formed are appropriate for the question. As with stability, and to some extent resistance, it is a problem that all communities constantly change. Entire recovery to a previous state is not a reasonable expectation, since the community would have changed anyway. The target moves.

4.8.5.1   Redundancy

Walker (1992) argued that maintaining function after a perturbation should be easier if there are several members of each guild, a situation he calls species/ecological/functional redundancy, since if a perturbation affects the dominant species in a guild, a subsidiary species will be available to assume its role.

This concept has not often been applied to real communities. Cowling et al. (1994) suggested that the high richness in Cape fynbos, South Africa, meant there was considerable redundancy within guilds. After the massive fires and droughts that are endemic to the region, some members of each guild could survive, so that the original guild structure was recovered: redundancy giving resilience. Presumably the missing species reinvade in time, restoring the richness.

In a second example, Walker et al. (1999) suggested that an Australian rangeland community was rather precisely organised this way: the most abundant species differed from each other functionally, the most functionally similar species differed in abundance, and when a species decreased under heavy grazing another species increased that was functionally similar but had previously been sparse. For example, using Walker et al.'s functional characters, when a large perennial grass with soft, thin leaves is reduced by grazing, another species with those characters should increase. This is not easy to imagine: if those characters affect palatability and grazing tolerance, all functionally similar species would decrease together. The relevant theory is that environmental changes (here, grazing) lead to dominance by species of different beta niches but with the same alpha niches still mostly represented, but the characters used to characterise alpha-niche function need to be chosen more carefully than Walker et al. were able to do. The concept of redundancy as espoused by Walker et al. falls all too easily into teleology, to quote: 'Why do the majority of species occur in low abundance? We propose that…They…confer resilience on the community with respect to ecosystem function'. We ask who knows to keep them in the community at low abundance for that purpose? The real answers to such questions come from the mechanisms of coexistence that we discussed in Chapter 3. In particular, insurance could come from migration rather than from redundancy (Loreau et al. 2003).

4.8.5.2   Predictors of Resilience

In a test of a hypothesis that higher richness gives higher resilience, Pfisterer and Schmid (2002) sowed Swiss BioDepth plots with a range of grass species richnesses and perturbed them with simulated drought, but contrary to theory the high-richness plots were less resilient in above-ground net primary productivity. In contrast, when Engelhardt and Kadlec (2001) set up aquatic mesocosms with one to five aquatic species and applied the perturbation by clipping off the above-substratum shoots, there was no effect of species richness on biomass and respiration resilience. Mathematical theory on resilience is sparse, but Haydon (1994) used community matrix models, which, after excluding unfeasible[2] and unstable communities, indicated that resilience

---

[2] That is, those that come to an equilibrium, but with one or more species having negative abundances, which is ecologically nonsense.

should increase with connectance and interaction strength, but decrease with species richness. The latter prediction matches Pfisterer and Schmid's finding, be it that Haydon's prediction was for species-composition resilience. Other models can give different results: Okuyama and Holland (2008) found that if the species interactions comprised mutualism, resilience increased with species richness, with connectance and (if symmetric) with mutualism strength.

There have been many studies of the relation between productivity (usually really standing crop, see Section 1.5.3) and resilience. Nearly all deal with grasslands, and generalisations might therefore be expected about the effects involved. Sadly, this is not the case. Moore et al. (1993), using randomly parameterised models, concluded that productivity and resilience are inextricably linked, but Stone et al. (1996), using a range of mathematical models, suggest it is far from simple, and depends on the model used. Rietkerk et al. (1997) indicated similar complications using a more mechanistic model, concluding that if plant growth is water-limited, the vegetation on sandy soils is more resilient, but if plant growth is nutrient-limited, the vegetation on clayey soils is more resilient, and this depends on the extent of the plant's reaction on the soil. Sometimes, it is not easy to see a deep meaning in differences in resilience. For example, Herbert et al. (1999) examined resilience in above-ground net primary productivity in Hawai'ian tropical rainforest after a hurricane. Plots that had been fertilised with P were less resistant but more resilient. Both seem rather simple results of lush growth and high growth rates with fertilisation.

The answer is probably that species' types differ, and simple conclusions are unrealistic. For example, in the trampling experiments of Cole and Trull (1992) and Cole (1995)'s experimental trampling, resilience in cover (which was objectively measured in the latter work!) was lower in communities containing more shrubs and chamaephytes.

4.8.5.3 Correlation of Resistance and Resilience

Community resistance and resilience are often inversely related. We saw this with Herbert et al. (1999) above. Taking our case of trampling, Cole and Spildie (1998) found that when montane understorey communities were subjected to hiker, horse and llama traffic at two intensities, forb-dominated vegetation was highly vulnerable but recovered rapidly, whilst shrub-dominated vegetation was more resistant but lacked resilience. Similar correlations have been seen at the between-species level: again with trampling, Cole (1995) found a negative correlation in mountain vegetation especially within the chamaephyte and graminoid guilds. In contrast, Pfisterer and Schmid (2002) found that species-poor plots were both more resistant and more resilient to drought. Again, simple generalisations are premature and probably wrong.

4.8.6 Reliability (Constancy)

Reliability is the reciprocal of variation in the ecosystem (e.g. Naeem and Li 1997; Rastetter et al. 1999). The concept of 'stability' was based on it up to 1975, though Orians (1975) described 'constancy' as one meaning of stability. It has tended to creep back, sometimes as the more honest 'temporal stability', apparently because true, Lyapunov stability is difficult to measure and reliability is an easy way out. Reliability tells us more about environmental variation, much less about community structure than do stability, resistance or resilience. There is more current interest in reliability than about the other three, but that is because it is much easier to measure, not because it is more interesting.

Species richness could increase reliability through the *portfolio, insurance, interaction* or *sampling* effects. In the *portfolio* effect, named after a portfolio of shares that varies less than individual shares, the community response to environmental fluctuation is dampened just because it is an average of the random changes in the biomass of plants of different species, say from year to year (Ives and Hughes 2002). The same random differences and averaging could occur with plants of the same species in a monoculture, so this is clearly a null model, not a community effect. The *insurance* effect is that species differ in their response to environmental fluctuations; when the environment disfavours one species, others will increase. The *insurance* effect could operate in the absence of species interactions[3] if species were in completely separate niches. However, interference and probably facilitation will be present in all communities, magnifying and perhaps reversing the differences in environmental response through differential apportionment of resources, and is the *interaction* effect[4]. A *sampling* effect is also possible: if some species have large biomass and are insensitive to yearly environmental variation, such a species is more likely to be present when there are more species (Allan et al. 2011). The *insurance* and *interaction* effects (sometimes jointly called the *covariance* effect because there needs to be a negative covariance between the abundances of different species to promote reliability) will always operate, since there are always species interactions and species are never identical in their environmental responses. Since averaging over minor species has little effect, these effects will increase with species evenness as well as with species richness. The negative covariance produced by the *insurance* and *interaction* effects may well be obscured by the overall effects of environmental favourability/stress on the growth of the vegetation. Presumably for this reason, literature surveys have shown that negative covariance across years is far from universal, though notably more common in plant communities than in animal ones (Houlahan et al. 2007; Valone and Barber 2008).

Some authors have suggested that overyielding can be a mechanism of reliability (e.g. Tilman et al. 2006; Isbell et al. 2009). This is not logical. No mechanism is ever suggested, except that: (a) reliability is often measured by the CV of biomass; (b) CV is variation/mean; (c) if there is overyielding, the mean is higher and (d) therefore the CV lower (Lehman and Tilman 2000). This is an artefact of calculation, not an ecological mechanism. It is true that niche differentiation will tend to lead to both overyield and

---

[3] That is, no facilitation and no interference

[4] Often called the *competition* effect. Mutualism is not a possibility here, because it would lead to variation, not reliability. The literature seems confused on definitions of *portfolio* effect, *insurance* effect and *interference/competition* effect.

reliability, but that is no logic. Thus, for all eight of the datasets analysed by Hector et al. (2010, appendix G) the SD increases in biomass with species richness, in most cases markedly, casting doubt on their interpretation of higher reliability. Cardinale et al. (2013) confirmed this by analysing the results of 34 experiments that had manipulated species richness: the strength of richness effects on reliability were independent across experiments of those on overyield.

However, we cannot assume that richness → reliability effects will be stronger when species richness is higher. Hughes and Roughgarden (2000) modelled a community using density-dependent Lotka–Volterra competition equations, and found that with diffuse competition the reliability of biomass decreased with richness, except in systems with few species and very low interaction strengths. With each species competing with only its two nearest neighbours, as in MacArthur and Levins (1967) model, the richness/ reliability relation was the reverse. DeWoody et al. (2003) used community models based on variations of MacArthur's 'broken stick' and found that whilst reliability increased with richness, it was, surprisingly, higher with more difference in competitive ability between the species. Hughes and Roughgarden (1998) found that in a two-species Lotka–Volterra community matrix model with the two species equal in competitive ability, biomass reliability was not affected by the strength of competition, but as the competitive abilities of the two species became more unequal, reliability decreased. Faced with this disagreement between models we shall have to descend to reality.

It has often been assumed that species-rich tropical forests are reliable in species composition unless disturbed. This is controversial, and evidence is hard to come by because of the lifespan of some of the tree species and because of a paucity of time-lapse data; it seems to be generally true (Richards 1996, pp. 67–68). On the other hand, records of long-term stasis are also found among stressed, species-poor communities, e.g. the dwarf pitch pine shrubland on Mt Everett, Massachusetts (Motzkin et al. 2002), the *Carex curvula* zone in the European Alps (Steinger et al. 1996) and the ombrotrophic mires of Quebec (Klinger 1996). Comparisons among field communities are fraught with confounding factors, and careful experiments with higher plants have generally shown the greater reliability in community biomass at higher richness, as expected from simple theory,[5] for example Tilman et al. (2006) in 168 plots at Cedar Creek sown at species richnesses of 1–16 and maintained thus by weeding; and van Ruijven and Berendse (2007) with 102 plots in an outdoor experiment in the Netherlands, with plots of 1–8 species each. However, no such pattern is seen in algal microcosms (Gross et al. 2014). Cadotte et al. (2012), using eight years of data from the Cedar Creek experiment, showed the same effect using phylogenetic diversity as a proxy for functional diversity, though Roscher et al. (2011) were unable to show a direct effect of the number of functional groups. The latter workers were able to pinpoint the richness → reliability mechanism in the Jena experiment to a considerable turnover of species of different functional groups, i.e. the covariance effect. On the other hand,

---

[5] The portfolio, insurance and interference effects.

Tilman et al. (2006), analysing 10 years of the Cedar Creek biodiversity experiment, and finding greater biomass reliability in more species-rich plots, found evidence of the portfolio effect but none of a covariance effect.

These experiments are the nearest there is to an answer so far.

4.8.7   Conclusions

The concepts of resistance, resilience and reliability are significant in community ecology, and data on these issues are important. However, it is a pity that for ease of determination reliability, under the name of 'stability', has driven determination of stability *sensu stricto*; Lyapunov stability, underground. The basic question about plant communities – are they real – is answered by whether they can restore themselves after press perturbations, and the processes that would do this, even if they never quite achieve it, are those that underly Lyapunov stability.

4.9   Diversity → Productivity

4.9.1   Concept and Interpretation

There are two major possible outcomes of niche differentiation between species: over-yield (Figure 4.12) and stability (*sensu lato*, Section 4.8); the conditions required for the two are closely related (Hector 2006). The concept of overyielding has a long history, with Darwin (1859, chapter 6, section 6) noting in *On the Origin of Species*, 'It has been experimentally proved, that if a plot of ground be sown with one species of grass, and a similar plot be sown with several distinct genera of grasses, a greater number of plants and a greater weight of dry herbage can thus be raised'. Despite Darwin's certainty on the process, investigation of overyield has been fraught with problems, notably the 'selection effect'. *If* the species with the highest interference ability and therefore the one that comes to dominate the mixture *also* has high productivity, there will be overyield, and with higher richness there is more chance that such a species will be present (if that species has low productivity there may be a negative selection effect). In terms of seeking the results of niche differentiation this is an artefact, with little relevance for the assembly of real communities. If the productivity of the mixture is greater than that of any of the monocultures ('transgressive overyielding': Figure 4.12c), the selection effect can be ruled out, and niche complementarity is the obvious cause. Loreau and Hector (2001) introduced a mathematical separation of the selection and niche complementarity effects. However, we suggest that if transgressive overyielding is not seen, the results are not of great ecological significance. Moreover, Carroll et al. (2011) have thrown considerable doubt on the ability of Loreau and Hector's (2001) much-used method to identify the fraction of overyield that is due to alpha-niche partitioning. Loreau et al. (2012) contest this to some degree, but conclude 'however useful, no post hoc analysis will ever be able to replace detailed knowledge of the biological mechanisms at work'.

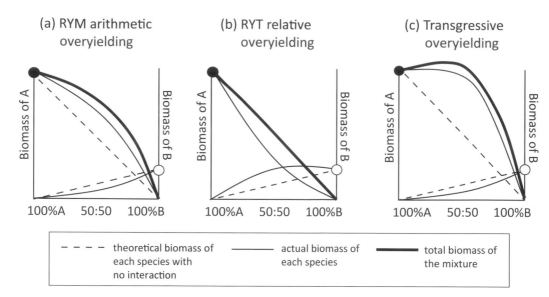

(a) RYM arithmetic overyielding

(b) RYT relative overyielding

(c) Transgressive overyielding

- - - - theoretical biomass of each species with no interaction

———— actual biomass of each species

━━━━ total biomass of the mixture

Figure 4.12 (a) There is arithmetic overyielding: the absolute gain by A in mixture is greater than the loss by B so Relative Yield of the Mixture (RYM) >1.0. (b) Relative gain by B is greater than the loss by A so RYT >1.0. (c) Some mixtures yield more than either monoculture. RYM is the yield of the mixture compared to the mean of the monocultures; RYT is the sum of each species' yield in mixture compared to its yield in monoculture: Wilson (1988c).

Experiments to detect overyielding have had mixed results. Tilman et al. (2001) planted field plots at Cedar Creek in 1994 in a range of species richnesses up to 16, from a pool of 18 species, and analysed results for 1999 and 2000. There was evidence that some mixtures of grassland species could outyield the highest-yielding monocultures, especially several years after establishment. Lambers et al. (2004) examined the same plots, analysed up to 2002. Over/underyielding species (i.e. those that yielded more or less as richness increased) were determined by regression of yield on species richness. Compared to the −1 slope on log–log basis expected under a null model, six species significantly overyielded and four underyielded. There was no time trend for overyielders to replace underyielders, which is strange. In early results from biodiversity experiments maximum overyielding effects usually occurred with only a few species, perhaps only two, and beyond this extra species had no effect. However, Marquard et al. (2009) reported that five years after sowing in the Jena biodiversity experiment, overyielding showed no sign of slackening from 8 to 16 species.

Using a different approach based on the analysis of the growth of individual trees within a 50-ha patch of tropical forest, Chen et al. (2016) observed increasing focal tree growth rates with increasing neighbour trait dissimilarity – a convincing demonstration of neighbourhood complementarity. Computer simulations confirmed that these local interactions were consistent with a positive biodiversity–productivity relationship at the community scale.

4.9.2   Time

There is a tendency in such experiments to find that if overyield is reached, it is only after several years' growth. For example, Reich et al. (2012) found this for Cedar Creek experiments and van Ruijven and Berendse (2009) for the Wageningen, Netherlands, experiment, though in the Wageningen experiment and in one of the Cedar Creek experiments there was no increase in the overyielding effect after five years. Cardinale et al. (2007), reanalysing 44 biodiversity experiments, found a significant tendency for the Loreau/Hector complementarity effect to increase with time.

One explanation for this is that time is required for the species to sort themselves into their respective niches, e.g. in the canopy or in rooting depth. Another possibility is that overyield is attained only with a certain optimum proportion of each species (as can be seen in many of the results of de Wit's group), and time is needed for the species to reach these proportions. Possibly, as in Whittaker's (1965) Niche-preemption concept, the dominant species takes all the resources it can, and subordinate species remain in minor amounts to pick up the crumbs that fall from the strong man's table. A further possibility is that character displacement may be occurring during such experiments, perhaps genetic selection, though it is difficult to exclude plastic effects when vegetative propagation is present, or even carry-over effects from the seed stage (Zuppinger-Dingley et al. 2014).

4.9.3   Mechanism

Overyield is not necessarily due to niche differentiation. Facilitation between species could cause overyield, as Vanelslander et al. (2009) observed in microbial microcosms, where substances leached from some diatom species were taken up by another. Legumes are probably the most common cause of overyield through facilitation. In the Jena biodiversity experiment, Marquard et al. (2009) found that by four years after sowing all the overyield (all of it being Loreau/Hector niche complementarity) was due to legumes. Van Ruijven and Berendse (2009) did report niche-complementary overyield with no legumes present in the Wageningen biodiversity experiment, though with no indication of transgressive overyield.

Interactions via heterotrophs (Section 2.7) are another possible mechanism of overyield, especially a reduction in pest pressure (Section 3.4). Thus, Maron et al. (2011), in a field experiment in Fort Missoula, USA, where 2–16 species were sown and with overyielding observed with up to 4–8 species, the overyield effect disappeared when fungicide was added.

Just as environmental fluctuation can be the mechanism of reliability (Section 4.8.6), so it can be of overyield, if different species can rise to the challenge in different years (Allan et al. 2011). There seems to have been little investigation of this.

Overyield is increasingly being reported, but the mechanisms warrant much further research.

4.9.4   Conclusion

Overyielding does not necessarily develop in a mixture. Hooper and Dukes (2004) planted field plots with local grassland species and followed them for an impressive eight years, giving the species time to reach their optimal niches and optimal proportions, but no significant transgressive overyielding was seen. Cardinale et al. (2007) in their literature survey/analysis found transgressive overyield in only 12 per cent of their analyses.

When overyielding is seen, it has long been known that it usually develops over time and is correlated with niche differentiation. Whittington and O'Brien (1968) planted field plots with *Lolium perenne* (ryegrass), *Festuca pratensis* (meadow fescue) and their triploid hybrid. Transgressive overyielding was seen in the simulated grazing treatment sporadically in the first two years, but almost constantly in the third. In two of the mixtures, the two components had diverged in their proportional contribution to the sward. The overyield in these mixtures can be attributed to niche differentiation in the depth of nutrient uptake. *Lolium perenne* and *F. pratensis* grown in monoculture took up very similar percentages of their P from 10, 30 and 60 cm depth (35/40/25 per cent versus 38/35/27 per cent, respectively) (O'Brien et al. 1967). Yet when they were grown together, *L. perenne* took up its P from deeper (33/22/45 per cent) and *F. pratensis* from nearer the surface (41/34/25 per cent). In contrast, when grown with the deeper-rooted hybrid (monoculture uptake 18/36/47 per cent), *L. perenne* utilised the surface P (56/26/18 per cent). This is circumstantial evidence that the overyield was due to niche differentiation in uptake depth. If the effect is in root system morphology, not just uptake activity, this is another example of plasticity in competition.

4.10   Conclusions

Many have tried to make generalisations about plant community change, but, none are generally applicable. Even general phrases such as 'Succession as a gradient in time' (Pickett 1976) can be misleading when a switch results in stasis or a mosaic, or when there is cyclic succession. There have been fashions. Before Egler (1954), and for some time afterwards, many ecologists dismissed Egler as an eccentric, and most ecologists held to Clements' (1916) concepts of succession-towards-climax in a much more simplistic way than Clements himself had. The movement towards other concepts began in earnest with Connell and Slatyer (1977), and progressed until some ecologists declared that the very word 'climax' should be avoided. Many have declared that Gleason was 'rejecting virtually all of Clements' ideas on succession' (Barbour et al. 1999), but in fact his concept of the process of succession (Gleason 1917, 1927) was almost identical (see Section 6.2).

The background to community processes (Figure 4.13) includes geology, speciation, biogeographic distribution through vicariance and dispersal, and environmental filtering (steps **A–C** of Section 1.5). In Chapter 2 we discussed the interactions between species that follow. It is clear that given an area initially without plants, certain species will

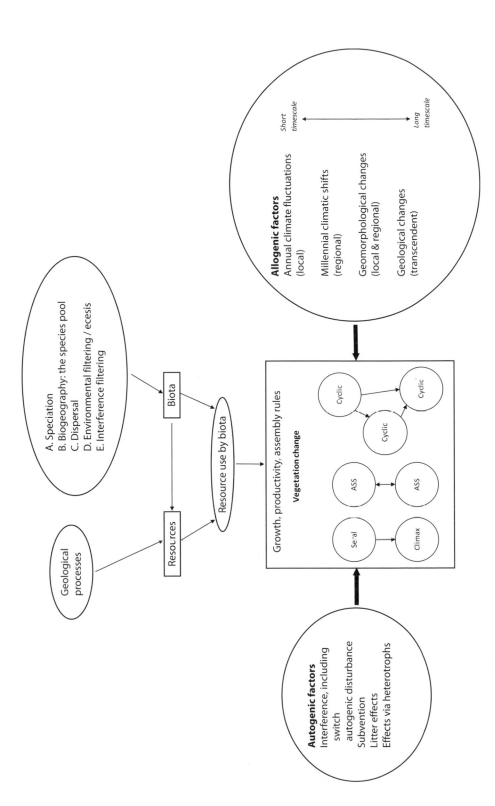

Figure 4.13 The background of vegetation change.

predominate first. Whether they do this through dispersal, fecundity or stress tolerance will depend on the environment, and there has been very little work attempting to separate those. Other species will enter later partly because, though not having those characteristics, they have greater interference ability, both through their ability to suppress other species and through their ability to tolerate the later-succession environment, especially lower P and light availability. The extent to which the later arrivals rely on facilitation by the pioneer species will again depend on the environment and on the species, but is very rarely known (but see Walker and Chapin 1986; Walker et al. 2003a). A single, stable 'climax' endpoint is present in one of the three possible pathways (Figures 4.1), and the concept should not be thrown out with the bathwater. On the contrary, succession-towards-climax should be the null model, not because Clements was almost always right (though he was), but because it happens often and it is more profitable to focus on the exceptions. The situations that Clements termed subclimaxes are another exception to succession-towards-climax, but his use of them was much more restrained than that of those peddling a caricature of his theory. He saw fire and grazing subclimaxes as human-induced, but these days ecologists are more likely to see fire *suppression* as human-induced and to advocate the reintroduction of grazers and their predators for restoration. The only environmental factor that Clements (1916) saw as retarding succession to give a subclimax was saline/alkaline soils; he thought that in some cases reaction would lead to the climax, but that other soils would be too intransigent. The replacement of communities (*sensu lato*) in a cycle is an obvious exception from the null model, but evidence for it is so hard to find that one can be fairly sure it is less common than Watt supposed.

Reaction is so widespread that altruistic facilitation should be part of the succession-towards-climax null model. However, there is every reason to expect that reaction is often in the opposite direction, giving a switch (a deviation from the null model), and this appears to be the case. Although we have emphasised that most putative switches are unproven, this is quite as true for succession-towards-climax and for cyclic succession. Moreover, switches are probably often part of relay-floristics succession, either speeding it up, in which case they will hardly be noticed, or slowing it, in which case the persistence of an earlier seral stage is noted (Egler 1954). It is when switches sharpen a gradient or magnify small initial differences to give a mosaic of ASSs in space or time that they produce patterns that are hard to miss. Clements did eschew the idea of alternative pathways of succession (Clements 1916, p. 67), but he allowed two types of reaction that retarded succession: what we have described as the *Sphagnum* bog pH switch (Section 2.4.3.1), which he described as flooding, and the Ericaceae (e.g. *Calluna vulgaris*) switch (Section 4.5.4.5), which he ascribed to 'acids and other harmful substances' (Clements 1916, p. 107).

Throughout succession-towards-climax, cyclic succession or the operation of a switch, there may be an allogenic environmental change (press perturbation; Section 2) or allogenic disturbance (pulse perturbation; Section 4.2), changing the rules of the game. Response to and recovery from pulse perturbation were the subjects of Section 4.8 of this chapter, but we concluded that it is hard to see generalisations, and probably always will be because of individualistic responses by species.

Having reached any of the endpoints of a climax or ASS, or one of the phases of a cyclic succession, the behaviour of that state under perturbation goes a long way towards telling us whether there had been a rather fragile assemblage of species, or whether there was, and will be, a plant community in any real sense. Such a community should show Lyapunov stability. This is difficult to test for, and as a result, very rarely has been. This should be the primary target of community ecologists. Resistance and resilience to perturbation are poor substitutes; in fact they have little meaning if the community is not Lyapunov stable. Resistance to invasion is a probe of community structure, and we consider it further in Chapter 5 (Section 5.12). Of these 'stability' *sensu lato* features, reliability is the least informative, but currently the best known, because it is more easily determined. If plant ecologists are serious about the nature of plant communities, considerable further work is needed.

5   Assembly Rules

5.1   Introduction

In Chapter 2, we outlined the processes that occur in plant communities: interference, facilitation, litter effects, autogenic disturbance, parasitism and interactions via heterotrophs. Many ecologists wish to go no further with plant communities than look at such processes, but we want to make generalisations at the community level. Descriptive ecology, or 'phytosociology', seeks to make regional vegetation inventories using the methods originating with Braun-Blanquet (1932), identifying and naming communities. This has value in conservation advocacy and an ecological tourist's guide, but ecologists must ask: where are the testable hypotheses? In our approach we look for the rules of engagement in plant associations that would arise from interspecific interactions. These are the assembly rules, which we define as 'restrictions on the observed patterns of species presence or abundance that are based on the presence or abundance of one or other species or groups of species (not simply the response of individual species to the environment)' (Wilson 1999a). This is close to Hubbell's (2005) definition of assembly as 'which species, having which niche traits, and how many species, co-occur in a given community'. Perhaps this is the true meaning of phytosociology. Take care when reading the literature, as at least two differing definitions of 'assembly rule' have been adopted (Booth and Larson 1999; Temperton et al. 1999). The first limits the definition to include only constraints from biotic interactions (e.g. Diamond 1975, and our definition above). The second is typically much broader and combines both biotic and environmental constraints on community assembly (e.g. Keddy 1992).

It would problematic if, as has been suggested, assembly rules do not operate after a disturbance until the community regains equilibrium (e.g. Drake et al. 1999; Stokes and Archer 2010). This idea was based only on speculation, and when Cash et al. (2012) tested the concept in alpine areas of southern New Zealand, they found there was as much evidence for assembly-rule structure in the areas disturbed by ski-run making as in the undisturbed ones. In any case, the concept of an undisturbed community is a mirage since allogenic and autogenic disturbance are both omnipresent in plant communities. Nevertheless, we shall tend to concentrate on what seem to be equilibrium communities. Another problem comes from Yodzis' (1978, 1986) distinction between founder control of community composition (priority, Section 5.4.2) and dominance/ niche control. Under the former, the species composition of a community will depend largely on which species arrives first and there will be no further predictability, no assembly rules. Ozinga et al. (2005) addressed this issue using a 20,000-quadrat database. On average among species the first four axes of a CCA ordination, constrained by six Ellenberg scores, explained only 7.7 per cent of species occurrences, though the value was 10.3 per cent for species with long-lived seeds and a mechanism for long-distance dispersal. This implies a role for founder control, though this conclusion relies on the completeness of the environmental characterisation: the same argument and the same problem we encountered with 'chance' as a mechanism of coexistence (Section 3.10). Another problem is that the species the taxonomist sees are not necessarily the entities that the plants see. We have forsworn, in general, consideration of within-species genetic (e.g. ecotypic) differences and plastic responses in this book, but both are important.

There is a widespread and commendable scepticism as to whether assembly rules occur at all (e.g. Ulrich 2004). This may not be our conclusion, but our reductionist aim demands that we start with null models and that we be especially careful in examining the evidence.

5.2   What Rules Should Ecologists Search for, and How?

5.2.1   Types of Assembly Rule

We shall consider a wide range of possible assembly rules (Box 5.1).

5.2.2   Inductive Versus Deductive

Inductive and deductive approaches both have value in community ecology (Dale 2002; Wilson 2003), and both will be seen below. An example of the deductive approach is guild proportionality in forest: differences between species in their mature height are well established, and we can reason that these represent different niches, with the species potentially capable of occupying each niche together constituting a guild. We can reason that a species will colonise more readily where few species from its own guild are already present, giving a rule of rather constant guild proportions (Section 5.7.1). If the appropriate null model is disproved, and if other explanations such as environmental effects can be ruled out, the existence of the rule has been proved, though not its exact mechanism. On the other hand, a search for intrinsic guilds (Section 5.7.3) is inductive in that no structure is being assumed save that guilds might exist, but so long as the guilds are formed and tested on independent data there is a strong pointer to the processes that are structuring the community. Finding a repeated pattern is the first step to finding its cause.

5.2.3   Null Models

To demonstrate assembly rules, is it essential to compare an observed pattern with that expected under a null model. However, the null model is often difficult to frame.

Box 5.1 Types of assembly rule

Zonation
        Boundaries in zonation
        Fundamental and realised niche
Species sorting
        Species associations in succession
        Compositional convergence and priority
Richness
Limiting similarity
        Limiting similarity in functional traits
        Limiting similarity in phenology
        Limiting similarity in phylogeny
        Guild proportionality
        Removal and invasion experiments
        Intrinsic guilds
Texture convergence
The fate of associations through time
Abundance
        Biomass constancy
        Species abundance distribution (SAD, RAD)
        Rank consistency
        Sparse species
Keystone species
Exotic species as community structure probes
        The nature of exotic species
        Exotic establishment and community assembly

What does a plant community look like when it isn't there? However, a prior question is what pattern to seek: what does a plant community look like when it *is* there?

In these comparisons, randomisation tests are often needed, in which a test statistic is calculated on the observed data, then on data randomised under a certain null model, and significance (i.e. the probability that the observed results would occur under the null model) is determined from the proportion (P) of randomised values that are equal to, or more extreme than, the observed one. There are traps here. Any test statistic can validly be chosen, though it obviously must be one that tests the ecological question asked. Selection of the null model is more crucial; many studies have come unstuck from choosing the wrong null model and demonstrating as a result an obvious fact such that species differ in frequency (Wilson 1995). The Tokeshi principle applies here: the null model must include all the features of the observed data except the one it is intended to test (Wilson 1999a). Lastly, tails: if it is conceivable that the observed data could differ from the null model in either direction, or results either way will be noted, a two-tailed test must be used. This comprises either doubling the P value obtained above or using, say, two 2.5 per cent tails for a 5 per cent test. The H_1 hypothesis to be tested in such frequentist analyses must be formed a priori. If the hypothesis is formed a posteriori, based on a pattern seen in the data being tested, the test is of one out of many patterns that might have been present. This is akin to multiple-hypothesis testing, and the P-values obtained are invalid.

5.2.4   The Confounding Effects of Environmental Variation

In seeking assembly rules, i.e. the repeated patterns of MacArthur (1972):

(a)   The rules sought will not necessarily depend on the identity of particular species. This contrasts with Diamond's (1975) original assembly rules, which specified that particular combinations of named species were forbidden. That approach has not proved useful because it tests no general theory and gives no inductive insight.

(b)   The rules have to transcend in their generality ones of the type: 'species x occurs at low/high values of environmental factor z' (the 'easy task' of community ecology: Warming 1909). Whilst such effects are certainly very important, for assembly rules the search is for species' behaviour that is not primarily environmental, but is based on species' interactions.

Although the search is for some generality, the rules cannot be expected to apply worldwide, in all habitats. For example, rules based on stratification cannot apply to the very few communities that have no stratification, and communities in deserts can be expected to be constructed quite differently from those in rainforests. For trait-based, limiting similarity rules, the traits involved will be different in different habitats, where different resources are limiting.

Environmental correlations ('b' above) are actually a huge problem in seeking assembly rules. Environmental variation occurs at all scales in all communities (Goodall 1954). Often, in seeking assembly rules, environmental variation acts as noise. More worryingly, the null model against which the observed pattern is tested often assumes no environmental variation, so that if the analysis disproves the null model, this could be either because there really is an assembly rule, or because environmental variation has mimicked the effect.

Take the simple case of testing whether variance in species richness differs from a null model. Suppose, unbeknown to us, there is environmental variation such that some environments have few species (just 'A' in Figure 5.1, Environment 1), but others have many ('A B C D', Environments 2–4, Figure 5.1) – the 'waterhole effect' of Pielou (1975) – with no between-quadrat variation of species richness within each of those environments. The community structure is in fact determinate, but will appear as greater variation in between-quadrat richness than expected at random if an overall randomisation – a 'site' model – is used. Under a site model, species are allocated to quadrats independently and at random, irrespective of the underlying environment.

Suppose the number of species is the same in each quadrat, and they are the same species in each quadrat within each of two environments (Figure 5.2). Site-level randomisations will include some quadrats with 0, 1, 3 and 4 species, and the observed

Environ. 1		Environ. 2		Environ. 3		Environ. 4	
A	A	A B	A B	A B C	A B C	A B C D	A B C D
A	A	A B	A B	A B C	A B C	A B C D	A B C D
A	A	A B	A B	A B C	A B C	A B C D	A B C D
A	A	A B	A B	A B C	A B C	A B C D	A B C D

Figure 5.1 Four environments containing different species assemblages, with constant species richness within the eight quadrats that comprise each environment.

Environment 1				Environment 2			
A B	A B	A B	A B	C D	C D	C D	C D
A B	A B	A B	A B	C D	C D	C D	C D
A B	A B	A B	A B	C D	C D	C D	C D
A B	A B	A B	A B	C D	C D	C D	C D

Figure 5.2 Two environments containing different species assemblages, but the same richness.

state will look like constant richness compared to this. The effect is real in that there is the same number of species in each environment. However, this is being tested 20 times in each environment (i.e. the 20 quadrats): habitat pseudoreplication. A test over several environments would be valid and interesting, but then one has to include each environment/community only once and one needs many environments.

Fuller and Enquist (2012) illustrate the problem by analysing a 14-ha plot of tropical dry forest in Costa Rica, showing that not taking into account spatial heterogeneity gave false estimates of species associations. The best answer to these problems is to use a patch model rather than a site model. This comprises making a prediction for each quadrat (the 'target' quadrat) on the basis of a limited number of adjacent or otherwise similar quadrats, which are more likely to be similar to the target quadrat in environment (Figure 5.3). The patch can be square (Figure 5.3) or linear, or a grouping of quadrats can be determined a priori as likely to be similar in environment. If the quadrats are not contiguous, quadrats can be grouped into patches, usually nearby groupings. Incorrect groupings are inefficient, but not invalid. Patch models tend to have lower power: a small price to pay for validity.

Since environmental variation is ubiquitous in nature, and very liable to confound assembly rules, we shall emphasise ways of dealing with this problem, and especially patch models.

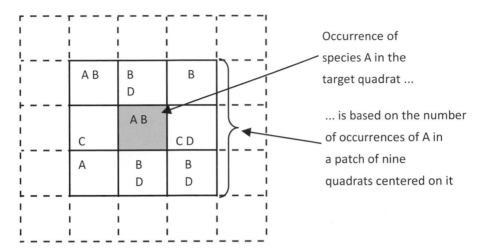

Figure 5.3 A patch randomisation model based on a grid of contiguous quadrats. The frequency of species A in the 3×3 patch is $3/9 = 0.333$, so in the randomisation species A has a 0.333 probability of occurring in the central square.

5.2.5 Taxonomy- and Phylogeny-Based Limiting Similarity

In animal ecology, membership of a genus/clade, or even a family, is commonly used to indicate similarity in alpha niche. However, in plants the alpha niche is commonly quite independent of taxonomy and phylogeny. For example, genus/clade *Hypericum* ranges in life form from annual herbs to 12-m trees in Africa (*H. revolutum* and *H. bequaertii*); genus/clade *Euphorbia* includes life forms from annual herbs through cactus look-alikes to long-lived trees 30 m tall. Indeed, sometimes membership of a genus is more representative of a species' beta niche (e.g. *Salicornia* spp. in saline areas). However, genera and clades are clearly a priori classifications, independent of the assembly-rule questions being asked, and if they are not a good guide to ecology, the result will be non-significance, not spurious significance.

5.2.6 Process versus Pattern

Ecologists often suggest that 'assembly rule' should mean the *process* by which the community is established. Whilst this is a logical thought, Diamond (1975) first used the term for the *results* of that process. Most later workers have used it to mean the pattern, i.e. the results, and we do so here.

5.3 Zonation

The study of changes in ecological communities along environmental gradients – zonation – has a long history in ecological research (see review by Keddy 2007). Indeed, the readily observable patterns evident in zonation provide opportunities for studying and testing a wide range of ecological hypotheses. Of particular relevance to our discussion of assembly rules is the detection of boundaries (Section 5.3.1) and the use of zonation patterns to explore species niches (Section 5.3.2).

5.3.1   Boundaries in Zonation

The ideal way to determine whether species are associated into discrete communities, as Robert H. Whittaker (1975a) pointed out, is to see whether their boundaries are clustered on an environmental gradient, e.g. to distinguish between the situations in Figure 5.4 a and b. Answering the question is much more difficult (Wilson 1994b).

The appropriate test here is to compare the observed pattern of upper and lower species boundaries against an appropriate null model; if species' boundaries tend to be clumped together, and if sets of species are coincident in their distribution, then this provides confirmatory evidence in favour of Figure 6.1a.

The first rigorous application of such a test was conducted by Pielou and Routledge (1976), using data collected from salt marshes at different latitudes in North America. They found statistically significant clustering, particularly for the upper boundaries, and a latitudinal pattern with evidence for a stonger trend at higher latitudes. Subsequent analyses in other environments also provided evidence for boundary clustering, though not always detecting the neat differentiation of discrete assemblages as depicted in Figure 6.1a. Examples include Keddy (1983) for a lakeshore zonation in Ontario, Canada; Shipley and Keddy (1987) for a marsh zonation on a gently sloping shoreline in Ontario, Canada; and Auerbach and Shmida (1993) for an altitudinal zonation on Mt Hermon, Israel. In a study of 42 wetland sites in Minnesota, USA, Hoagland and Collins (1997) also found evidence for clustered boundaries in 10 of their 42 sites. These results are heartening, but perhaps not entirely surprising. After all, many zonation patterns are visually identifiable, and thus should be amenable to statistical verification.

Whilst studies such as these clearly demonstrate the reality of non-random species distributions along environmental gradients, exactly how to interpret the results in the context of species assembly rules is problematic. First, our search for assembly

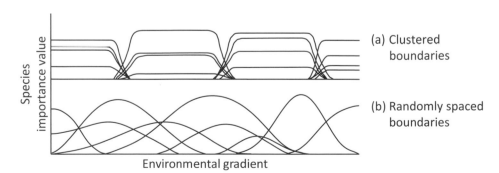

Figure 5.4 (a and b) Whittaker's concept of different distributions of species along an environmental gradient (see also Figure 6.1).

rules requires removing (or accounting for) the effects of environmental variation (Section 5.2.4); however zonation, by definition, involves changes along an environmental gradient. Any search for assembly rules along such a gradient therefore runs the risk of confounding environmental filtering with limitations imposed by biotic interactions (which are likely to be present and important, and to be acting in concert with environmental filtering). Pielou and Routledge (1976) provided evidence for a role of biological interactions in shaping some of the patterns in their data, and we note that such boundaries could be maintained via a switch (Section 4.5). As is often the case, careful observational and experimental work is required to unambiguously identify the underlying mechanisms.

A second complication is that zonation studies typically define gradient axes a priori, such as elevation, wave exposure or temperature. Such gradients are a proxy factor for those actually affecting the plants and may not be related to any of them. In general the true factors and the scale on which to express them are both unknown, so that any clustering might not be on the true factor. Observing bimodality of species distributions would be a mildly interesting feature because it is less likely to be affected by factor scaling, but evidence for it (e.g. Whittaker 1960, 1967) is weak (Wilson et al. 2004).

The problem of defining the scale of environmental gradients was solved by Dale (1984) by abolishing scales. He took up a previous implication that, looking at the sequence of top- and bottom-boundaries along a gradient (an intertidal shore in his case), the top boundary (T) of one species would be immediately followed by the bottom boundary (B) of another (the one replacing the other in the same alpha niche): a TB pair. Therefore, overall there would be an excess of TB pairs compared to random expectation. This test is non-parametric, in that it is absolutely unaffected by any monotonic rescaling of the axis. However, the non-null (H_1) hypothesis assumes very precise replacement of one species by another, with a small gap, or it will not be possible to identify the TB order. Such a gap is hard to envisage in the real world (Wilson 1994b). It is surprising that Dale himself found excesses of TB pairs significantly often. Thomas et al. (1999), using Dale's method on three coastal zonations in South Island, New Zealand – sand dune, saltmarsh and cliff – did not.

Perhaps valid answers can be obtained only by changing the question. Wilson and Lee (1994), examining an altitudinal gradient in the Murchison Mountains, southern New Zealand, formed a null model in which the number of species, their frequency patterns and positions along the gradient and the number of species in each genus were all held as observed, but species were assigned to genera randomly. The test statistic was the amount of overlap along the gradient between species in the same genus. The basis for this is an assumption that members of one genus will tend to be similar in alpha niches (if this assumption is false, it will lead to non-significance, but not to a spurious positive result: Section 5.2.4). They will compete with each other in either ecological or evolutionary time (the 'ghosts of competition past'), and hence be spaced-out in beta niche (altitude), with less overlap than expected for a random selection of species. The results are complicated because testing several genera separately comprises making multiple significance tests. Some genera are known to have altitudinal biases (Pielou

1978 showed that this was true overall for the distribution of algal congeners along a latitudinal gradient) and others have too few species to give significance. However, taking all this into account Wilson and Lee concluded that there was evidence that the species of a genus were more spaced-out in altitude than expected at random.

5.3.2   Fundamental and Realised Niches in Zonation: Experimental Evidence

5.3.2.1   Niche Shift and Contraction

Niche shift (including 'habitat shift') is a change in mean/modal resource usage by a single species in different areas and/or when its associated species are different (Schoener 1986). It is the difference between fundamental and realised alpha niches, or between realised niches with different associates. Such differences have long been recognised (Gleason 1917). Niche contraction, and its opposite niche expansion, are similar concepts, except that under contraction the niche width changes, not the mean/ mode. There is disagreement in the literature, sometimes even within one paper, as to whether these responses are plastic/behavioural or genetic.

5.3.2.2   Beta Niche

A species' realised niche is certainly related to its fundamental one (Section 1.4), but it is not clear just how. Generally, when two species with largely overlapping fundamental niches meet in the field, their realised niches are different.

For example, Kenkel et al. (1991) grew three species, one a facultative halophyte, in a range of rather low salinities in sand culture. In monoculture, they all grew best with no added NaCl, but in mixed pots they sorted themselves into three realised-niche optima along the gradient (though there was little sign of niche contraction). In most situations, one species shifts further along the gradient than the other. A well-known example is the work of Grace and Wetzel (1981) growing two *Typha* (cattail) species on a gradient of average water depth. In monoculture, both had the same optimum depth of 50 cm. In mixture they hardly overlapped in the depths at which they grew: *Typha latifolia* contracted its niche and shifted its optimum to 15 cm depth and *T. angustifolia* to 80 cm. Similarly, Pennings et al. (2005) investigated a southeast USA saltmarsh, where *Juncus roemerianus* grows higher up on the marsh and *Spartina alterniflora* grows lower, with a sharp boundary between them. The lower limit of *J. roemerianus* is set by the physical environment (salt and/or waterlogging), but the upper limit of *S. alterniflora* is set by interference, for when *J. roemerianus* was removed from its zone, *S. alterniflora* grew, if anything, slightly better there than in its own zone. Niche shift and niche contraction are clearly common in nature; the experiments simply clarify the process.

Can we generalise? Austin (1982) grew five grass species in a greenhouse sand culture with a range of nutrient solution concentrations, both in monoculture and in a five-species mixture. Performance was calculated as shoot dry mass relative to the highest-yielding species in those conditions. He found that in most concentrations the performance of a species in mixture was correlated with its performance in monoculture, but the shape of the relation depended on the nutrient level and was often markedly non-linear. Thus, a species' realised niche could generally but not always be predicted from its fundamental niche. Similarly, Pickett and Bazzaz (1978) grew six species along an experimental soil moisture gradient in a greenhouse, in monoculture and in a six-species mixture. The optimum soil moisture remained the same for four of the species, but for most species the niche widths had contracted in the mixture. Fascinating results came from Wilson and Keddy (1985), who examined a field gradient in organic content along a lakeshore. The vegetation gradient is probably caused by wave action, and it is correlated also with soil mechanical composition, nutrients and water depth. Twelve of the species were also grown in pots out-of-doors containing various mixtures of sand and highly organic soil collected from the field. The shape of the response to the gradient, field versus experimental, was:

- Not or hardly related: five species.
- The opposite skewness: three species.
- Related or vaguely related: four species. (The response was sharper in the field in one of these, less sharp in another, equal in a third, and the relation was too vague to see in the fourth.)

One possibility is that species with weaker interference ability are pushed towards the less favourable end of the gradient. This can be seen in the work of Pickett and Bazzaz (1978), where one of the two species most suppressed by interference, *Polygonum pensylvanicum*, was pushed in mixed stands to the dry end of the gradient, where overall growth was less. This also seems to be the situation in the work of Pennings et al. (2005) where *Spartina alterniflora* was restricted by interference to the lower marsh. Further work is needed to determine the generality of these rules.

5.3.2.3 Alpha Niche

The examples discussed above relate to the beta niche, but alpha-niche shift could also be realised, for example in soil depth, canopy height or potentially time. Niche shift has indeed been found in the dimension of rooting depth. For example, Nobel (1997) found that rooting depths for the three codominant species in a site in the Sonoran Desert were 9–10 cm for isolated plants, but roots for interspecific pairs in close proximity contracted to an average of 2–3 cm more shallow for *Agave deserti* and expanded to 2–3 cm deeper for the other two species. The results of O'Brien et al. (1967; this volume Section 3.2.3) are similar. Miller et al. (2007) found niche shift in N source: not technically zonation, but still a type of niche shift.

For the time dimension, i.e. phenology, Veresoglou and Fitter (1984) suggested that when *Holcus lanatus* was growing with certain species (their Area III), its nutrient uptake peaked earlier than in other communities. However, this was true for only one of the two nutrients they examined. Even then, Area III could have been different in other ways.

5.4   Species Sorting

5.4.1   Species Associations in Succession

Greig-Smith (1952) suggested that the strength and direction of species associations would change through succession, and Gitay and Wilson (1995) synthesised these suggestions with the terms of Watt (1947) to suggest three phases in succession:

1. Pioneer phase: Initial colonisation will be essentially at random, with weak associations between species, those tending to be negative. There will be low consistency in the ranks of species between patches (Section 5.10.3 below; Watkins and Wilson 1994).
2. Building phase: As further dispersal removes the effects of chance, some positive and negative association will appear due to microhabitat sorting, but the removal of founder control will lead to higher rank consistency.
3. Mature phase: Species will sort themselves by microhabitat and assembly rules, especially at a larger scale, giving stronger associations, with negative ones predominating if different communities have approximately equal species richness, the microhabitat differentiation giving lower rank consistency again.

In Greig-Smith's (1952) original work in Trinidad tropical rain forest he found little interspecific association in 1.5×1.5 m plots during the pioneer phase (site A, cutover five years earlier). There was considerable association (indicating spatial heterogeneity) in secondary forest that had recovered over a longer period, which we can see as the Building phase (sites B and D). In the mature phase (site E, primary forest) species associations were effectively absent. This fits the Gitay–Wilson model only imperfectly.

More precise estimates of successional status were available to Gitay and Wilson (1995) through known times since burning in tussock grassland sites. The pattern expected under the above model was seen, i.e. association was low and rather negative for the first 10 years, negative and positive associations balancing at 10–20 years, and more negative beyond 20 years. Aarssen and Turkington's (1985) study of three pastures of different age in western Canada showed partial support for the model. They claimed consistently stronger and more negative associations between grass species in the older pastures, though the relevant information presented shows that the total number of significant associations (positive plus negative) was *lower* in the oldest pasture. They do give figures to demonstrate that the number of associations were more consistent over seasons and years in both direction and significance in older pastures. Turkington and Mehrhoff (1990) interpret this as 'transition from an essentially unorganised assemblage of species to a more organised community'. However, in an analysis of Ontario sand quarries of various ages, O'Connor and Aarssen (1988) expected to see what we have called the Mature phase developing; in fact the frequency of negative species associations decreased with time.

The analyses discussed thus far were based on 'space-for-time substitution', with the implicit assumption that the only material difference between sites of different ages is time; an assumption that is very difficult to test. However, the Gitay–Wilson model was confirmed using actual records of succession taken through time from a restoration experiment at Monks Wood, eastern England, where over 13 years, rank consistency increased during the Pioneer phase, was maximal in the Building phase and then decreased markedly in the Mature phase (Wilson et al. 1996a); an identical but non-significant trend was seen over the six years of a restoration experiment in the western midlands of England. del Moral et al. (2010) analysed another genuine record of succession through time, in this case primary succession after the eruption on Mt St. Helens, and found that the sorting of species into habitats increased through succession, from 3.4 per cent of the species composition explained 10 years after the eruption, 7.9 per cent (significant from then on) at 16 years, to 12.5 per cent at 26 years.

The Greig–Smith model as developed by Gitay and Wilson fits well in a number of successions, but not all.

5.4.2 Priority Effects

Ever since Clements (1916) there has been an expectation that the species composition of sites that were initially different for some reason would converge with time. However, more recently it has been suggested that in some situations initial differences would persist, recalling one of the ideas in Egler's (1954, 1977) Initial Floristic Composition proposal: 'priority effects' (Drake 1991). Priority effects have been implicated in the initiation of ASSs (Section 4.6); however, convincing evidence for either convergence or priority effects in natural communities is vanishingly sparse. Convergence is never complete, and failure to converge could be due to habitat differences, and often demonstrably is so. On the other hand, how long are priority effects supposed to last? Meso-/microcosm experiments are therefore very valuable, where near-uniform conditions can be achieved, and several to many generations observed.

Chase (2010) explored priority effects by setting up pond mesocosms, adding N and P at three levels for years 1 and 2 of the experiment. He added, annually, different random selections of pond animals and plants (including green algae and cyanobacteria) into different ponds, such that by year 3 all species were present in all ponds. Other animal species appeared as volunteers. In years 4–7 the species composition of each pond changed little, and Chase suggested this indicated priority effects leading to ASSs. Similarities between replicate ponds in the presence and abundance of plants (including filamentous algae) were consistently higher in the low-nutrient treatments. Chase suggested this was because the species found there comprised only a subset (37 per cent) of those found in low-nutrient ones – a smaller species pool giving less opportunity for priority effects.

Most experiments have been done using just microorganisms. Dickie et al. (2012) inoculated 10 fungal species into experimental wooden discs, but with one species put into its plug three weeks before the others, giving it priority in colonising the disc (some species took advantage of their headstart more than others). The discs were then put into soil in a New Zealand *Nothofagus solandri* (mountain beech) forest for 6 or 13 months. Generally, all species were affected by priority given to others, but the results were somewhat idiosyncratic. Notably, *Phlebia nothofagi* disappeared from the discs after six months, unless given three weeks priority when it remained present in all 10 replicates, though with what abundance was not reported. This can presumably be called a priority effect. Other species showed similar, though less dramatic, effects. *Trametes versicolor* and *Daldinia novae-zelandiae* were the two fungi having most effect on the colonisation of other species after six months. However, by 13 months most of their dominance had disappeared. Some effect of three weeks advanced colonisation would be expected under any null model due to temporal inertia (Section 3.12.1), though some of the effects seem disproportionate. Yet, overall most of the priority effect had disappeared after 13 months. Fukami et al. (2010) performed a very similar experiment, but in the laboratory. The results were in some cases more marked, and possibly remained for longer, though with the same tendency to decrease.

Again with a microorganism, Tan et al. (2012) grew red and white clones of the freshwater bacterium *Serratia marcescens* on agar plates in an incubator, one clone introduced seven days after the other. Examined 42 days later, when the white clone had been introduced second, it comprised 47 per cent (by density) of the mixture, but when it had been introduced first, it came to comprise 82 per cent. This is priority between genotypes of one species, extremely similar genetically. Their interference abilities would have been very similar, and thus replacement slow, and there is no indication from their results how long-lasting the effect of priority would have been beyond the 42 days. Mixtures of other species within and between genera showed much subtler, and in some cases inconsistent, effects. These results do support Vannette and Fukami's (2014) suggestion, arising from their 5-day-long, 2-day delay experiment with yeast species in artificial nectar, that priority effects will be stronger among entities with higher resource use overlap. Since almost all species–species interactions, notably competition, are driven by reaction, their second suggestion of stronger priority effect for species with greater impacts on the environment and response to it is almost a truism.

Any ecological effect may differ with the external environment, and Clements et al. (2013) in an experiment with three bactivorous protozoan species in petri dishes found convergence at some temperatures and divergence at others, and even this depending on the species involved.

Waiting for several generations in an experiment is boring. Mergeay et al. (2011) overcame this by using a past natural experiment, analysing microfossils to reconstruct communities of *Daphnia* in a freshwater lake that had changed between low and high water levels 16 times over the previous 1,800 years. The correlation between the species composition at the end of a low period and that 10 years into a low→high change was highly significant, with 70 per cent Bray–Curtis similarity, and similarity remained c. 50 per cent for 40 further years, i.e. > 500 generations since the change. This was independent of reconstructed environment, and is evidence of past priority effects. A similar study with plants would be interesting were a suitable study system found.

It would be fascinating to see how similar species assembly was in identical conditions. Truly identical can never be achieved, but Crawley et al. (1999) approached it by sowing a mixture of 80 forbs into six replicate blocks in an experimental field. After seven years, *Tanacetum vulgare* (tansy) predominated among the sown species, varying across five of the blocks from 10 per cent to 72 per cent of the standing crop (Table 5.1). However, it comprised only 0.1 per cent in block 3, where four of the other five sown species present exceeded it. There is no convergence here. Among the volunteers the most abundant was *Alopecurus pratensis* (foxtail), which reached 86 per cent in one block but was absent in block 6, then *Holcus lanatus* (Yorkshire fog) with a 64 per cent maximum but absent from four of the six blocks, and *Arrhenatherum elatius* (oat grass) varying from 0 to 31 per cent. Again, huge ranges. Recall that these plots had been made as similar as possible. Crawley et al. describe this variation among them as a 'quite remarkable degree of similarity', but we would describe it as quite remarkable dissimilarity. Crawley et al. (1999) went further and described the blocks as remarkably similar in species diversity, but in fact variance in species richness between blocks was three times greater than expected under a null model, and significantly so (by the method of Wilson et al. 1987).

Fukami et al. (2005) reported an experiment in which outdoor plots[1] were sown (unreplicated) to five different combinations of four species out of a pool of 15. Natural colonisation was also allowed. One year after establishment, the quantitative species composition of the mixtures was considerably different,[2] and those differences remained eight years after sowing with no sign of convergence (Figure 5.5a). The authors called this a priority effect, which might imply a switch, but if interference abilities were rather similar, inertia might be involved. But in spite of the persistent differences in species composition, the different five-species mixtures converged in terms of composition as defined by 14 guilds (Figure 5.5b; an example guild being 'Autumn-germinating annuals, typically tall with semi-rosette form and wind-dispersed seeds').[3] This is the most impressive example of a priority effect: the convergence in texture indicates that considerable species sorting is going on, yet differences in species composition remain.

How large and long-lasting must an effect be before it can be declared a 'priority effect'? The species that is abundant at first will, depending on the length of life of its ramets and their reproduction, continue to dominate in abundance simply by temporal inertia. Moreover, resource levels may have been depleted in the initial colonisation, slowing its replacement by other species. For us, a priority effect is interesting only if it transcends such temporal inertia and establishes ASSs. A true demonstration of ASSs arising from priority must test that each state has Lyapunov stability, and none of the above have made such a test (we dealt with studies that have more rigorously tested for ASSs in Chapter 4). For an ASS, a switch must have been operating, but in none of the above studies has the switch-mediating factor even been discussed.

---

[1] In the results we are presenting.
[2] Using Euclidean distance based on cover values that were 'visually estimated'.
[3] The danger that the clearer trend with guilds was because they averaged over several species was disproved with a randomisation test.

Table 5.1 Dry mass (g m^{-2}) of selected species in six replicate plots in the experiment of Crawley et al. (1999)

		Replicate					
		1	2	3	4	5	6
Tanacetum vulgare	Sown	10	32	<1	72	12	27
Alopecurus pratensis	Volunteer	86	33	61	28	46	–
Holcus lanatus	Volunteer	–	–	–	–	41	64
Arrhenatherum elatius	Volunteer	4	31	3	<1	–	9

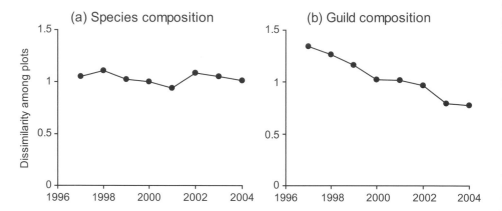

Figure 5.5 Euclidean dissimilarities between field plots sown with different mixtures of species in 1996. (a) Species competition; (b) guild composition.
Redrawn from Fukami et al. (2005).

5.5   Richness

A basic question in community ecology is whether there is a limit to the number of species that can coexist locally, a hypothesis that follows simply from the limiting similarity concept of MacArthur and Levins' (1967). Under this concept, there can be no more species present in a sample (e.g. quadrat) than there are niches, so the number of species in a quadrat should be limited by the number of niches and be rather constant across quadrats in the same habitat. In practice a niche may be empty in a patch, or split between two species, or two species could coexist transiently, but there is constraint (of course, this simplistically ignores the other 11 mechanisms of species coexistence: Chapter 3). There should therefore be lower variance in the species richness of quadrats than would be expected under a null model in which the number of occurrences of each species is held at that observed, but those occurrences are scattered across the quadrats, independently of other species (Wilson et al. 1987). It is often difficult to see such an effect because of overlain environmental variation and disturbances, and perhaps the presence of empty niches. Possibly for these reasons, Wilson et al. (1987) failed to show variance lower than expected under the null model in two communities at scales of

5×5 m and 2×2 cm respectively, and Wilson and Sykes (1988) at 10×10 m. However, Watkins and Wilson (1992) found lower variance than expected under the null model at the scale of 13×13 mm, and this remained for 6 of the 12 lawn communities when analysed with a patch model (Section 5.2.4). There may be remaining doubts that the limitation at this scale is due to geometric packing of individuals, but variance in richness remains a basic question.

Niche limitation should also be seen in constancy in species richness through time. Wilson et al. (1995a) claimed this for limestone grassland (alvar) on Öland, Sweden, though it was contested. Elmendorf and Harrison (2011) found similar constancy in Californian (USA) grasslands. However, experimentally adding species did not lead to others going locally extinct, nor did removals lead to other species colonising. They therefore concluded that the relatively constant richness was caused by environmental heterogeneity (Section 5.2.4), species negative covariance (Section 4.8.6), or because in species-rich plots there was low colonisation (because most species were already present) and high extirpation (because it was sparse species that made those plots species rich).

In an experimental approach to niche limitation, Levine (2001) sowed seeds of several native and exotic plant species into tussocks of *Carex nudata*. He found even the most diverse tussocks were colonised, and concluded that they had not been completely saturated with species. However, some of these species might not have persisted in the long term.

Behind all this is the question of species saturation. Wilson (1961) concluded that most of the ant faunas of the Moluccas-Melanesia are saturated, using as evidence a close correlation between the size of the fauna and the area of the island. Cornell and Lawton (1992) suggested that it would be possible to identify niche saturation from the relation between regional and local richness. If niche saturation exists, then as the regional species pool increases, local richness will increase proportionally at first, but level off to a maximum. If there is no saturation, the relation will continue to be linear. It is easy to show in models of community assembly that such saturation will occur (Fukami 2004), but will it in the real world? Although it is easy to determine richness at the site level, the estimation of regional species pools involves too many arbitrary and subjective decisions. There is also a problem of circularity: local richness is affected by the regional richness, as Cornell and Lawton reasoned, but regional (gamma) richness is a combination of local (alpha) richness and between-site (beta) richness and so not independent of it. Wilson and Anderson (2001) concluded that comparisons between habitats are not appropriate for such tests because of overlap of the species pools and because ecological processes differ between habitats. Only comparisons between equivalent habitats in areas with different floras, such as different continents, are valid, and they cannot be made because there are too few continents for a statistical analysis. A wooden light bulb is beautiful and interesting, but of little use (Wilson and Anderson 2001); likewise the concept of saturation from the species pool concept is stimulating, but because of circularity between alpha and gamma richness and because of the difficulty of finding independent species pools, it is probably operationally impossible to test.

There have also been simple comparisons between different continents in florula size and/or in quadrat species richness. As Orians and Paine (1983) say, 'Implicit in community convergence in species richness patterns is the notion that assemblages eventually reach some saturation level'. However, such comparisons have generally found that the areas compared differ in richness at both area and quadrat scales, e.g. annual grassland in California and Chile by Gulmon (1977), the floras of California and Israel by Shmida (1981) and in the brown intertidal algae in various points around the Atlantic, Pacific and Southern oceans by Orians and Paine (1983). Richness convergence would have implied niche saturation; divergence does not disprove saturation, because the habitats may not be as similar as hoped, or there might be niche straddling/splitting.

Robert H. Whittaker travelled the world recording species diversity in a standard way and in plots whose exact location was carefully selected, attempting to find patterns and thus predictability. In Whittaker (1977) he had reached the conclusion, which he put in a more straightforward way in seminars, 'We once thought species diversity was the one fixed, predictable feature of plant communities. But it isn't'.

5.6   Limiting Similarity

Hutchinson (1959) instigated this topic, as he instigated so much in ecology, by observing that in some mammals and birds of Britain, Iran and the Galapagos Islands, the morphological size ratio between each species and the next larger one was about 1:1.3 for a linear measure. He actually reported a range of 1:1.1 to 1:1.4, but this is usually forgotten. Hutchinson implied that this was partly due to within-species genetic character displacement, and there is some evidence for that in his data. So far as we know this concept has not been applied to plants.

MacArthur and Levins (1967) put the idea of a limiting similarity between the niches of coexisting species on a solid mathematical foundation, be it with some assumptions. A consequence of their analysis is that if two species are too similar in resource use patterns, one will be excluded (Abrams 1990), a reassertion of the Principle of Gause (1934). This is the basis of several of the topics included in this chapter: constant richness (Section 5.5), guild proportionality (Section 5.7) and texture convergence (Section 5.8). It 'underlies current discussions of biodiversity' (Pianka and Horn 2005). The concept has also been referred to as 'community-wide character displacement' (confusingly, because 'character displacement' means genetic change: Section 1.5.4.3) or 'ecological character displacement' (Strong et al. 1979).

It is sometimes suggested that although interference can lead to even trait spacing (i.e. character spacing) in traits that are related to niche differentiation, in traits determining interference ability it will lead to aggregation, because species of lower interference ability will be eliminated (e.g. Herben and Goldberg 2014). This is true, but unrealistic on a local scale. It implies that some patches will comprise species of lower interference ability than others. Is that likely? If the patches are alike in other ways, surely species of higher interference ability will invade. It may be true on a large scale: short-statured species may not be able to persist in a tall grassland (though many do). Yet the tall grass species must be there for a reason. This is environmental filtering.

The quantitative predictions of the MacArthur and Levins theory have not been tested, but even qualitative testing has been fraught. It is even difficult to know what test statistic to use – e.g. minimum trait distance, even distances, greater range – or which traits are appropriate (Stubbs and Wilson 2004). It is usually unclear what is being examined: plastic responses, exclusion by interference between species, character displacement or the coevolution of species. Hubbell (2005) concluded: 'The empirical evidence, in general, has not borne out these [MacArthur and Levins, etc.] predictions..., particularly in plant communities'. We wish to look further, and with plants at that.

Terminology has been a problem. When co-occurring species are closer in trait space (i.e. more similar) than expected, the terms used have included 'clumped' and 'aggregated'; when they are less similar, terms have been 'evenly spread', 'evenly spaced', 'spaced-out', 'staggered' and 'regular'. These terms are self-explanatory. 'Overdispersed' and 'underdispersed' have also been used. This is unfortunate because overdispersed is the mathematical term for aggregated and underdispersed for evenly spaced (Greig-Smith 1983), whereas in the ecological literature the meanings are often transposed, with 'overdispersion' used to describe evenly spaced. They are therefore ambiguous in usage, and best avoided.

As elsewhere, species have usually been used as units, ignoring polyploids, other within-species variation (except when measuring traits), within-plant somatic variation and generally dioecy. In all these studies there will be plasticity in traits. The aim is not to examine how plants react plastically to their neighbours, so ideally some intrinsic measure of traits should be used, such as their traits in conditions near-optimal for RGR. Yet the plastic response to competition could be part of their ability to coexist. This is difficult.

5.6.1   Limiting Similarity in Functional Traits

Limiting similarity has generally been sought in 'functional' traits, i.e. those thought to be related to a species' alpha niche, since it is similarity in the alpha niche that determines species' ability or inability to coexist. Morphological traits have generally been used for their ready measurement or availability in existing databases, though in more thorough studies traits that are more closely related to physiological function, such as nutrient and chlorophyll contents, have been used.

In a fascinating and pioneering study, Cody (1986) reported a number of pieces of evidence for limiting similarity among woody plants of desert and of South African fynbos. In the Granite Mountains, Mojave Desert, California, he demonstrated that the *Opuntia* species, which are shallow-rooted, were negatively associated among themselves, but all were associated positively with the somewhat deeper-rooted *Yucca schidigera*. At four fynbos sites, he showed even spacing of species of the major proteaceous shrubs in a morphological space of leaf shape and leaf length, with little overlap between species. Positions in morphological space were occupied by different species in different sites, and the position of some species changed between sites, both making the spacing that was observed even more notable. He made no probabilistic test against a null model, and a null model would not be easy to frame, but the patterns are compelling. The one exception to the morphological sorting was overlap between *Protea eximia* and *P. nitida*, and they occurred in different aspect microhabitats, so exclusion by interference would not operate. Most remarkably, in some species, notably *Leucadendron salignum*, plants of the two sexes overlapped considerably on each of the axes, yet were largely separate in the two-dimensional morphological space. For *Leucadendron*, Cody offered evidence that species pairs that are more similar in the two-dimensional space co-occur less often than expected at random. He also found indication that the 80 species of *Leucadendron* in Cape Province, South Africa, were more spaced in morphological space than expected by chance, though with only 20 randomisations the probability cannot be accurately determined, and details of the null model are not clear, especially the treatment of the edges of morphological space. Cody's work is fascinating, and it would be wonderful for some of these leads to be followed up in more detail, but Potts et al. (2011) failed to replicate Cody's results with more rigorous sampling and analyses, including a null model.

There are traps for the null modeller. Weiher et al. (1998) tested for limiting similarity in herbaceous riverside vegetation, with quadrats placed to deliberately give a range in environment (soil fertility and disturbance) and vegetation ('from cattail marshes, to wet sedge meadows to sandy beaches'), measuring 11 vegetative traits. They found a significant tendency for the minimum nearest-neighbour distance in 11-trait space to be greater than expected under their null model (though other test statistics did not give significance). Four of the individual traits showed even spacing. They concluded that there are morphological assembly rules that constrain wetland plant community composition. The main problem with this work is that there was no attempt to avoid environmental heterogeneity, or to allow for such heterogeneity by a patch model (Section 5.2.4) or the like, so the null model they used combined species from several species pools. This means that the departures of the observations from their null model are likely to reflect species habitat preferences, rather than community structure resulting from limiting similarity, as discussed above. To put it another way, there was habitat pseudoreplication, as there would be if we analysed the data of Figure 5.2 by randomising across habitats. It is as if we saw one person's garden with uniformly two species, one a dicotyledon and one a monocotyledon, and a second person's garden also with one dicotyledon and one monocotyledon. Is this limiting similarity in the plants that gardeners choose? Perhaps, but significance could never be obtained from four species in two gardens. Suppose we sampled each garden 100 times, with of course identical results, and analysed the whole dataset. We would get significance, but it would be spurious because we had pseudoreplicated. That is in effect the trap into which the brave attempt of Weiher et al. fell, an illustration of the traps that await those who are less wary.

Stubbs and Wilson (2004) attempted to avoid such problems when they tested for limiting similarity in a New Zealand sand dune community. Twenty-three functional traits were measured on each of the species, covering the morphology of the shoot and root systems and nutrient status, and intended to represent modes of resource acquisition. Since it is not clear at what scale limiting similarity would occur, sampling was at four spatial scales, from a single point up to a scale of 50 m^2. These multiple scales allowed patch models to be used. A carefully selected range of test statistics was used, for example excluding any that were affected by the range of trait values. A test over all traits found that the mean dissimilarity between nearest-neighbour species in functional space and the minimum dissimilarity were both greater than expected under the null model at the 0.1 × 0.1 m scale, supporting the MacArthur and Levins (1967) limiting similarity concept. Limiting similarity effects were seen in separate root and leaf traits when within-species variation was taken into account to calculate measures of overlap – the test most closely aligned to MacArthur and Levins' original theory. The traits showing limiting similarity were mainly those related to rooting patterns and leaf water control and thus probably reflected the acquisition of nutrients and/or water. The implication that competition for water and nutrients limit coexistence seems reasonable for a sand dune. The main problem with this work is the number of tests made – four spatial scales, 23 traits and different test statistics. This seems inevitable when analysis of limiting similarity in plant communities is still in a relatively early stage and it is not yet known at what scales, in what traits and how it will operate. However, almost all the significant results were in the direction of limiting similarity, and the overall results are convincing.

The same workers sampled three zones on a southern New Zealand saltmarsh (Wilson and Stubbs 2012), using very similar sampling methods, traits and methods of analysis. There was evidence of limiting similarity in the two upper zones: the rush community (dominated by the restiad *Apodasmia similis* (jointed rush)) and the shrub community (dominated by the shrub *Plagianthus divaricatus* (Saltmarsh Ribbonwood, Malvaceae). These were both dense-canopy communities, with leaf traits showing significant limiting similarity, implying that canopy interactions were the cause.

An impressive demonstration of limiting similarity comes from the analysis of Kraft and Ackerly (2010) in the most different type of community imaginable from the lawns, saltmarshes, grasslands, etc. of other studies: trees in a 500 × 500 m plot of tropical rainforest in Ecuador. Highly significant even spacing was indicated for SLA, leaf [N] and leaf area, seed mass and some bole characteristics. The best evidence was at the smallest grains (20 × 20 m and especially 5 × 5 m), as seen in many other assembly-rule studies, and as expected from local exclusion by interference. Some of the results are odd, for example effect sizes were reported that were in the opposite direction to those expected if limiting similarity were present, yet significance (from one-tailed tests) suggested, conversely, the presence of limiting similarity. In such testing significance can be severely inflated by spatial autocorrelation, which would surely have been present in such a closely sampled area (a problem by no means limited to their study); but another analysis (reported in supplementary material of Kraft et al. 2008) used a null model that preserved the spatial structure by rotating and shifting the observed grid for each species (thus retaining the spatial autocorrelation structure), and the results were similar to their main analysis. The major reservation is the effect of the environmental and related species-composition variation, notably between ridgetops and valley bottoms and presumably intermediates. Kraft and Ackerly (2010) innovatively dealt with this by using two test statistics that they believed reflected environmental filtering and four that they believed reflected assembly rules, though they admitted that the even-spacing statistics were somewhat influenced by environmental filtering. In an earlier version of the study (Kraft et al. 2008) a very basic patch model, randomising separately within the ridgetops and the valley bottoms, reported that the evidence for even spacing was generally stronger. The use of a local patch model (Figure 5.3) would have strengthened their conclusions considerably, and any loss of power would hardly matter with up to 10,000 quadrats (i.e. at the critical 5×5 m spatial focus). Kraft and Ackerly attributed the limiting similarity that they found to stratification, which is very reasonable, but also to patches of different disturbance age within quadrats. However, it seems unlikely that differently aged disturbance patches would occur within a 5×5 m quadrat (which is where their strongest evidence was found), nor that this should occur quite consistently across the area.

None of these studies have made full use of abundance, but Mouillot et al. (2007) introduced a method to determine whether abundant species tended to be close in functional space because of environmental filtering, or far apart because of limiting similarity. The latter was found in one site out of eight, a shrub fen, but bizarrely only when photosynthetic biomass was replaced by presence/absence.

Most studies of limiting similarity in functional traits have used vegetative traits. In contrast, Armbruster (1986) used traits related to pollinator usage at 12 sites in central and northern South America with unique combinations of *Dalechampia* species (reduced from 26 populations observed in the field). In the ecological sorting ('pure assemblage') null model, the *Dalechampia* species richness of each site was fixed at that observed and the species frequencies, whilst not so fixed, were taken as probabilities of occurrence. As with most assembly-rule work, environmental differences between sites are potentially confounding, no less and probably no more so than in work on a micro scale. Armbruster coped with this by using five different species pools taking into account climatic and geographical ranges. This is a patch model on a grand scale. The test statistic was the number of cases where two species similar in pollinator usage co-occurred (within 50 m) at a site, pollination vectors being determined by observation and flower morphology. After this careful work, *P* was 0.16, not significant. Twelve sites are really too few for a good test. Another model, with character displacement, does not strictly concern us here since we are limiting ourselves to ecological assembly eschewing ecotypic differentiation, but the results were significant, though only using a one-tailed test which is debatable. Almost a decade later at 25 sites in Western Australia, Armbruster et al. (1994) performed a similar study on *Stylidium* species, another genus with complicated floral organs. The test statistic was overlap in the morphological similarity in the flowers of species co-occurring at a site, and again there was a large-scale patch model based on habitat and geography. Only one site with overlap was observed, compared to an average of 4.38 expected under the null model, but this result was not significant ($P = 0.055$, and perhaps it should be doubled to 0.11 for a two-tailed test). Again there was significant character displacement. An important question here is whether to base analyses on vegetative or reproductive traits? Armbruster (1995) suggested that limiting similarity due to ecological sorting would operate more readily in vegetative traits than in reproductive ones (in the latter genetic character displacement would occur), and comparison of his own ecological sorting results with the results of, for example, Cody and Stubbs and Wilson supports this.

Various null models are possible, closely based it is to be hoped on the ecological question being asked. de Bello et al. (2012) introduced an alternative type, where the null species pool for a quadrat (c. 1 ha in their example) comprised species present in the quadrat plus other species that were close (± 1.5 Ellenberg habitat preference scale units) to the mean environmental position in beta niche space (calculated as the mean Ellenberg values of species present in the community). Tested on 27 grassland and forest sites in Estonia this gave more evidence for trait convergence (notably in lateral spread) than by simply randomising the species present. The approach is superficially like a patch model, where species are selected for inclusion in the null model based on ecological similarity from an hypothesised regional pool, rather than spatial proximity. The validity of the approach rests on ensuring these selected species have trait values that are representative of the species that *could* potentially occur at the site. The problem is that this is difficult, if not impossible, to test.

An alternative approach to a spatial-patch model is an environmentally restricted null model. Cornwell and Ackerly (2009) analysed 14 traits for 54 woody species in 44 20×20 m plots along a gradient in soil water content on a ridge in California, USA (average 652 mm/year precipitation, but with soil moisture affected by topography). They included in the null model for each plot only species whose distribution along the soil–water gradient included that plot (but a gradient based on mean trait values gave very similar results). Out of 11 leaf and stem traits, SLA showed significantly even spacing. This is one significant result out of 11, and significance at $P < 0.05$ in a one-tailed test might be non-significant in a two-tailed test. Soil water content is less meaningful than water potential, and it was measured on only one occasion. However, the approach is novel and useful. Using a trait-based variant of Cornwell and Ackerly's (2009) null model, May et al. (2013) sought to control for habitat filtering in semiarid, Mediterranean shrubland and grassland. Over three sites with precipitation ranging from 291 to 420 mm year^{-1} there was evidence for trait divergence only at the smallest sampling scale (0.25×0.25 m), for canopy height and seed mass (but not for SLA nor seed number, correcting to a two-tailed test).

Hubbell (2005, p. 170) wrote: 'Does a limiting niche similarity for species in functional groups exist?...I believe the answer to [this] question is *no* (at least in plants)'. Grime (2006, p. 257) wrote: 'there can be little doubt that the Darwin-Diamond model of competition as the mainspring of trait variation within communities is not supported by the empirical study of plant communities'). Both were too dismissive. Limiting similarity exists in plant communities and can be demonstrated.

5.6.2   Limiting Similarity in Phenology

5.6.2.1   The Concept

The simplicity of time as a niche axis has led to several attempts to ask the question – are the flowering or fruiting times of the species in a community evenly spaced? That is, is there a constraint on the phenology of species that can co-occur? In such work, either the position of species flowering/fruiting peaks can be compared, or the time span of flowering/fruiting, or quantitative measures such as the number of flowers/fruits open at any time. The ecological and evolutionary selective pressures against species that are too similar in flowering time would come from several interactions discussed in Chapter 2 such as competition for pollinators/dispersers, pollen wastage, interference on the stigma and maladapted hybrids. On the other hand, aggregation of reproductive events could be an adaptation to attract pollinators/dispersers, for predator satiation, or a response to seasonal pollinator/disperser availability (Thompson and Willson 1979). Because even spacing and aggregation are both likely, it is essential that any significance test be two-tailed.

The same general concept applies to vegetative phenology too.

5.6.2.2   Flowering Phenology

Investigation was sparked when Stiles (1977) claimed to find evenly spaced flowering for hummingbird-pollinated plants in a Costa Rican tropical forest. Statistical analysis of this dataset, and of such datasets in general, has proved difficult and controversial; an excellent summary is given by Gotelli and Graves (1996). In general, the more recent studies use appropriate randomisation tests, and are valid.

Ashton et al. (1988), examining the six species of *Shorea* section Mutica in tropical rainforest in Malaya, found even spacing of flowering 'at the 4.6 per cent confidence level', but it is not clear whether this was a two-tailed test. Wright and Calderon (1995) tested separately 59 genera from Barro Colorado Island. Flowering times were aggregated in some genera, but evenly spaced in six genera (so far as one can tell converting the two one-tailed tests into a two-tailed one and with the limited number of randomisations used). Thies and Kalko (2004) found that eight forest *Piper* species flowered within a short period and at random within that, but fruiting was evenly spaced. The *P*-values were not adjusted to give a two-tailed test, though the results may have been significant anyway, again with few randomisations.

In French calcareous grasslands, Bernard-Verdier et al. (2012) sampled the vegetation in 12 plots on a soil gradient (primarily soil depth), using point quadrats. Randomising species across all 12 plots, the abundance-weighted variance in flowering onset was greater in most plots than expected at random, significantly so across the 12, but not for seven other traits, and when randomising only species abundances only plant height at reproduction was significant.

Not all niche differences in pollination are via phenology, and interesting conclusions can be made bringing in other information. Pleasants (1980) calculated from flowering time overlap and flower densities the potential for competition for pollinators between bumblebee-pollinated species in some Rocky Mountain Meadow

species; he found that such competition was negatively correlated with presence/
absence association between the species.

5.6.2.3 Fruiting Phenology

Similar tests have been made for an even spacing of fruiting times. For example, Burns
(2005), among 10 woody angiosperms common below the canopy of conifer forest in
an area of British Columbia, Canada, found no evidence for a significantly even spacing
of fruiting times. Poulin et al. (1999) examined phenology in central America, recording
the fruiting times of *Miconia* (Melastomataceae). *Miconia* species' fruiting times from
Barro Colorado Island were not significantly different from a null model, but those from
the same genus in Trinidad and Colombia showed significantly even fruiting times,
though again with few randomisations. In the same study, *Psychotria* (Rubiaceae)
fruiting times were aggregated.

Conclusions on Flowering/Fruiting Phenology
Overall conclusions are difficult, especially with the danger that non-significant results
or aggregation are underreported, but it seems that even spacing of reproductive
phenology sometimes occurs.

There are major problems with all such studies:

It is difficult to know whether to compare overlap between the most similar neighbours, or between all possible pairs of species (Pleasants 1990). Probably species are
affected by the cumulative competitive pressure from several, but not all, species.

Flowering times are usually aggregated on a seasonal scale. In temperate areas, few
species flower in winter, but there is normally aggregation in the tropics too, corresponding to wet/dry seasons (Stiles 1979; Wright and van Schaik 1994). There can be up
to three clumps per year (Parrish and Bazzaz 1979). It is very difficult to demonstrate
even spacing when it is laid over aggregation.

Even within the flowering season (or within a clump), there is usually variation, with
fewer species flowering at the beginning and end. Although it would be possible to
estimate this variation from the data, incorporation of it into a null model starts to
involve circular reasoning. This problem is difficult, but Aizen and Vazquez (2006)
overcame it for hummingbird-pollinated plants in southern South America by basing
their null model on the distribution of flowers pollinated by other vectors, finding good
evidence of even spacing.

There will probably also be variation in pollinator and disperser availability, so
pollination competition will be more intense at the two ends of the season with few
pollinators around (Hanya 2005). This will actually tend to mitigate the problem of
imprecise flowering time within a clump.

The patterns in flowering/fruiting could be caused by any of four processes: (1)
exclusion by interference between preadapted species or ecotypes (i.e. ecological
sorting), (2) coevolution of species, (3) evolution of coadapted ecotypes within species
(i.e. character displacement), or (4) plastic responses (i.e. niche shift). Rarely is it clear
which process particular workers have been intending to test. Most recent studies have
been based on *in-situ* observations of phenology. Although this sounds commendable, it

would actually be preferable to use data on flowering/fruiting times of the species generally, even from deliberately outside the area, in order to exclude '3' and '4' and narrow the possible explanations. Coevolution of species ('2') seems unlikely here because most species occur in several different communities, with different neighbouring species, and could not adapt their flowering times to each community. Ecotypic differentiation ('3') would be difficult when species associations are constantly changing. Plasticity, ('4'), at first sight unlikely, is possible since fruit removal from a plant often causes its flowering period to be extended. Analysis with multiple null models (as performed by Armbruster (1986; Armbruster et al. 1994) would be needed to distinguish between these possibilities. Also, relative flowering time may not be consistent from year to year, because species are responding to different signals (Rathcke and Lacey 1985).

5.6.2.4   Vegetative Phenology

Vegetative phenology might also constrain the coexistence of species. For example, Parrish and Bazzaz (1976) commented that among the six oldfield species that they examined only one pair was similar in the time of peak root growth. Comparison with a null model would have been useful. Veresoglou and Fitter (1984) found differences in vegetative phenology (growth and nutrient uptake) between co-occurring grasses, suggesting that this helped permit coexistence between them, but again there was no comparison with a null model. Rogers (1983) examined sorting of species by vegetative phenology among the vernal guild of herbs in North American deciduous forest. Effects of environment in producing negative correlations were potentially removed by excluding species pairs with negative correlations at a larger scale (50×100 cm), though in fact none were found, an approach conceptually related to the method of Dale (1985). Associations between species in the same guild (ephemeroid, summergreen, annual) were no more or less frequent than between species in different guilds.

Cody and Prigge (2003) made the curious observation that individual shrubs of *Quercus cornelius-mulleri* affect each other's phenology of leaf replacement. Late and early timing alternated annually within individuals and between large or close individuals in space. In response to the question: 'How do the plants decide which is to go early and which later?' the authors proposed that these phenomena could be due to resource depletion or the cost of early bud break. Cody and Prigge do not suggest how staggering of leaf replacement affects fitness. This is an interesting case that could be considered as either interference or facilitation. The results suggest temporal niche partitioning and thus may be implicated in an assembly rule, but without mechanistic insight it is difficult to know how to characterise it.

5.6.2.5   Conclusion

Limiting similarity in phenology is an interesting approach to community structure. It is mainly restricted by difficulties in specifying a null model in which the test focuses on possible assembly rules, but some evidence for phenology-based assembly rules has emerged.

5.6.3 Limiting Similarity in Phylogeny

If functional traits are phylogenetically conservative, it should be possible to use phylogenetic relatedness as a proxy for functional similarity. Using a proxy is liable to offer a weaker test, so why use a proxy when one can use the traits themselves? One advantage is clearly convenience: hypothesised phylogenetic trees derived from DNA are quite widely available, and it is much easier to look this up than to measure hundreds or thousands of plants. The other, theoretical, advantage is that it is never possible to be sure what the crucial functional traits are, and phylogenetic relatedness could encapsulate traits the ecologist had not thought of. Indeed, Liu et al. (2012) found in a subtropical forest in south-east China that species that co-occurred at spatial grains up to 50×50 m tended to be less phylogenetically related, and an associated experiment suggested the effect occurred via host-specific fungal pathogens. A study based on morphological traits would not have found this.

However, several studies that have found significant evenness in functional traits have found that effects were weaker, or lost, using phylogeny as a proxy for ecological similarity (e.g. Kraft and Ackerly 2010; Soliveres et al. 2014). This is not surprising. There are often considerable differences in habit and habitat between closely related species, such as congeners (Section 5.2.5), and this lack of phylogenetic conservatism can be seen in responses to environmental factors such as light and nutrients (Bennett and Cahill 2013).

5.7 Guild Proportionality

Guild proportionality is based on the concept of Pianka (1980): species that are in the same alpha guild will tend to exclude each other. It is akin to some of Diamond's (1975) rules in which closely related species could not coexist. The process would be:

1. Species arrive at a point and some establish (Sections 1.5.1–1.5.3).
2. A further species arrives (the challenge):
    2a. The species may fail to establish. Failure is more likely if the new species is similar in resource use to the majority of the species already present, i.e. it is a member of the same alpha guild (Figure 5.6); or
    2b. If the new species does establish, and species previously present are excluded, the excluded species are more likely to be from the same alpha guild as the newly established species.

Care is needed here, because in the simple scenario above invasion will be determined by the total abundance of each guild, not the number of species in it, so the hypothesis must be of within-guild differences. Note that the mechanism is assumed to operate at a small enough scale to allow the constant possibility of arrival of disseminules and thus challenges. The result would be a tendency towards a relative constancy in time or space in the proportion of species from each of the guilds – guild proportionality (Wilson 1989b). In reality we would not expect exact constancy, but less variation

Patch 1   Patch 2

Figure 5.6 In Patch 1 there is only one tree species, so further tree species will find it easier to invade; in Patch 2 there is only one understorey species, so further understorey species will find it easier to invade. This process will produce more similar guild proportions between the two patches.

than in a null model. The appropriate null model here is one that holds both quadrat richnesses and species frequencies equal to those observed. The finding of guild proportionality would mean that: (1) there is constraint on species presence and (2) it is at least partially related to the traits used in the guild classification. These must be alpha guilds since, to quote Pianka (1980), they refer to niche in the 'narrow sense of resource utilization'. Usually the guilds have been determined by a priori criteria so most of our account concerns them, but it is also possible, and perhaps better, to determine them from within the data: intrinsic guilds (Section 5.7.3).

5.7.1   Evidence: Constancy in Space

5.7.1.1   Local Scale

The first application of the concept to a plant community was by Wilson (1989b) in a New Zealand rainforest, sampled at 10×10 m and analysed with synusial guilds (strata, liancs and epiphytes). There was significant proportionality for canopy trees. In a similar forest sampled with 2-m diameter circular quadrats by Wilson et al. (1995c) the ground and herb strata showed significant proportionality when coastal broadleaved forest and *Nothofagus* forest were combined, which is not ideal. In both studies a site model was used which casts doubt on the results. At the other end of the plant physiognomic spectrum Bycroft et al. (1993) found guild proportionality at the scale of 1×1 m in the herb stratum of a New Zealand *Nothofagus* forest, but it was significant only with a site model, not with a patch model, reinforcing the doubts on the earlier analyses. Working at an even finer scale Wilson and Watkins (1994), sampling 11 lawns at a scale of c. 13×13 mm and using a 3×3-quadrat patch model

(Figure 5.3), found guild proportionality between graminoids and forbs in three of the lawns, but only in the more species-rich quadrats. This was promising: more significances than expected at random,[4] none in species-poor quadrats where there might be empty niches, but significant in the high-richness category where limitations to species packing would be expected to appear. In a follow-up study Wilson and Roxburgh (1994) sought guild proportionality in one of those three lawns, the University of Otago Botany Lawn, using point quadrats. Again there was a significant guild proportionality using graminoid versus forb guilds (we shall synthesise the Botany Lawn results in Section 5.13). Similar results were found in a salt meadow (similar physiognomically to mown lawns), where Wilson and Whittaker (1995) found highly significant guild proportionality for two, though related, a priori guild classifications: narrow versus broad leaves and monocotyledons versus dicotyledons, analysing with a patch model.

Evidence for guild proportionality has also been sought in taller grass/herb-lands. Thus, Weiher et al. (1998) analysed their rivershore data (Section 5.6.1) for guild proportionality. They reported significant proportionality for three guilds, but discounted them after Bonferroni correction. The use of Bonferroni is problematic here since the tests include complementary guilds and are thus far from independent. However, the much greater problem is the deliberate combining of different habitats, which could have led to spurious significance due to habitat variation. Other studies that have sought evidence for guild proportionality in herblands include Wilson and Gitay (1999) who found significant guild structure at a 10×10 cm scale in the intertussock vegetation of 21 sites in a New Zealand tall tussock grassland, and Kikvidze et al. (2005) who analysed subalpine meadows in Georgia (Caucasus), using 4×4 cm quadrats. In this latter study the index of guild proportionality RV_{gp} for the proportions of monocotyledons and dicotyledons was 0.64, impressively below the null-model value of 1.0 and highly significant. A site model was used, but the reality of the result was reinforced by an interference experiment, where the yield of a monocotyledon +dicotyledon mixture was greater than for either monocotyledons or dicotyledons alone.

Most tests for guild proportionality have examined the proportion of occurrences in a guild. Wilson et al. (1996c) examined this for the Park Grass experiment, but then performed a test examining the proportion of *biomass* in each guild, with a null model that kept the occurrences as observed and randomised the biomass among them. In neither test was there significant indication of guild proportionality. However, the approach is sound, and it would be interesting to apply tests based on biomass more widely.

5.7.1.2   Puzzling Results

Great care is necessary with evidence for guild proportionality, partly because community structure is so elusive, and partly because it is so easy to obtain artefacts from habitat variation. The danger is that with habitat variation the null model may be

---

[4] $P = 0.015$ by one-tailed binomial test.

inappropriate. In the case of guild proportionality, if A and C in Figure 5.2 are in one guild and B and D in another, each observed quadrat has guild proportions of 0.5:0.5, with zero variance between quadrats. If occurrences could be randomised (i.e. with somewhat different quadrat and species totals), constant guild proportions would be concluded, given the considerable variation in the null model. This would be guild proportionality that was highly significant but spurious, arising not from species interactions but from environmental control. It is a real result that each environment has one species from each guild, but the difference between the environments is being multiplied 20 times – habitat pseudoreplication again.

Some studies have reported highly significant guild proportionality, yet with results that are puzzling ecologically, and we have to wonder whether they have fallen into one of those traps. Thus, Klimeš et al. (1995) recorded for five years 30×30 cm permanent quadrats in two meadow communities that differed in fertilisation and mowing regimes. There were many cases of guild proportionality using a wide variety of guild classifications and fewer cases of variance excess. Yet, to be frank, plant community structure is often so elusive that caution is needed when it is found. Using a site model, there could possibly be problems with environmental heterogeneity even within the 1.5×1.5 m area, but more worrying is that many of the guilds that showed significance were in traits typically of beta-niche differentiation, not in traits expected to relate to alpha niches, i.e. differences in resource use at one location. Light response could relate to stratification in the community, but how could there be alpha-niche differentiation, i.e. at one point, in pH and soil nitrogen? The winter-green guild is more convincing, suggesting phenological guilds, and with that guild there were significant differences in the fertilised meadow in four years out of the five recorded.

Similarly Bossuyt et al. (2005) analysed 52 1×1-m quadrats, each in a different dune slack in western Belgium and northern France, using forb versus graminoid versus shrub guilds. They found highly significant guild proportionality with forbs. The sampling of 52 slacks differing in age from 5 to 45 years makes us worry about environmental artefacts. Using C-S-R (Section 6.5) they found significant guild proportionality with ruderals but not with C or S species. This is difficult to understand. There could well be disturbed patches for ruderals within each 1×1-m quadrat, but a proportion more constant than expected at random? How would this arise?

5.7.1.3   Biogeographic Scale

The concept of guild proportionality can be seen at a biogeographic scale in the conclusion of Gentry (1988) that the familial composition of tropical rain forests is remarkably constant. For example, members of Fabaceae virtually always dominate neotropical and African 'lowland primary forests'; the plant families represented are 'almost entirely' the same in the New World as the Old. He saw similarity at the generic level too, for example between the New World and Madagascar. These are fascinating observations. Gentry commented that it 'can hardly be due to chance', but made no comparison with a null model. The finding is relevant to guild proportionality only if

families occupy particular niches, Gentry's 'familial-specific niches'. We would generally doubt these, but how else could the result arise? As with taxonomic guilds in
general, non-significant results would be unsurprising, but significant ones are valid
(Section 5.2.5).

In fascinating work, Mohler (1990) examined co-occurrence patterns between species
of *Quercus* (oak) from different subgenera at various sites across the USA. For 12 of the
14 regions that he examined (apparently with a variety of quadrat sizes) there was a
significant tendency for the two most abundant oak species present to be from different
subgenera. This was not related to consistent pairing of particular species. His null
hypothesis was a 0.5 chance of each subgenus, which assumes they are equal in size, but
this would bias the test *against* the situation he found. The data were collected in
various ways, but his consistent result is in spite of this. It was apparently an a posteriori
test (i.e. he thought he saw an interesting effect so he tested it), but the consistency of
the effect over several regions largely overcomes this problem. Mohler considered
various explanations: disease-driven pest pressure, differences in the cues for mast
fruiting, dispersal differences, etc., but could not find any clear single explanation.
The study was considerably extended in careful work by Cavender-Bares et al. (2004).
They examined several *Quercus* spp. in three reserves in central Florida, USA. Traits
that tended to be similar in more frequently co-occurring species included bark thickness, radial growth rate, seedling absolute growth rate and rhizome resprouting. These
are traits that probably adapt to water stress, fire tolerance and soil fertility: i.e.
predominantly beta-niche traits. Habitat preferences were more scattered across a
phylogeny inferred from ribosomal DNA than expected at random, suggesting that
the three *Quercus* subgenera occupied different alpha niches and that the species within
the subgenera had evolved to cover the beta-niche range, mainly in moisture availability. In the phylogeny, the traits indicated as changing less within a clade included acorn
maturation time, embolism due to freezing, wood density, second-year vessel diameters,
and non-significantly seedling leaf lifespan and perhaps specific leaf weight (SLW).
Traits that tended to be dissimilar in co-occurring species, indicative of different alpha
niches, were acorn maturation time, embolism due to freezing, leaf life span and first
year vessel diameters and, non-significantly, SLW and perhaps seedling leaf lifespan.
Because of the tendency for species from far parts of the phylogeny to co-occur, the
latter list should be similar to the list of conservative traits, and it is almost identical.
These should be traits that are related to alpha niche, but it is less easy to see how they
are. Cavender-Bares et al. suggest that acorn maturation time might be related to
phenological niche differentiation in masting and seedling regeneration, and they imply
that frost tolerance might be related to year-to-year weather variation and leaf lifespan to
timing of nutrient uptake. The crucial correlation ($P < 0.034$) is that species that co-
occur more often are more distant on their 'phylogenetic tree'. However, this is
essentially a test between habitats and therefore their 74 plots were not all independent.
Again the ugly head of habitat pseudoreplication rears, via what we might call environmental autocorrelation: testing between three reserves using 74 plots.

Work on guild proportionality at the biogeographic scale is fascinating, but even after
the work of Cavender-Bares, remains puzzling.

5.7.1.4   Patch Models in Guild Proportionality Studies

We have referred repeatedly to the problem of spurious guild proportionality due to environmental differences and consequent habitat pseudoreplication. The solution, as mentioned above, is not to randomise over all the quadrats. Wilson and Roxburgh (1994) made some attempt by having their points arranged in 10 24 × 24-cm plots, randomising occurrences only within each plot, and accumulating the departures from the null models over the 10 plots. Wilson and Gitay (1999) used a similar technique creating separate null models for each of their 21 sites and then combining the results to give an overall test, and Wilson and Whittaker (1995) applied the same method over six sampling lines. An even better technique is to form a separate null model for each quadrat, randomising over a few quadrats adjacent to it: the local patch model technique described above (Figure 5.3). Bycroft et al. (1993) did this by using a linear 7-quadrat patch based on the target quadrat, with the result that the proportionality that had been seen with a site model was reduced in size and no longer significant. Although the loss of significance could be due to the reduced power of patch model, the effect size was less too – only half. This was in vegetation selected to be uniform, and warns us to be careful about any study that does not use some kind of patch model. Wilson and Watkins (1994) used a patch of nine quadrats centred contiguously on the grid. This is probably the ideal, and in their work some significant guild proportionality was seen.

5.7.2   Evidence: Removal and Colonisation Experiments

It should be possible to see guild effects in perturbation experiments. If member(s) of one guild are removed, the resident species that increase should be from the same guild. Indeed, when Herben et al. (2003) removed the dominant grass species, *Festuca rubra*, from a mountain grassland, it was grass biomass that increased more than that of dicotyledons, with the particular species that increased dependent on the year in which the removals started.

It is also possible to see whether colonisers tend to be in the same guild as those removed. Symstad (2000) removed three guilds – forbs, C_3 graminoids and C_4 graminoids – from existing grassland at Cedar Creek, Minnesota, USA. After three years of growth, seeds of 16 native prairie species were added: legumes, non-leguminous forbs, C_3 graminoids and C_4 graminoids. There was only weak evidence that resident species repelled functionally similar colonisers. Fargione et al. (2003) used plots at Cedar Creek that had been planted with 1–24 species in 1994. Then in 1997, 27 species were added that occurred in the area but had not been planted in 1994. Multiple regression of cover 'visually estimated' in 1999 of four coloniser guilds on the resident guilds indicated that each guild as a resident had a greater inhibitory effect on colonisation by its own guild, though all coloniser guilds were inhibited most by C_4 grasses. Hooper and Dukes (2010) performed a similar experiment using experimentally established plots on serpentine soil. Early-season annuals had least invasion success into plots comprising the same guild, and late-season annuals had least success in any mixtures containing that guild. Petermann et al. (2010) sowed species into plots of the Jena (Germany) biodiversity experiment. The number of colonisers that were sown into the plots decreased with species richness, as it had to (a species cannot be recorded as 'invading' if it is already present). However, invasion was much more likely, giving higher coloniser biomass, when the coloniser was of a functional type (grass, legume, small forb, tall forb) not already represented. However, the results of such experiments do not always support guild theory. When Von Holle and Simberloff (2004) planted into a floodplain 10 species commonly found in that habitat, and weeded particular a priori guilds from some, there was no tendency for species to survive better or grow more when planted into a plot from which their guild had been removed.

These removal experiments do give evidence for guild-based assembly rules. However, field experiments on the response to removals are prone to high experimental error and the intensity of work required limits replication, so the statistical errors are usually large and any interesting effects are likely to be non-significant.

5.7.3   Intrinsic Guilds

The majority of guild investigations have used a priori guilds (Wilson 1999b). Sometimes, the guilds have been chosen directly (e.g. MacNally 2000). Occasionally several traits have been chosen and multivariate methods have been used to classify species into guilds (e.g. Landres and MacMahon 1980; Willby et al. 2000), but this begs the question of whether the traits measured are the appropriate ones and whether they have been weighted correctly. Tests for the reality of a priori and multivariate guilds (e.g. Hairston 1981; Hallett 1982; examples above) can indicate that some guild structure has been found, but not that it is the true guild structure of the community. Wiens (1989) summarised the problem:

There is an arbitrariness to guild classification and the determination of guild membership, which is especially evident in *a priori* classifications. This raises the prospect that the guild 'patterns' that emerge from studies based on such classifications are consequences of imposing an arbitrary arrangement on a community that is actually structured ecologically in some other way altogether (or is not structured at all). Using multivariate statistical procedures does not grant immunity from this problem.

A solution to Wiens' dilemma is to ask the plants what guilds they belong to, i.e. to select an index of guild structure and to find the guild classification that maximises this index. This classification is the intrinsic guild structure (Wilson and Roxburgh 1994). As someone said to us, 'Why don't you interview the plants to find the guilds?' We did.

5.7.3.1   Distributional Data

The approach was initiated by Wilson and Roxburgh (1994), using distributional information from the University of Otago Botany Lawn. To avoid circularity they divided the data in two, optimising the guild classification on one half of the point quadrat records in each of the 0.24 × 0.24-m quadrats – the optimisation subset –and testing it on the other half – the test subset. With field data it is impossible to examine every possible two-guild classification, the number is generally astronomical ($2^{(\text{number of species} - 1)} - 1$), so they took their a priori graminoid versus forb+bryophyte classification, and swapped species iteratively to reduce guild proportionality index RV_{gp} (variance in guild proportions relative to the null model). This showed that some forbs were better assigned to the 'graminoid' guild, perhaps because of the role of their laminae in the upper canopy, and vice versa. After many iterations the process converged to intrinsic guilds that gave an even stronger tendency towards guild proportionality, not only in the optimisation subset, but also in the independent test subset. Searches for intrinsic guilds starting from two random initial configurations resulted in classifications quite similar to the optimised 'graminoid' versus 'forb+bryophyte' guilds, and with further optimisation using the whole dataset the three optimised classifications converged to become identical. It is important to remember that these intrinsic guilds are alpha guilds, not beta ones. That is, there is a tendency for the species of one guild *not* to occur together. Presumably the reason is that they are too similar in resource use, and exclusion by interference occurs. Rather, for example at a 2-species point, e.g. there will tend to be one species from one guild and one from the other.

Applied to a Welsh saltmarsh by Wilson and Whittaker (1995), three random-start searches for intrinsic guilds using the Wilson–Roxburgh method produced very similar classifications, which converged to become identical after further whole-dataset optimisations, indicating that real guilds were occurring in the saltmarsh. Intrinsic guild membership could subsequently be correlated with leaf morphology; all the monocotyledons were in one guild together with other narrow-leaved species, as in a lawn examined by Wilson and Roxburgh (1994), suggesting the importance of canopy interactions in controlling species' coexistence.

Wilson and Gitay (1999) performed 100 random-start searches (computer processing power had increased in the interim) on the tussock grassland data (Section 5.7.1). A guild classification that showed significant guild proportionality in the test subset was found in a significantly greater number of searches than expected by chance (28 out of 100), and the 10 classifications that gave the lowest RV_{gp} comprised three groups. Further optimisation of representatives of these groups using the whole dataset confirmed that the community contained at least two genuinely independent, alternative guild classifications. It seems that two or more guild classifications can exist within the same set of species in a community, orthogonal in the sense that they are unrelated to each other and operate simultaneously. This is not surprising; the true guild relations are probably quite complex. The intrinsic guilds showed some relation to growth form/height, suggesting that the local community tends to comprise even representation from different height strata.

The general impression from these results is that guild membership in these grasslands depends on canopy relations, especially vertical stratification as affected by leaf morphology. However, this may be partly due to the traits considered; other traits, correlated with them, may be the real determinants.

5.7.3.2   Interference Experimental Data

Wilson and Roxburgh (2001) used an interference experiment to seek intrinsic guilds. Seven species from the Otago Botany Lawn had been grown in boxes in all possible two-species mixtures (Section 3.5). They argued that when a species from one alpha guild was grown with a species from another alpha guild, by definition differing in resource use, then by the Jack Spratt[5] effect the yield of the mixture should be considerably greater than the mean of the two monocultures, as measured with index RYM (Relative Yield of the Mixture, Wilson 1988c). With only seven species it was possible to test all possible two-guild classifications to find the one that maximised the mean RYM of mixtures, and this resulted in guilds very similar to those obtained from distributional data described in the previous section.

5.7.3.3   Experimental Removals Data

In theory the guilds intrinsic to the community could be discovered by removing one species at a time and observing which of the other species increased. Clements et al. (1929) had experimented with removing species from communities, and Fowler (1981) took this approach in removing single species from a North Carolina grassland. For all removals, at least one other species was affected significantly. Often several species were affected. Usually removal effects between a pair of species were not reciprocal. There was no sign of guilds of species that especially affected each other, and it was hard to predict which species would be affected when one was removed. A few negative effects were seen, in which removal of a species decreased the yield of another; if these effects were real, they could have been due to disruption of facilitation or to indirect interactions via a third species ('My enemy's enemy is my friend'). The conclusion is that species interactions in that grassland were complicated, often indirect and diffuse. Intrinsic guilds were not found. Similar experiments, with similar conclusions, were performed by Allen and Forman (1976) on a New Jersey oldfield, Abul-Fatih and Bazzaz (1979) on an Illinois oldfield, Silander and Antonovics (1982) on North Carolina dune, slack and saltmarsh, del Moral (1983) in Washington subalpine meadows and Gurevitch and Unnasch (1989) on a New York oldfield.

These results seem to exclude a simple model of community structure, with distinct guilds. However, the many indirect effects known from analysis of the community matrix (Section 4.8.2), together with the necessary compromise that the community is disturbed by perturbation of species removals, and the low power in removal experiments (Section 5.7.2), will always make intrinsic guilds difficult to find by this method.

5.7.3.4   Conclusion on Intrinsic Guilds

A major advantage of the intrinsic guild approach is that it can fail. Approaches such as multivariate classification of traits must give guilds, whether any exist in reality or not.

---

[5] 'Jack Sprat could eat no fat and his wife could eat no lean, and so between the two of them they wiped the platter clean'.

In contrast, a search for intrinsic guilds by minimising RV_{gp} can result in all the species being in one guild, or in a guild structure that is non-significant, as it did for Wilson et al. (2000a) and in several other datasets where the failure remained unpublished. In experimental studies searches via maximising RYM can find that there is no overyield. Thus, if there is no guild structure, the intrinsic guild approach can indicate this; no other method offers this possibility.

Although functional-trait relations between species are often expressed in a classification, ordinations have also been used to see trends and continuous variation. It would be good to have an intrinsic equivalent to ordination, placing the species on guild gradients according to their distributions or their responses in experiments. Attempts to describe species trait distributions in continuous space are being developed in the context of large-scale dynamic global vegetation models (DGVMs) (Pavlick et al. 2013), and analogous approaches could be adapted at local scales in studies seeking evidence of community structuring by alpha guilds.

5.8   Texture Convergence and Trait-Based Analyses

Vegetation texture was defined by Jan Barkman (1979) as: 'the qualitative and quantitative composition of the vegetation as to different morphological elements, regardless of their arrangement'. Ecologists have now caught up with Barkman's approach, be it using the term 'functional characters' or 'traits', and ideally (but rarely in practice) extending it beyond morphology into physiological traits and regeneration traits (Larson and Funk 2016), not necessarily including all the species in a community. However, the aim remains to describe communities not by the names of the species, but by plant traits that, it is hoped, represent function. Two types of convergence have been examined, due to environmental filtering and to biotic filtering, respectively.

Environmental filtering (Section 1.5.2) might cause texture convergence due to parallel adaptation to the physical environment, i.e. beta-niche filtering. The null model for this, even if rarely applied, is greater similarity between areas than a random selection of species from other (all?) habitats. Fuentes (1976) did this for lizard communities in California and Chile from sea level to 2,000 m a.s.l., testing for greater similarity between communities at the same altitude (in different countries) than between those in the same country (at different altitudes). Cowling and Witkowski (1994) compared five Mediterranean-climate sclerophyllous shrubland sites in south-western Australia with five edaphically-matched sites in South Africa, and concluded 'strong convergence' between continents on the basis of lack of significant difference in many soil/trait comparisons, such as growth form (shrub/graminoid/forb), leaf consistency (sclerophylly and succulence) and SLW. However, convergence cannot be concluded from non-significant difference, and there was no null model using other areas. Jacobsen et al. (2009) extended such texture convergence work into the more directly functional xylem-cavitation resistance, finding some overlap between species from Californian and South African Mediterranean-climate areas, but again with no null model that included species from other regions.

Continent 1   Continent 2

Figure 5.7 The concept of texture convergence. A similar range of traits is present on the two continents, even though the species involved are different.

The opposite result is texture divergence, which would be expected in different environments. For example, Fortunel et al. (2014) found that in two South American regions stem wood density was greater than expected at random when comparing seasonally flooded and clay-soil forests, and lower than expected when comparing white-sand forests.

Biotic convergence, an assembly rule, almost requires environmental convergence as a prerequisite, but then supposes that interactions between species as they fight (or cooperate) for niche space leads to texture convergence. To take the example of plant height (stratum), the null model here is that the species in the joint (environmentally filtered) pool occur at random in the two (or more) sites. The H_1 non-null model, the assembly rule, is that there is a greater range of heights within each site, leading to greater similarity between sites in the range of heights, i.e. each site has a good representation of canopy trees, subcanopy trees, shrubs, herbs, etc. (Figure 5.7; Wilson and Smith 2001). In these comparisons, species can be weighted by their abundance, or only by their presence. The basic concept is the same as limiting similarity and guild proportionality. It is convergence due to species interactions, not environmental tolerances. To put it in anthropomorphic terms, it is as if a community were saying: 'I have an empty niche for a tree here; hey you, tree X, you'll do. Sorry shrub Y, I can't take you, I have enough shrubs already' (Figure 5.7).

It is sometimes asserted that such convergence cannot be demonstrated unless the evolutionary history of the species is known. This is not true, because the test is against an ecological null model, of assembly of species from the existing joint-site pool. The problem is rather that biotic convergence may be impossible to see because the sites are too different in environment, so that prior environmental filtering has not occurred. For example, Mediterranean-climate regions, the focus of almost all plant convergence studies, are far from identical in either soils or climate (Cowling et al. 2005).

Biotic convergence might be: (a) species-for-species matching, or (b) even distributions in niche space. Cody (1974) introduced the concept of species-for-species

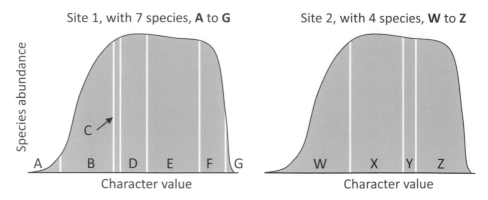

Figure 5.8 Site 1 has the same texture as Site 2 with respect to the trait, even though they differ in the number and abundances of species.

matching in his 'one-for-one matching of individual species', i.e. convergence because species fit into fixed niches that are the same in each area. However, that would not necessarily be expected, nor that there would be the same number of species since a niche filled by one species in Site 1 could be split between three species in Site 2. The requirement is only that the same niche space is occupied fully, with an even distribution in niche space due to limiting similarity in both communities (Figure 5.8). The result in terms of community-level convergence is the same (Smith and Wilson 2002).

It is possible that the mean texture might converge, but not the distribution of traits (Figure 5.9a), or the distribution could converge, but not the mean (Figure 5.9b), or both, or neither. Any convergence could be due to sorting in ecological time or convergence of the species pools in evolutionary time, but the latter is just a genetic fixation of ecological convergence (Smith and Wilson 2002).

Wilson et al. (1994) compared convergence between two carr (i.e. wooded fen) communities in Britain and two in New Zealand, in five functional traits related to light capture, such as SLW and photosynthetic unit (PSU) support fraction. The null model comprised random assignment of species from the pool to the four communities. In fact, the texture of the four carrs diverged when weighting species equally. However, weighting the species by their photosynthetic biomass, convergence was seen for PSU width and possibly for PSU area. This is biotic convergence, resulting from coevolution and or ecological sorting – niches filled either by evolution or by immigration.

The first texture–convergence studies compared continents, but comparisons can be made between nearby sites, or between patches within sites with slight adjustment to the null model (Watkins and Wilson 2003). This is close to an intuitive question that ecologists have when looking at different patches within an area of vegetation: is the texture similar, i.e. does a species in one site substitute for a different but functionally similar species in a second site? As Clements (1907, p. 294) wrote: 'a species found in one area may be replaced in another by a different one...essentially alike'. One just has to realise that it is ecological, not evolutionary, convergence. Smith et al. (1994)

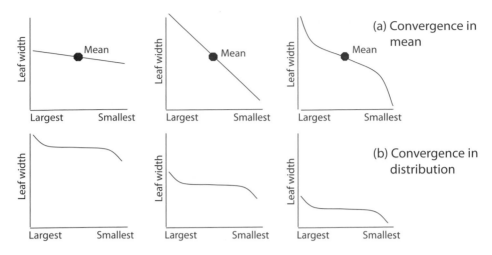

Figure 5.9 Texture convergence can be in (a) mean or (b) shape.

investigated sites in conifer/broadleaved forest in southern New Zealand, recording similar traits to those used by Wilson et al. (1994) and found convergence in all traits. As in the Wilson et al. study, trends were apparent only when traits of the species were weighted by the abundance of the species, but such weighting makes ecological sense. Matsui et al. (2002) conducted the same type of investigation but more locally, within three sites, and evidence of convergence was found for a subalpine grassland: each patch (quadrat) tended to comprise a mixture of small-leaved species and large-leaved species, a more constant mixture than expected if the species were being assigned to quadrats at random. Watkins and Wilson (2003) took this approach further by examining replicate quadrats within 12 herbaceous communities, measuring 11 traits that were intended to reflect the functional above-ground niche of the species and laboriously obtaining the biomass of each species in each quadrat. Convergence was seen in chlorophyll content, indicating a significant tendency for each patch in a community to comprise a rather constant mixture of species types in terms of their different chlorophyll contents, though other results were non-significant or showed divergence. In these local convergence studies it is necessarily explicit that the question is of ecological assortment, not of evolution. If species interactions are significant in structuring plant communities, texture convergence must occur, even if, as is so often the case, environmental differences act as noise, perhaps preventing convergence from being demonstrated (Schluter 1990).

If the traits of species control the occurrence of species in a community, it should be possible to invert the problem and predict the abundance of each species from its traits. In a much-discussed paper, Shipley et al. (2006) used an empirical model, Maxent, to predict the abundance of a species in a community from its functional traits. They applied it to 12 French ex-vineyards estimated to be 2–42 years since abandonment, in which the above-ground biomass of 30 species out of those present had been determined. Half of the species were present in only one field, and only two of the 30 were present in more than three fields, so the species differences were largely confounded with field (/successional) differences. Eleven functional traits had been measured on these species (Vile et al. 2006); they used eight, omitting three reproductive traits. Of the eight, height, vegetative mass, stem mass and leaf mass per plant were, not surprisingly, highly and significantly correlated ($P < 0.001$ in each case). Their Maxent model successfully ($r^2 = 0.94$) predicted from the functional traits of a species its biomass in a field relative to the total biomass there.

There is always a problem calibrating a model on a set of data and testing its predictivity on the same set: the model parameters may be fine-tuned for those particular data. The solution, which should be standard practice, is to use a jackknife, leave-one-out procedure, and using this Marks and Muller-Landau (2007) showed that the predictivity that Shipley et al. had shown disappeared. Roxburgh and Mokany (2007) also explored the circularity in the Shipley et al. method, demonstrating the method had high predictivity even applied to random data.

In spite of the statistical circularity of the original Shipley et al. (2006) analysis, there may be some real trait/biomass structure present in these vineyards. Roxburgh and Mokany (2010) introduced a randomisation test that they validated with random data, i.e. it gave the correct proportion of type I and type II errors. It still indicated significance in Shipley et al.'s (2006) original data. But what is this really telling us, ecologically? Yes, the woody species in a community often have higher biomass, and are taller (maximum height was one of Shipley et al.'s eight predictive traits), that is well known. In this case, four of the traits were highly correlated with estimates of plant size. It seems one possible interpretation here is that species with higher mass per plant tend to have higher mass per quadrat. This should not be a surprise to any ecologist.

5.9   The Fate of Associations through Time

Time has done natural assembly-rule experiments for us. When the climate has changed, e.g. in the c. 15,000 years since the last glaciation, species have migrated. But have they moved as whole communities, i.e. reassembled into the same communities, or have many of our present-day communities no analogues in the past communities, and vice versa?

Clements (1936) wrote that 'climaxes have evolved, migratcd and disappeared under the compulsion of great climatic changes from the Paleozoic onwards, but [the student of past vegetation] is also insistent that they persist through millions of years in the absence of such changes'. He continued: 'The prairie climax has been in existence for several millions of years at least, and with most of the dominant species of today'. Clearly his concept of the community as a complex organism led to a conclusion that there were only a limited number of combinations in which species could assemble. He envisaged that in the very long term new communities could 'evolve' and some disappear, but the changes in climate since the last glaciation would result largely in the migration of existing combinations.

However, several palaeoecologists have suggested that many of the communities that were extant earlier in the Holocene, usually as seen in pollen assemblages, are not found anywhere on earth today: they are 'no-analogue' communities. For example, Birks (1993) concluded, from several studies that reconstructed past communities from the pollen record, that many communities that were present in previous interglacials up to the late Holocene have no convincing analogues among present-day communities. Veloz et al. (2012) formed species distribution models from palaeo-climate estimates from Global Circulation Models (GCMs) and combined them with plant genus/species palaeo-distributions derived from pollen. The models, applied to climates of 21–15 ka BP, predicted species distributions of that time well. However, applied to modern climates the models predicted the distributions of some genera poorly, especially those common in areas with past climates that had no climate analogue today.

Thus, many of the notable associations of the present day, for example the *Tsuga canadensis* (hemlock)–*Fagus grandifolia* (American beech) forests of northeast USA have existed since the last glaciation for only c. 6,000 years, for 500 years in some places. Before that, the species had little overlap (Davis 1981; Graham and Grimm 1990). These results challenge a simplistic interpretation of Clements' concept of the plant community, and indeed the existence of assembly rules. However, more rigorous testing using appropriate null models based on paleocommunity occurrence, as used by Bennington and Bambach (1996) for marine faunas, remains to be done for plant assemblages.

A more local picture of vegetation change can be obtained from macrofossils, and DiMichele et al. (2002) examined coal balls as a record of the vegetation of mires in Illinois, USA, through several intervals of glacial advance and retreat, covering a period of up to 650,000 years. Essentially the same community seemed to be forming many times, though we ask how same is 'same' and whether this was just a case of beta-niche filtering (Section 1.5.2). However, this approach avoids the problem with pollen reconstruction that a past 'no-analogue community' may not be a real, local community itself, but comprise the sum of pollen dispersed inwards from several surrounding communities.

Where do the species come from for a community with no precedents? Barrington and Paris (2007) concluded from fossil records and palaeo-distributions of habitat that the current New England flora has been synthesised by immigration from a variety of sources: the arctic, the west, the now-submerged North Atlantic coastal plain, the southeast coast and the lower Mississippi valley. Such movement of species could occur again (and already is with exotic species), and they concluded that the present-day plant communities of the region 'are most likely transient'.

There are actually many possible explanations of no-analogue communities, and Jackson and Williams (2004) evaluate them carefully. They discuss the problem of how different, and by what criterion, a 'no-analogue' community has to be. They reject, as major explanations of no-analogues, artefacts such as differential pollen preservation, mixing of sediments, different pollen production by some species under the CO_2 levels of the past and a different juxtaposition of communities over the landscape (though the use of macrofossils avoids this problem). It is remarkably difficult to find exact matches

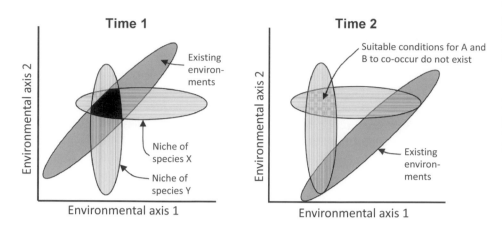

Figure 5.10 At Time 1, the realisable niches of Species X and Y overlap in an area of environmental hyperspace that exists. At Time 2, the combination of environmental conditions where they overlap does not occur.
Inspired by Jackson and Williams (2004).

between any two current climates, and this is undoubtedly true for the past as well, and differences in CO_2 levels will be present too, making matching impossible. They suggest that the most likely explanation for no-analogue communities is that whilst similar ranges of climatic variates occurred, often the combinations of such variates that are present today did not (Figure 5.10). This interpretation means that no-analogue communities are a function of no-analogue climates.

This interpretation is supported by comparing (a) the degree of mismatch between reconstructed past plant communities and the closest modern analogues, with (b) the degree of mismatch between reconstructed past climates (from GCMs) and the best modern fits. Using this approach Williams et al. (2001) found community misfits (no-analogues) tend to occur in the same place/time as climate misfits.

The realisation that many of the communities that we see around us are of quite recent origin is the most significant discovery since Clements' synthesis, and contrasts with his understanding of constant communities moving around the landscape. However, it does not distinguish species reacting individualistically to the climate, as suggested by Gleason in some of his writings, from a model in which the occurrence of a species is determined by the identity of other species present, a view attributed with some truth to Clements (Section 6.2).

5.10   Abundance

5.10.1   Biomass Constancy

The constancy of biomass per unit area, compared to null models in which species abundances are random, has been used as an assembly rule (Wilson and Gitay 1995a). Similar biomass constancy can be seen on a regional extent (Culmsee et al. 2010; no trend with elevation 1,050–2,400 m a.s.l. in tropical rainforests in Sulawesi, Indonesia) and even on a global scale (Enquist and Niklas 2001; no trend with latitude or elevation across tropical and temperate forests). This may be an inevitable consequence of the limited input of energy (Section 1.1). However, it is also a demonstration from the field that interference is occurring and causing community structure. It has the ability to distinguish between communities (Wilson et al. 2000a). Of course, this constancy will not hold across different biomes, with their associated large differences in environmental range.

5.10.2   Species Abundance Distribution (SAD, RAD)

Various models of community construction give predictions for the relative abundance distribution between species, 'abundance' ideally being determined as biomass, but sometimes as cover, density, etc. (SAD = species abundance distribution, RAD = relative abundance distribution, dominance/diversity; Wilson 1991; Figure 5.11). SADs are one of the very few types of evidence on community structure available for one point in space and time.

Various models of community construction give predictions for the shape of the SAD (even if they were originally empirical or arbitrary constructs), so that recording the SAD might ideally enable identification of the way the community was constructed. The Geometric (Niche-preemption) model is based on competition and the Zipf–Mandelbrot can be interpreted as succession/facilitation. The Broken Stick and the General Lognormal (Sequential Breakage) models are alternative models of the random

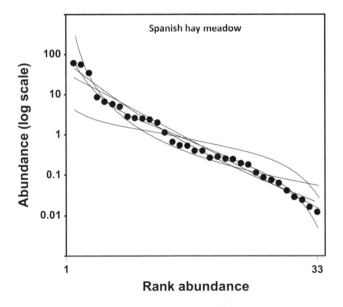

Figure 5.11 A SAD (species abundance distribution) plot for biomass in a Spanish hay meadow, showing observations (circles) and fitted curves for six SAD models.
Reprinted from Wilson (1991), with permission from John Wiley and Sons Inc.

assignment of resources (i.e. alpha-niche widths) between species. Several other subtly different models of the latter type can be constructed (Tokeshi 1996). All of them are null models, i.e. alternative models of what a community looks like when it is not there. This means that the analyses are liable to end up testing between null models, not against them. It is also a problem that some of the distributions, notably the General Lognormal, can be derived from alternative assumptions.

The major SAD models for plants, with their continuous scales of abundance, are Broken Stick, General and Canonical Lognormal, Geometric and Zipf–Mandelbrot.

In MacArthur's (1957) Broken Stick model, species' abundances reflect the partitioning of resources among competing species by random divisions along a one-dimensional gradient. Unfortunately, other ecological assumptions can give the same distribution, including models with no restrictions on niche overlap (Cohen 1968).

Preston (1948) proposed the use of a lognormal distribution for empirical reasons, though it might express community structure in three ways. Firstly, plant growth will be affected by several environmental factors and by the Central Limit Theorem this will give a near-normal distribution. Since plants have intrinsic logarithmic growth, the distribution will be lognormal (May 1975). Secondly, a lognormal distribution can arise, again by the Central Limit Theorem, from the summation of SAD distributions in many subareas of the sample (Šizling et al. 2009). Thirdly, a variant of MacArthur's Broken Stick model with occupation and subsequent division of niches, i.e. the breaks made sequential and breakage probability being equal for all existing segments independent of their length, gives a lognormal distribution of species abundances (Pielou 1975). Preston (1962) proposed further that the distribution was a reduced-parameter subset of lognormal distributions that he called 'Canonical Lognormal', defined by the mode of the individuals curve coinciding with the last point on the species curve (i.e. gamma = 1). The Canonical hypothesis was empirical; there is no ecological basis for it (Caswell 1976). Whether it is a mathematical artefact is controversial (May 1975; Connor and McCoy 1979; Sugihara 1980; Connor et al. 1983).

The Geometric Model (Whittaker 1965) suggests that the 'most successful species' takes fraction k of the resources and therefore forms approximately k of the abundance. The second most successful species takes k of the remainder (i.e. $k(1-k)$), etc.

The Zipf–Mandelbrot model originated in information theory. Applied in community ecology it implies that the entry of a species into a community is dependent on prior changes (as in the 'facilitation' model of succession: Wilson 1991). It tends to fit frequently, partly because it is highly parameterised and partly because it can fit a concave-downwards relation. The latter implies strong dominance.

With so many different models, and with sampling variations, there is a worry that it would be impossible to discriminate between them. However, simulations show that in a 15-species community, for example, one can identify the correct model with reasonable correctness given 10 or more quadrats; success depends on the model and the number of species (Mouillot and Wilson 2002).

Models can be tested by comparing their SAD predictions with those observed, and after many years of parallel presentation of theoretical curves of theory and data, Wilson (1991) showed how the two could be compared. The results have been frustrating.

Wilson and Gitay (1995b) found that in four dune slacks in west Wales the best fit was given by either Geometric or General Lognormal Models, but there was no consistency between the two subsites within each slack as to which gave the better fit. Wilson et al. (1996a) fitted SAD models to plots from three fertiliser/succession experiments in the English midlands; basically there were no trends except those reflecting the higher evenness in plots to which P had been applied, and it is hard to see much ecological meaning even in this. Watkins and Wilson (1994) sought a relation between the level of vertical complexity in a community and the SAD model that fitted, but could find none. Cheng et al. (2012) compared predictions of tree density SADs from four models with those observed in an evergreen subtropical forest in eastern China. The observed SAD could be distinguished only from a Poisson distribution over a homogeneous area. In particular, they could not distinguish whether the SAD reflected environmental hetero-geneity or dispersal limitation. Spatharis et al. (2011) analysed marine macroalgal communities at two sites in Greece. At a shallow, sheltered site, cover values from most subsites fitted the General Lognormal distribution (described as 'random fraction' from its stick-breakage graphical illustration and ecological explanation). At a deeper, more exposed site, most subsites fitted a variant of the Geometric Model, in which the value of k (niche size at each sequential step) is random. The ecological meaning of the difference in SAD between the two sites is not clear.

Particular interest has been whether the SAD resulting from a niche-based model of community assembly could be distinguished from one assembled according to Hub-bell's (2001) Unified Neutral Theory (UNT, Section 6.4). The general conclusion is that they cannot, especially since one or both models will involve chance. Moreover, Chisholm and Pacala (2010) demonstrated that for high-diversity communities a niche model can produce SAD predictions mathematically equivalent to UNT ones.

Determining the SAD of a community is ideally an easy way to determine its mode of construction, the assembly rules. In practice the similarity of predictions from different models (including several other variants that the four described above), the problem that the predictions of several models are stochastic by the nature of the models, that there will be stochasticity and heterogeneity in nature anyway, problems of multiple eco-logical interpretations of some models, and that the model fitting best can be dependent on the spatial grain of sampling (Wilson et al. 1998) all mean that the approach has given minimal insight in practice.

There has more recently been interest in comparing community assembly models with spatial distributions. Brown et al. (2011) found in simulations that although the model used to assemble a community could not be consistently identified by the resulting SAD, the spatial pattern resulting from it could be. Perhaps a combination of SAD and spatial structure would be informative.

Species diversity can be split into richness and evenness, and the latter summarises in a single value some of the information in SAD curves. Caswell (1976) compared evenness to that expected from a null model. He found that tropical rain forests tended to be less even than predicted from the null model, but temperate deciduous forests of eastern North America were significantly more even than the null model. The contrast was the opposite of what he expected from previous theories. Other attempts to obtain evidence on community assembly from evenness have not been fruitful. We conclude that the information analysed here is potentially useful. Fits to an abundance distribution model based on ecological theory are the most interesting, though usually ambiguous, and no conclusions of real ecological value have emerged yet. Any regularity, such as adherence to Preston's Canonical hypothesis, would be evidence that the structure was deterministic.

5.10.3   Rank Consistency

In the Geometric Model of SAD 'most successful' can be interpreted either as the first species to arrive when competition is cumulative so that it has a big advantage (Section 2.2.1.2), or as the species that has the highest intrinsic interference ability (Watkins and Wilson 1994). Similarly, niche boundaries in the Broken Stick and General Lognormal models can be seen as occurring as species disperse to a site ('founder control') or as the species evolve their interference ability ('dominance control'). In the dispersal case there is no reason to suppose that the identity of the species occupying each abundance category will be the same in each patch/site. A measure of this is rank consistency (Watkins and Wilson 1994) – the tendency for the abundance rankings of species (either across space or through time) to be more similar than that under a null model. For example Wilson et al. (1996a) used such an index to examine changes in community structure through succession (Section 5.4.1).

Collins et al. (2008) introduced graphical displays of rank consistency, 'rank clocks', and used them to examine temporal rather than spatial consistency in USA desert grassland and shrubland, old-field and prairie. There was considerable variation in abundance ranks in spite of more constant species richness (cf. the Carousel Model, Section 4.2).

5.10.4   Sparse Species

Species that are sparse within the community (locally 'rare') are a puzzle (Rabinowitz et al. 1984). Firstly: are they filling special niches that exist for sparse species? Zobel et al. (1994) investigated this in a wooded meadow in Estonia by removing 10–17 species from certain plots, all with a cover of 1 per cent or less (a different list for each plot), repeating the removals for five years. There were no visible gaps and they say very little biomass was removed, but species richness was reduced by 25–33 per cent. Species did not immigrate to fill the niches: the number of immigrants was no higher than in control (i.e. no-removal) plots, actually non-significantly lower. They concluded that there were no special niches for the sparse species.

Perhaps the sparse species have a distinct effect on the major species? Bracken and Low (2012) performed a rare-species removal experiment on a rocky shore, removing some sparse species (mainly algae), which had no significant effect on cover (presumably 'estimated by eye') but reduced the richness of motile animals (herbivores, carnivores and omnivores) by 30 per cent and their estimated biomass by 43 per cent. Lyons and Schwartz (2001) in a meadow in the mountains of California manipulated species richness by removing either: (a) all plants of the least abundant species, thus reducing species richness to between two and seven species, or (b) an equivalent biomass of the most common species (to control for possible disturbance by the removals in treatment 'a'). The exotic grass, *Lolium temulentum* (darnel) was then introduced. Its establishment was higher when more sparse species were removed, indicating a role for the sparse species in invasion resistance. It is not immediately obvious how this result squares with that of Zobel et al. Generalisations in these questions are far off, especially since the latter study involved only one coloniser. Myers and Harms (2009) took the opposite approach of adding seeds of 38 'mostly rare' species to a Louisana, USA, longleaf pine savannah. This increased the species richness at the 0.25 m^2 grain 7–22 months later by 53 per cent. However, it is unknown whether any of that increase would be permanent.

We have some answers to the question of why rare species are there in the explanations of the 'Paradox of the Plankton' – mechanisms of coexistence (Chapter 3). However, their effects on the more abundant species and on the whole community remain one of the greatest puzzles in plant community ecology.

5.11   Keystone Species

The term 'keystone species' was coined by Paine (1969) for a single native species high in the food web that, whilst perhaps unimportant as an energy transformer, is vital for the maintenance of the community, e.g. a top predator causing a trophic cascade. This cannot be applied literally to plants, but some have redefined 'keystone species' as the one in a community with the greatest effect on others, for example a herbivorous insect in some herbaceous communities in eastern USA (Carson and Root 2000; Schmitz 2004), or the greatest effect relative to its biomass (Jordán et al. 1999). The term has been applied to plants with litter that is high in polyphenols and of low pH, effecting a switch that maintains the current state (e.g. *Empetrum hermaphroditum*; Mallik 2003; Section 4.5.4.5). The contribution of plants as furniture for birds has been seen as keystone (arborescent succulents by Midgley et al. 1997), and this may also operate as a dispersal switch (Section 4.5.4.7). Plants are relied on by many frugivores (Diaz-Martin et al. 2014). Bond (1993) proposed a broad definition of 'keystone species', 'If loss of a species results in a large effect on some functional property of the ecosystem, that species may be called a keystone', a concept similar to that of 'foundation species'. An excellent example of a Bond-keystone/foundation species is the brown alga *Undaria pinnatifida*, which has invaded in Patagonia (and elsewhere), dramatically reducing native marine algae as well as affecting fish and the benthic macrofauna (Casas et al. 2004).

Carnivores can affect plant communities indirectly. For example, Green et al. (2008) describe how *Anoplolepis gracilipes* (yellow crazy ant), invading Christmas Island, has drastically reduced the density of *Gecarcoidea natalis* (the native red land crab), possibly by spraying formic acid into the crabs' eyes when it is disturbed and frightened. Since the crabs are herbivorous, consuming seeds and seedlings selectively, the diversity and density of plant seedlings is markedly higher in ant-infested areas. Other indirect effects of ant invasion have included changes in the litter insect fauna, probably because crab consumption of litter has decreased. There has been an increase in populations of the honeydew-secreting scale insect *Tachardina aurantiaca*, which has in turn increased sooty mould populations. The ants themselves, or the scale insects, or the sooty mould, seem to have interfered with frugivory and killed shoots of canopy trees. The changes have apparently allowed invasion by an exotic snail *Achatina fulica*, caused increases in the densities of some invertebrates but decreases in others, and changed bird densities and behaviour. It has been suggested that the high densities of red crabs when scientific investigations began were due to extinction of the endemic rat *Rattus macleari* by c. 1903. All these comparisons have been between invaded and non-invaded sites that may have differed in other ways; plot placement has not always been random, and most of the mechanisms are unknown or speculative. However, it is clear that there have been ecosystem-wide effects. The red crabs can be described as Bond-keystone/foundation species, affecting many other species. The yellow crazy ants are almost keystones *sensu* Paine, except that whilst they kill the herbivorous red crabs they do not consume them. *Rattus macleari*, in its speculative role as a consumer of red crabs, might have been a genuine keystone species, thus causing a trophic cascade.

But we are concerned with plants, and a plant species with a strong reaction on the environment will either change the current state, in which case it would not be called a keystone, or it will reinforce the current state, in which case it is a keystone because it operates a switch. Examining mechanisms, i.e. the factor mediating a switch, is more helpful than labelling species 'keystone' or 'foundation'.

5.12   Exotic Species as Community Structure Probes

It is not always easy to define 'exotic species', but the native/exotic status of most species is clear, and in some parts of the world exotic species have displaced much of the native cover (e.g. the Seychelles, Hawaii, New Zealand: MacDonald and Cooper 1995). Exotic species are sometimes practical/management problems, but they also represent opportunities for the theoretical community ecologist, as natural experiments. In this section we explore exotic species through the lens of community assembly, starting with an overview of exotic species, and the associated process of invasion.

5.12.1   The Nature of Exotic Species

In one way invasion by exotic species is surprising: the native species have presumably evolved to meet the local environment, physical and biotic. Moreover, exotic species cannot be intrinsically different because all species are native somewhere (except species of garden origin and a few species of recent origin such as *Spartina anglica*, cord grass), and therefore the concept that 'exotic species' in general have distinct properties is deficient in logic. Leger and Rice (2003) found the exotic (Chilean) ecotype of *Eschscholzia californica* to be more vigorous in California than the native genotype. Would the Californian genotype, as an exotic, be more vigorous in Chile than the native one? How would that situation arise? It is far from clear that exotics are consistently different, e.g. Kissel et al. (1987) found no consistent difference in water relations between the three major native woody species and four exotic ones of the most semiarid area of New Zealand, and in a nearby area King and Wilson (2006) found no difference in experimental water stress tolerance or nutrient response, though the exotic species did have a greater RGR_{max}. We suspect that often generalisations are made from special cases, especially from invasive species with a practical/management impact.

There are a number of possible reasons why the exotic species might be different from the native ones. Many hypotheses have been put forward for individual species, relying on the properties of those species, but we restrict ourselves to explanations that might apply to any species, just because it is exotic.

5.12.1.1 Diversity and Species Identity

The role of the species diversity of the target community in providing resistance to invasion was discussed in Section 4.8.4.1. Here two additional aspects are considered: the question of species saturation, and the potential for lower interference abilities of the species in the target community.

i.   *Species Saturation*
The flora of the invaded area might be 'depauperate' (Section 4.8.4, point 1.ii). Islands are often given as examples of invasion into areas with a depauperate flora (e.g. New Zealand: Dulloo et al. 2002). Niches could be empty, which no species in the depauperate native flora was able to fill, but which an exotic invader could. For example, Shimizu and Tabata (1985) explained the invasion of *Pinus lutchensis* into the shrublands of the Ogasawara Islands, Japan, by postulating that there had been an empty niche for an emergent tree. Cappuccino and Carpenter (2005) comparing nine exotic species in natural areas in Ontario, New York and Massachusetts for which they had evidence of invasiveness, found that invasive species were more taxonomically isolated than non-invasive plants, belonging to families with 75 per cent fewer native North American genera. Strauss et al. (2006) found the same with grasses of California, this time using a phylogeny inferred from both morphological and molecular data, rather than taxonomy. This is superficial evidence for empty niches/missing guilds in the native flora: that species that were less related and therefore perhaps less similar in niche more readily found empty niches and invaded. Davies et al. (2011) found in a field survey of a serpentine site in Northern California, USA, that exotic species were less phylogenetically related to natives than expected under null model at small spatial grain ($16 \, m^2$), but with no such significant effect at large grain ($10,816 \, m^2$). They interpreted this in terms of niche limitation: exotics that were less related to the existing community might be adapted to different niches and more readily find an empty niche locally. However, all these results – Cappuccino and Carpenter (2005), Strauss et al. (2006) and Davies et al. (2011) – could also be explained in terms of natural enemy release (Section 5.12.1.4 and ii below).

Tropical rain forests are an interesting case, since they are generally less invaded by exotic species. It would be helpful to conservationists to ascribe the lack of exotics to the saturation of available niche space through high diversity of species or guilds, but some species-poor tropical forests also have no invaders (Rejmánek 1996). A more likely explanation is that most of the exotic species that are transported to tropical countries lack the specific life history traits, most importantly shade tolerance, necessary for successful invasion of undisturbed tropical forests (Fine 2002), especially since fast growth in tropical rain forest results in rapid canopy closure after disturbances (Rejmánek 1996).

Empty niches are possible in theory, and are clear in a very few cases, but evidence for them as a general phenomenon and explanation of the success of exotic species is elusive.

ii.   *Interference Ability*
The other possible consequence of a depauperate flora is that the native species might not be vigorous enough, not competitive. MacDonald and Cooper (1995) wrote 'an individual island's biota is based on too small a sub-sample of the global gene pool to have generated robust competitors for every available niche. ...Insular species are frequently outcompeted by species that have been honed in much more exacting biotic communities of the mainland. ...[suggesting] superior competitive ability of mainland species'. For New Zealand, Dansereau (1964) wrote: of 'apparently vacant space', occupied only by 'weaker' species. Is this really true? Perhaps super-species, once limited by dispersal (e.g. to the Old or New worlds, Northern or Southern hemispheres), are now able to spread everywhere. In that case, homogenisation of the flora is set to change the world (which it is). Still, these super-species do not seem to have been that super in their original hemisphere. In Britain, there are many yellow composite herbs with rosette leaves, leaving one to key out between a number of quite likely possibilities. In New Zealand *Hypochaeris radicata* (cat's ear) is present in a huge range of environments and often quite frequent within them. An exception may be *Ammophila* spp. It has been suggested, with some truth, that when high coastal dunes are built it is always by species of *Ammophila*. They seem to operate a switch, trapping sand and tolerating sand burial.

Acer platanoides (Norway maple) has been suggested as a super-species, invading NE North American deciduous forests and suppressing the native species (Paquette et al. 2012). The many hints as to its invasive/interference ability make it a useful case study, mostly via comparisons with the species of the forests that it invades; hints that include its continued shoot extension into the autumn (Paquette et al. 2012), greater winter survival and earlier leaf emergence of seedlings (Morrison and Mauck 2007) and higher use efficiencies of N, P and water, all compared to *Acer saccharum* (sugar maple) (Kloeppel and Abrams 1995). An increase in nutrient availability (pH, Ca, Mg, K, N) under *A. plantanoides* trees compared to native species has also been observed, though benefitting other species as much as itself (Gómez-Aparicio et al. 2008), as has a higher soil water content beneath its canopy than under that of *Pseudotsuga menziesii* (Reinhart et al. 2006). During early establishment the species is favoured by facilitation by the native soil biota, and initial natural enemy release from soil enemies (Reinhart and Callaway 2004). Release from natural leaf-eating insects and fungi when in USA compared to Europe has also been documented (Adams et al. 2009). *A. platanoides* can also be more plastic; in a greenhouse experiment its shoot:root ratio in an illuminance and R:FR regime typical of native forest was comparable to that of three native species at 1.1:1, but under the lower illuminance and lower R:FR typical beneath an *A. platanoides* canopy it increased to 2.3:1 (Reinhart et al. 2006).

Acer platanoides grows fast in the high light of gaps, but it can also invade slowly into undisturbed native forest, suppressing native understorey and tree species (notably *A. saccharum*) by creating deep shade and tolerating it (Niinemets 1997; Martin and Marks 2006; Paquette et al. 2012). The best evidence for this is the study by Reinhart et al. (2006), comparing an uninvaded, native reach of riparian forest where the canopy reduced PAR to 22 per cent of sunlight and R:FR to 37 per cent, with an invaded reach (with a largely *A. platanoides* canopy) where PAR was reduced to 1 per cent and R:FR to 16 per cent (with the common problem that the two reaches may have been different before invasion). These differences in light regime decreased the seedling survival of *A. platanoides* less than it did for three native species, and its biomass was reduced to 24 per cent of that in sunlight, compared to natives' 8–13 per cent. However, the models of Martin et al. (2010) indicated that the mortality of *A. platanoides* in low light would be higher than of most of its native associates.

Comparisons of *Acer platanoides* with other European tree species that might have become invasive in North America, but have not, are also informative: such as the ability of *A. platanoides* to shift its leaf distribution upwards through a shading canopy, its high SLA, and its ability to increase chlorophyll and N per leaf mass in low light (Niinemets 1996, 1997). Further comparisons with European species, similar to those conducted in North America, would be very interesting.

Whether these short-term investigations of *A. platanoides* mean its eventual dominance in the forests is less clear. Martin et al.'s (2010) simulations of mixed deciduous-angiosperm/evergreen-gymnosperm forest comprising tree species typical of Connecticut suggested eventual dominance by *Fagus grandifolia* (American beech), though with considerable presence of *A. platanoides*.

Fridley and Sax (2014) tried to find general reasons why exotics might have greater interference ability. They suggested three: (a) species from extensive regions have greater genetic variation, (b) evolutionary lineages from old, stable environments have had more opportunity to be honed by selection, (c) species from species-rich areas have been subjected to selection for interference ability against a wider range of plant types. These ideas are interesting, but are yet to be tested. Moreover, they may be trying to explain a super-species phenomenon that applies to only a few species. Dostal (2011) examined 12 species exotic to and invasive in Central Europe, and 23 species native to the region; some of the natives taxonomically related to an exotic, some not. Growing them in various three-species mixtures in pots in an experimental garden there was no tendency for exotics to have higher interference ability, nor for the outcome to be different if they were related to their associate.

5.12.1.2   Disturbance

Perhaps exotics can invade because they are *r*/*R*/ruderal/pioneer species, short-lived, more readily dispersed and rapidly reproducing in ephemeral habitats. Why should there be more *r* species among exotics? Probably disturbed habitats were sparse and small before humans changed the landscape. This has been suggested as the origin of arable weeds: once they were only in local disturbed areas such as riverbanks and with cultivation they expanded their geographical range into arable fields. In some floras the number of *r* species may have been very small. For example only c. 2 per cent of the New Zealand native flora is annual (Wilson and Lee 2012). Perhaps a similar situation was true of many areas before humans appeared.

5.12.1.3   Reaction

Many studies compare the soil in patches invaded by an exotic species with those not invaded. Like all observational studies of reaction they have the problem that the soil differences may be the cause of the invasion patchiness, not the result. However, the potential clearly exists for a switch to cause niche construction, e.g. *Ammophila arenaria* (marram grass) causing and tolerating fast dune growth, or *Genista monspessulana* (French broom) facilitating and benefitting from fire (Pauchard et al. 2008). Why should exotic species be especially capable of reaction? Perhaps because were they natives the change would already have taken place; perhaps because they are drawn from a wide species pool (the 'depauperate flora' concept, Section 5.12.1.1 ii), or perhaps because of the factors suggested by Fridley and Sax (2014; Section 5.12.1.1).

5.12.1.4   Novel Weapons

A possible reason for a higher interference ability of exotics is that they are using unfair means of interference, ones not familiar to the residents. This is hardly possible with competition because all plants know about N, P, K and the microelements, and about using them in a different way (e.g. different form of N, different soil depth). However, an exotic might produce an allelopathic toxin to which natives have no evolved resistance: a 'novel weapon'. Callaway and Aschehoug (2000) suggested this when they found in a greenhouse experiment that *Centaurea diffusa* (white knapweed), an exotic in Montana (USA), had greater interference effect on Montana grasses than on related species from within its native range in Georgia (Caucasus), and the difference was removed by adding active carbon. Similarly, *Centaurea stoebe* (incl. *C. maculosa*; spotted knapweed) has invaded western USA grasslands, and He et al. (2009) found in a greenhouse experiment interference from *C. stoebe* reduced the growth of several American species more than that of several European species, and in turn the European species inhibited its growth considerably more than the American species did. The flavenoid (±)-catechin is exuded from *C. stoebe*'s roots, and this inhibited the growth of the North American species more than that of related European species, suggesting that it was the novel weapon. Thorpe et al. (2009) found a similar effect of *C. stoebe* on plants of North American and European species in extant communities in the field in their respective regions. Similar novel weapon effects have been proposed for *Alliaria petiolata* (Jack-by-the-hedge), a Eurasian species highly invasive in forest understories in northeast USA. Its leaves contain a range of toxic chemicals, notably glucosinolates, not found in at least some related North American species (Barto et al. 2010).

Effects are not necessarily directly plant-to-plant. There have been suggestions of effects via the speed of litter decomposition, inhibition of mycorrhizal mutualists (*Alliaria* species, being in the Brassicaceae, are not mycorrhizal) and inhibition of the soil-nitrifying bacteria of the invaded range (but why should a reduction in nitrate availability impact more on the native plants species than on the invader?). A variant occurs when an exotic introduces diseases to which the native species have less resistance, perhaps by increasing the population of insect vectors (Malmstrom et al. 2005), or by root leachates increasing the population of a fungal pathogen (Mangla et al. 2008). We might call these novel biological weapons. Many other sorts of indirect effects are possible.

The invader might evolve to lose some of its novel weaponry, and Diez et al. (2010) and Lankau (2012) found some evidence for this in *Alliaria petiolata*: long-established populations post-invasion, comprising almost a monoculture of *A. petiolata* with few other plant species, produced less glucosinolates. Probably the more toxic genotypes invaded most readily, but in an almost-monoculture of *A. petiolata* there was self-toxicity and a biochemical cost to producing the toxin, and thus no advantage. The native species could evolve tolerance to interference from the invader and its novel weapons; Lankau (2012) found evidence for this in *A. petiolta* and Callaway et al. (2005) in *Centaurea stoebe*. However, all these studies comprise comparisons of different areas – old invaded vs more recently invaded – with the problem that the areas may have been different before invasion.

5.12.1.5 Natural Enemy Release

Perhaps exotics can invade because their vigour is enhanced by having fewer troubles with pests, having escaped from their enemies that were specific to them in their native range: natural enemy escape. Indeed, the general pattern, whether the pests are insects, crustaceans, fungi or viruses, is a lesser impact on populations in the exotic range of a species than in its native range, presumably because the pests specific to the species are missing (Mitchell and Power 2003; Vilà et al. 2005; Bossdorf et al. 2005). For example, Vasquez and Meyer (2011) selected somewhat arbitrarily four invasive exotic species, five non-invasive exotics and 30 natives from oldfield and prairie sites in Wisconsin, USA. Leaf damage, estimated 'visually', was generally less on invasive exotics than on natives. Below-ground effects can be found too: Callaway et al. (2004) found that the microflora in soil collected from the exotic range of *Centaurea maculata* in North America depressed its growth considerably less than that from its native range in Europe. However, the type of pest showing such an effect – viruses, bacteria, fungi and herbivores – differs between species (Agrawal et al. 2005). In their comparison of two invasive *Acacia* tree species in Portugal with their native Australia, Correia et al. (2016) found escape from predispersal predation combined with larger seeds, and a higher production of fully developed seeds per fruit, are likely contributors to increased seedling growth and the production of abundant soil seedbanks in the introduced area.

A similar contrast can sometimes be seen between exotic species that are 'invasive' and those that are not. For example, in Cappuccino and Carpenter's (2005) comparison (Section 5.12.1.1) there was much lower herbivore leaf damage in the invasive exotics than in non-invasive ones. The same trend was reported by Jogesh et al. (2008) using no-choice feeding experiments with adults of *Schistocerca americana* (American grass-hopper) and *Melanoplus femurrubrum* (red-legged grasshopper) feeding upon seven to nine naturalised exotic species (Asteraceae and Brassicaceae); the results showed less leaf area was eaten of the highly invasive species. In Vasquez and Meyer's (2011) survey, leaf damage was not significantly greater on invasive exotics, but when a subset of the species were released from pests with a systemic insecticide and a foliar fungicide, the leaf area and vegetative biomass of the one invasive exotic included in the experiment (*Pastinaca sativa*, parsnip) increased, compared to the control, more than most of the non-invasive exotics and natives officially listed as 'invasive'. Similar results were seen in a garden choice experiment.

Another argument is that closely related species are more likely to have common pests, so that exotic species that are less related to those in the area are liable to suffer less herbivory/disease damage. Indeed, when Hill and Kotanen (2009) grew 32 exotic plant species from an area in Ontario, Canada, in eight separate experimental plots, herbivore damage to plants, estimated by both counts of damaged leaves and 'visually estimated' area-per-leaf damage, was lower in species less phylogenetically related to native species in the same family (or close family) growing in the area.

How general is such natural enemy release? van Kleunen and Fischer (2009) found for 140 North American plant species naturalized in Germany there were records of 58 per cent fewer leaf and flower fungal pathogens than in their native range. However, including fungal pathogens of the native North American host range that had so far been reported only on other plant species in Europe, the estimated natural enemy release was reduced to 10.3 per cent, and most of the missing pathogens were rare on its particular host in North America. There was little escape from native enemies there. Also in contrast to natural enemy escape, cases have been reported of native herbivores preferring exotic plants to native ones (Parker and Hay 2005).

Presumably the enemies will catch up in dispersal time, as has happened with the invasion of *Lupinus arboreus* (tree lupin) on sand dunes in New Zealand, the invader later largely suppressed by the lupin anthracnose fungus *Colletotrichum gloeosporioides* (Molloy et al. 1991). Or in evolutionary time: native pests may evolve to utilise, and impact on, exotic plant species, and there is a little evidence for this, in disease (e.g. Gilbert and Parker 2010) and in invertebrate herbivory (e.g. Prokopy et al. 1988).

If a species has escaped from natural enemies, there will be no selective pressure on it to maintain its defences. It is far from clear that this normally happens (Felker-Quinn et al. 2013), but proceeding from the possibility that it does the 'Principle of allocation' (Cody 1966) indicates that resources (assimilates) will be available for other purposes. One possibility is that resources previously used for specific resistance may be used to increase general herbivory resistance. For example, in *Senecio jacobaea* (ragwort) native to Europe but invasive in North America, Australia, New Zealand and elsewhere, the specific defence against specialist herbivores *Tyria jacobaeae* (cinnabar moth) and

Longitarsus jacobaeae (ragwort flea beetle) has been lost in its invasive range, but some of the resources saved seem to have been put into increased protection against generalist lepidopteran herbivores via pyrrolizidine alkaloids (Joshi and Vrieling 2005; Stastny et al. 2005). Or in other cases it seems that the invasive genotype exhibits greater induced resistance but lower constitutive resistance as in *Lespedeza cuneata* (bush-clover, Fabaceae) in its introduced southeast North American range (Beaton et al. 2011). Alternatively, the spare resources may go into increased growth rate and/or interference ability (EICA, for Evolution of Increased Competitive Ability, except that other forms of interference may be involved too). Such faster growth can been seen comparing ecotypes from the native and exotic ranges grown in a common garden (e.g. Blair and Wolfe 2004), though in other studies the effect has been absent (e.g. Bossdorf et al. 2005).

There is, as in other native-range/exotic range comparisons, a problem that the genotype pool of the invaders may not have been that of the native-range population (s) used in the experiment. Beaton et al. (2011) minimised this problem by comparing, for *Lespedeza cuneata*, the ancestral genotype introduced to North America in 1930 with modern-day invasive (North American) and native (Japanese) genotypes, showing that the invasive genotype had higher interference ability than either the native or the ancestral genotype. Is the EICA effect restricted to a few populations? Flory et al. (2011) compared 10 native (China) and 10 introduced (USA) populations of the invasive annual *Microstegium vimineum* (stiltgrass) across 22 common gardens, planting seeds in a wide range of habitats and environmental conditions. Populations from the exotic range produced on average 46 per cent greater biomass and had 7.4 per cent greater survival, outperforming native-range populations in every experimental site.

However, in a broader survey Thébaud and Simberloff (2001) used data from floras to compare the heights of species native to the USA with the heights of the same species as invasives in Europe, and similarly for species that had invaded in the other direction. In some comparisons populations were no different, and in some species populations were taller in their native range, the opposite of the effect expected under the enemy release hypothesis. This study has the advantage of surveying many species and avoiding the possible bias of choosing problem weeds; the drawbacks are that it is not clear whence the flora writers obtained their height information, nor whether maximum height was defined consistently between different floras. And Felker-Quinn et al. (2013) found in a meta-analysis little evidence for increased growth or interference ability in species' introduced range than in the native range. The jury is out.

5.12.1.6   Invaders as Drivers Versus Passengers

It has been suggested that exotics could either be the cause of changes to native communities (the 'drivers') or whether they just take advantage of a disturbance (the 'passengers') (Section 4.8.4). Corbin and D'Antonio (2004) addressed this for the grasslands of California, which 200 years before had been dominated by native perennial grasses with associated annual and perennial dicotyledon species. These were almost completely displaced by European and Asian species. Under the 'passenger' hypothesis the change came about due to tilling for agriculture, introduction of livestock and a severe drought in the 19th century, leaving disturbed conditions, which the exotics could tolerate. Corbin and D'Antonio experimentally removed the vegetation, then sowed plots with three native perennial grass species, with three exotic annual grass species, or with both. Over time, the native grasses reduced the productivity of the exotic annuals, whilst the impact of the latter on the native perennials was minor and decreasing. The 'passenger' concept was supported for these species in this ecosystem.

Further south in California, Lambers et al. (2010) found that native and exotic grassland annuals did not differ consistently in competitive ability in terms of R* for monoculture in soil moisture, available N and P, or light below the canopy. This was supported by their approximately equal abundance in ungrazed plots. However, in plots open to grazing by cattle, exotics strongly predominated, as in the surrounding grass-lands. The conclusion can be that it is heavier grazing by different, introduced species that has allowed the exotics to invade. This is another exotics-as-passengers situation.

Stylinski and Allen (1999) compared almost undisturbed sites of chaparral and sage shrublands with nearby areas disturbed by vehicles, excavation or agriculture. Percent cover of shrubs was measured by canopy intercept, but of that herbs and seedlings only estimated 'by eye'. The vegetation of the disturbed areas comprised mainly exotic annuals (60 per cent cover), whilst the undisturbed areas had 68 per cent cover of native shrubs. This situation remained essentially unchanged in a site disturbed 71 years earlier, and the authors concluded that after invasion by exotics, the vegetation reached an ASS. Presumably a switch was operating, so that the passengers took over driving the vehicle, but what factor mediated the switch?

The drivers versus passengers issue is really a non-debate. Some exotics are drivers, many are passengers, for many we do not yet know their eventual fate so we cannot classify them either way.

5.12.1.7   Conclusion

The success of exotic species in many communities is clear, but the reason for this success is not. Although a less-than-satisfying conclusion, probably all the factors above are at work for different species.

5.12.2   Exotic Establishment and Community Assembly

The most fascinating way to use exotics as probes into community structure is to ask how they assemble when they reach new territory. Wilson (1989a) examined the native and exotic plant guilds of the Upper Clutha catchment, New Zealand. The two guilds produced classifications of the quadrats that were no more different than those using random groups of species, suggesting that the two guilds follow the same vegetational boundaries. However, there was some evidence that the guilds differ in which environmental factors controlled their distribution.

The roadsides of New Zealand generally comprise exotic species that have reassembled into communities. Wilson et al. (2000b) examined an area of southern New Zealand containing 152 exotic species, introduced mainly from Britain for

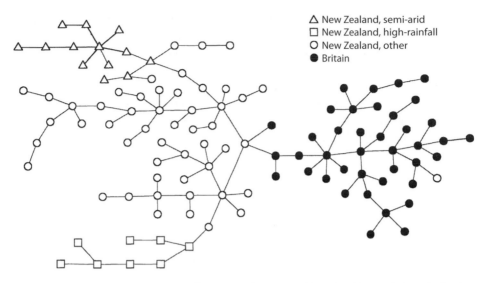

△ New Zealand, semi-arid
☐ New Zealand, high-rainfall
○ New Zealand, other
● Britain

Figure 5.12 A minimum spanning tree for the species composition of British and New Zealand roadside vegetation.
Reprinted from Wilson et al. (2000b), with permission from John Wiley and Sons Inc.

environmental and cultural reasons. Quadrats from these New Zealand roadsides were fitted to the British National Vegetation Classification (NVC). After excluding species that are not present in New Zealand, the fit was 61 per cent. Randomising the species/quadrats occurrences of the New Zealand data gave on average a 59 per cent fit to the NVC, so the fit of the real quadrats was only slightly, though significantly ($p < 0.001$) better than the random ones. British roadside communities were also compared to the NVC, as a control; they gave a 66 per cent fit. Thus, the New Zealand communities bear little relation to NVC communities in Britain (though the British communities did not fit brilliantly either). Comparing the New Zealand and British quadrats directly using a minimum spanning tree to connect similar quadrats, the two formed two almost distinct groups (Figure 5.12). The conclusion must be that the British species have reassembled into communities in New Zealand, most of which are new, i.e. distinct from those that occur in the native range of the species in Britain. The evidence points to community assembly by preadaptation. Wilson et al. (1988) reached similar conclusions comparing an area in southern England with one in southern New Zealand, which had considerable overlap in flora but no similarity in species associations.

Lord et al. (2000) studied in a similar way to Wilson et al. (2000b) the reassembly of species introduced from Britain in New Zealand calcareous-soil grasslands (4–24 per cent $CaCO_3$). Analysed as per the roadside study, the fits for six sites to NVC communities ranged from 48 to 77 per cent. Two of the six sites fitted British calcareous grassland communities. These two sites are on thinner soil (<10 cm depth), under lower rainfall, more likely to be influenced by the base rock. For these sites the soil and climate given for the community in Britain matched well that of the New Zealand site.

Together, these three reassembly studies suggest that only strong environmental filtering is able to reassemble communities. Even though the roadside dataset spanned a wide and very comparable environmental range in the two countries (e.g. rainfall 345–3,460 mm and mean temperature in the warmest month 12–17°C in New Zealand versus 485–1,777 mm and 14–17°C in Britain), it appears that environmental filters and assembly rules were not strong enough to reassemble the same communities. Instead, alternative states have been reached. We could not tell whether they are stable, and if so what switch is responsible, but the consistent separation in Figure 5.12 is remarkable.

5.13   Case Study: The Otago Botany Lawn

It is difficult to draw conclusions on assembly rules. Plants interact (Chapter 2) and plant species differ (Chapter 1), so there must be limitations to coexistence. However, the difficulty of finding assembly rules, and the difficulty of ensuring that tests for them are valid, combine to make it difficult to confirm that this is so in the real world.

In most studies just one or a few assembly rules have been tested on a site, and these might not have been the ones operating. The Botany Lawn of the University of Otago (Figure 5.13) has surely been more intensively studied in this way than any other community and offers a case study. We have already been introduced to some of the Botany Lawn results in the sections on guild proportionality (Section 5.7.1.1), intrinsic guilds (Section 5.7.3.1) and stability (Section 4.8.2.3). This section expands upon this work and presents the case that the Botany Lawn has, thus far, yielded the best evidence for the existence of assembly rules for any single community.

The lawn was established c. 1965 with the sowing of *Agrostis capillaris* (bent) and *Festuca rubra*. The former is still prominent, but the bulk of the 36 species present within the current community have arrived through natural dispersal, the commonest

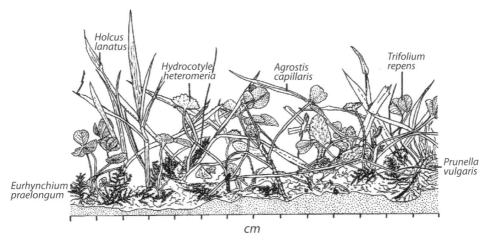

Figure 5.13 Profile through a part of the Botany Lawn.
Reprinted from Roxburgh et al. (1993), with permission from John Wiley and Sons Inc.

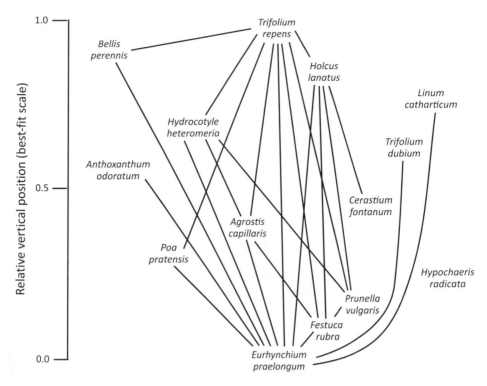

Figure 5.14 Stratification in the Botany Lawn after 14 days' regrowth. Lines connect species pairs that are significantly different in vertical position. Rare species are omitted.
Reprinted from Roxburgh et al. (1993), with permission from John Wiley and Sons Inc.

being the grass *Holcus lanatus* (Yorkshire fog), forbs *Trifolium repens* (white clover) and *Hydrocotyle heteromeria* (the only New Zealand native) and mosses *Eurhynchium praelongum* and *Acrocladium cuspidatum*. Since its establishment, the lawn has been maintained under a consistent regime of cutting to a height of c. 2.7 cm, fortnightly in the growing season and monthly in winter. There has been no application of fertilizer, herbicide or irrigation (the average annual rainfall is 784 mm year^{-1}). This constant management, together with the short lifespan of individual ramets in the lawn, has created the opportunity for the community to come to equilibrium, and indeed the species composition of the lawn is quite constant over time. There are seasonal changes on the lawn, but there is little evidence of directional change between years, and the abundance ranks of species have remained almost constant (Roxburgh and Wilson 2000b).

There is considerable stratification of species in the lawn (Figure 5.13). Even when the sward is only 2.7 cm high after cutting, there is significant evidence for three strata, and when the species have regrown 14 days later there are many more significant vertical relations between species, with evidence for up to four strata (Figure 5.14).

The variance in species richness across the lawn has been demonstrated to be lower than expected in a null model. This is seen at the scale of 13 × 13 mm (Watkins and

Wilson 1992) and the effect at that scale does not seem to be an artefact of environmental variation since its significance remains using a patch model. In fact, it was one of three out of the 12 lawns in that investigation to show a significant deficit of variance, greater than 20 per cent. A similar deficit of variance in richness can be seen at the scale of a point (Wilson et al. 1992b). The possibility has been raised that the effect is due to a physical limitation in packing plant modules at that scale. However, up to five species can be found at a point in this lawn, and on average only 1.45 species are, so space does not seem to be a limitation. Plants do not compete for space (Chiarucci et al. 2002; Wilson et al. 2007a; Section 2.2.1.1), and the profile diagram (Figure 5.13, drawn from life) confirms that the canopy is largely empty.

The restrictions on species coexistence can probably be seen better by analyzing guild proportionality. This removes us from questions of the number of modules that can be physically packed, by using a null model in which the numbers of species in each quadrat do not differ from those observed, and indicating restrictions in terms of types of species. Wilson and Watkins (1994) analysed thus at the 13×13-mm scale. Testing over all richness categories there was no significant guild proportionality for graminoid versus forb guilds ($P = 0.074$), but examining four-species quadrats alone there was ($P = 0.005$). This was true for one other New Zealand lawn and one Fiji lawn. Likewise, grass versus legume guild proportionality was significant in the Botany Lawn in three-species quadrats. Wilson and Roxburgh (1994) found significant guild proportionality at a point using graminoid versus forb guilds, whether or not the two bryophyte species were included with the forbs. There was no evidence that the rule was based on grass/legume interactions. There was also guild proportionality using as guilds the species that tended to be in the upper stratum of the lawn versus those that were basal (Figure 5.14), but only if the stratum assignments were based on species' positions at the *end* of the 14-day mowing/regrowth cycle. The constancy of the graminoid versus forb proportions increased as the number of species at a point did. All these results indicate that when there are few species present at a point, there is less constraint on which types, but as the species start to pack in, their ability to enter the community depends on their traits.

The a priori guilds that were used are not necessarily the true ones. At the scale of 13×13 mm, although two of the three grass–grass associations are negative as one would expect, so were those between *Plantago lanceolata* and two of the grasses (Wilson and Watkins 1994). It is possible to determine the guilds as perceived by the plants using the intrinsic guild approach (Section 5.7.3). With distributional data (minimising guild proportionality index RV_{gp}) the intrinsic guilds that resulted from the optimisation process generally confirmed both the particular role of graminoids and the importance of leaf position in the canopy (Table 5.2; Wilson and Roxburgh 1994). For example, *Trifolium repens* (white clover) with its horizontal laminae is often in the canopy fighting with the grasses (Figure 5.13), and it appeared in the same intrinsic guild as four of the five grasses (Table 5.2). Some other forbs were also better assigned to the 'graminoid' guild, again apparently because of their role in the upper canopy. All this suggests that there is one niche for species that occupy the upper canopy towards the end of the mowing/regrowth cycle, based on the interaction of lamina shape and

Table 5.2 Intrinsic guild classifications of species of a lawn obtained from (a) distributional data (Wilson and Roxburgh 1994) and (b) interference-experiment data (Wilson and Roxburgh 2001)

Species	Type	Guild from distributional data	Guild from interference experiment data
Agrostis capillaris	Grass	A	A
Anthoxanthum odoratum	Grass	A	
Bellis perennis	Dicotyledon, rosette	A	
Holcus lanatus	Grass	A	A
Hydrocotyle moschata	Dicotyledon, horizontal lamina	A	
Linum catharticum	Dicotyledon, upright	A	
Poa pratensis	Grass	A	
Ranunculus repens	Dicotyledon	A	A
Trifolium dubium	Legume, horizontal lamina	A	
Trifolium repens	Legume, horizontal lamina	A	A
Acrocladium cuspidatum	Moss	B	
Cerastium fontanum	Dicotyledon, erect	B	
Cerastium glomeratum	Dicotyledon, erect	B	
Eurhynchium praelongum	Moss	B	
Festuca rubra	Grass	B	B
Hydrocotyle heteromeria	Dicotyledon, horizontal lamina	B	B
Hypochaeris radicata	Dicotyledon, rosette	B	
Prunella vulgaris	Dicotyledon, creeping	B	B
Ranunculus repens	Dicotyledon, creeping	B	
Sagina procumbens	Dicotyledon, creeping	B	

position, and another for the basal species that may absorb the light that reaches further down just after mowing. Strong, almost surprising, support came from the intrinsic guilds obtained from the interference experiment by maximising the RYM, the yield of mixtures relative to the two monocultures (Wilson and Roxburgh 2001). The guilds formed gave, for the seven species included in the experiment, perfect agreement with those obtained from the distributional data (Table 5.2). These intrinsic guilds are real community ecology, because the ecologist is allowing the species to tell us what is happening in the community. This is inductive science, and made deductive for the distributional data by testing the guilds on independent data and for experimental data by confirming the results with the distributional data.

Mason and Wilson (2006) examined the traits of the seven most common species in each intrinsic guild. The intrinsic guild approach does not make any assumptions about the traits that determine coexistence, but the two guilds differed in Mowing Removal Index (MRI), calculated as the proportion of a species' mass typically removed during

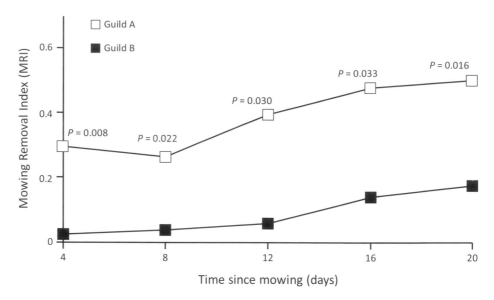

Figure 5.15 Mean Mowing Removal Index (MRI) of each guild at each sampling date. The *P*-values are from *t*-tests for differences between guilds in mean Mowing Removal Index. Redrawn from Mason and Wilson (2006).

mowing (Figure 5.15), though not in other traits related to light capture, such as specific leaf area, leaf area ratio and six photosynthetic pigment traits. This confirms the importance of canopy interactions, but sheds doubt on whether they involve light capture.

Mason and Wilson (2006) tested the limiting similarity concept directly by examining the traits of the species co-occurring at a point. Greater variance among the traits of co-occurring species than expected at random would indicate limiting similarity: a tendency for species that were alike not to co-occur. MRI and leaf length showed significant limiting similarity at all five times since mowing, as did two correlated traits, leaf area and length:width ratio. However, none of the other traits gave more than sporadic indication of limiting similarity. PSU length:width ratio showed significant limiting similarity for three of the dates, but it is related to MRI. Anthocyanin/dry mass demonstrated limiting similarity for the first two samples after mowing, and marginally ($P = 0.072$) after 20 days. None of PSU width, PSU thickness, PSU dry mass, SLW, ratio of lamina area or mass to shoot mass, chlorophylls a or b per dry mass, chlorophyll a:b ratio or UV pigment content were significant for more than one period out of five.

How can the restrictions on coexistence be due to canopy interactions yet not be related to light capture? One possibility, by analogy with the apparent importance of NPK and water resources in the results of Stubbs and Wilson (2004), is that although the guilds are canopy-related, the basic effect is below-ground. After defoliation there is generally 'root growth stoppage' (Hodgkinson and Baas Becking 1977). Species with a high MRI would be affected by this because more leaf is removed. The temporary cessation of root growth would affect P uptake, which is dependent on exploration of the soil by new roots. Species with a low MRI could carry on growing, not only absorbing light temporarily available by canopy removal, but with a continuing P supply. However, some support for the role of light comes from the local texture convergence study of Watkins and Wilson (2003), who found overall convergence between quadrats in chlorophyll, mainly due to strong convergence in two of the 12 sites, one of which was the Botany Lawn. It is simplistic to expect one process to be limiting coexistence.

Why is the evidence for assembly rules stronger in the Botany Lawn than anywhere else? Firstly, it has been more intensively studied than any other community. The short stature probably contributes to the ease of finding assembly rules. The canopy is in some ways like a forest canopy in miniature, but the relations are easier to see: in a forest it is hard to determine just which part of the canopy a ground herb is influenced by. However, following suggestions that assembly rules will be found in equilibrium communities (Bartha et al. 1995), the major factor may be that the lawn has reached equilibrium. It has been undisturbed for 30–40 years with little management apart from a constant mowing regime, with no allogenic disturbance to the soil and even few wormcasts. The lifespan of a ramet in the lawn is probably about a year, giving 30–40 generations of turnover. For forest trees, with lifespans of say 300 years, the equivalent would be 9,000–12,000 years. In temperate areas the forests have not been around that long since the glaciation, and in tropical areas there would almost certainly have been major disturbance. Ramet turnover in constant conditions has possibly been as great as in any community, and possibly no plant community anywhere is closer to its equilibrium than the Botany Lawn. If the community is close to equilibrium, we can ask about its stability, and as we discussed in Section 4.8.2.3, the Botany Lawn community has also been analysed for stability more intensively than any other community (Roxburgh and Wilson 2000a) and found to be on the borderline of stability, a conclusion confirmed by its response to perturbation (Roxburgh and Wilson 2000b). This stability is probably both the cause and the result of the assembly rules demonstrated.

5.14   Conclusions

Some types of assembly rule have been insufficiently tested (e.g. the fate of associations through time), or are difficult to test (e.g. zonation). In others, it is impossible to see trends in the limited work done so far (e.g. species abundance distributions, SAD). However, there are types for which good evidence is available: species sorting in succession, guild proportionality, limiting similarity and texture convergence. We note that the last three are those based on functional traits (characters) or on guilds (functional types), probably because these are more directly related to community functioning. Guild proportionality, limiting similarity and texture convergence are different ways of testing MacArthur and Levins' (1967) limiting similarity concept. Guild proportionality does this with discrete categories, texture convergence via similarity in traits between assemblages, limiting similarity looks directly for minimum trait distances/overlaps within communities. Texture convergence was first used between continents but can be applied very locally. Guild proportionality has been applied locally and, in an informal way, globally. Limiting similarity has been applied only locally but could be applied globally. 'Hard' traits are needed (Díaz et al. 1998), that is to say traits more closely related to plant function, for testing texture convergence and limiting similarity. The difference between fundamental and realised niche has to be considered, especially when there is niche construction. Plasticity due to associates in the community (Section 1.2.2) should be incorporated, but the task would be hugely complex.

There will still be many factors masking assembly rules: disturbance, microenvironmental variation, disease, the existence of assembly rules more subtle than those being testing for, and perhaps others. The failure to find an assembly rule in a particular community does not mean that it does not exist, and certainly that no assembly rules exist there.

The concepts behind the rules described here are implicitly or explicitly the niche and competition. This ignores the possibility of rules based on other types of interference, on facilitation, plant-to-plant interactions via heterotrophs, and the other interactions reviewed in Chapter 2. Assembly rules based on plant-to-plant interactions via herbivores and pathogens (Section 2.7) would be very interesting. Exploration of such rules is rudimentary, to the point of being non-existent. Some of the rules based on morphological/physiological traits, such as guild proportionality and texture, that have been found could reflect herbivory/disease effects, but it is unlikely the most appropriate guilds/traits for the purpose were used. The intrinsic guild approach is particularly appropriate as a tool, making no assumptions on trait choice, and allowing the community to 'speak for itself'. Perhaps the greatest need is to investigate the mechanisms underlying assembly rules in terms of plant physiology, and plant–plant interactions. Work continues on this.

6   Theories and Their Predictions

6.1   Introduction

Beyond Warming's (1909) 'easy task' of recording the environmental filtering of species, indirectly or directly by their traits (Section 1.5; Section 3.1), plant community ecology is based on plant interactions (Chapter 2). Most of these interactions operate via reaction on the environment (*sensu* Clements). Physical reaction causes interference, facilitation and various litter effects (Sections 2.2–2.4). Biotic reaction comprises modification of the surrounding heterotrophic biota, including interactions via the soil microflora, diseases, pollinators and herbivores (Section 2.7; Section 4.5). Direct interactions comprise autogenic disturbance and parasitism (Sections 2.5–2.6). Almost all of these interactions can be intraspecific, but it is interspecific interactions that structure the species composition of communities (Chapter 5).

In this chapter we now turn to general theories of plant community structure, and review the evidence for and against them. A general theory of plant ecology should apply to every vegetation type and to all species, and link species interactions with their reactions and resource requirements, and thus provide an overview of the topic of our book. Although a number of such theories have been proposed, our ultimate conclusion is they have been inconsistent in their focus on our core topics of enquiry (interactions [Chapter 2], coexistence mechanisms [Chapter 3], community processes [Chapter 4] and patterns of species assembly [Chapter 5]), and are thus necessarily incomplete. They have, however, provided many useful insights into different aspects of the functioning of plant communities.

The chapter starts with a broad classification of community ecology theories based on Whittaker's (1975a) oft-reproduced diagrams of species distributions along an environmental gradient (Figure 6.1).

There are issues with such graphical models of community structure:

1.   Gradients: Austin and Gaywood (1994) distinguish between environmental and resource gradients, but we believe many of the issues are similar for both. We agree with them that the proxy gradients used by Whittaker (e.g. altitude) cannot be used. Resource gradients are light, water and macronutrients, and theoretically CO_2, micronutrients, radiant heat and oxygen (Section 2.2.1.1), all with their

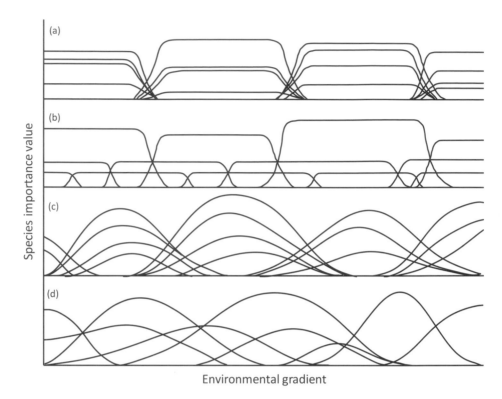

Species importance value

Environmental gradient

Figure 6.1 Representation of community structure gradient theories. The panels show the progression from discrete communities with low overlap in species composition (a), through to species arranged randomly along the gradient (d).
Reprinted from Whittaker (1975a), by permission of Pearson Education, Inc., New York, New York.

various modes of input and transience/storage. Environmental gradients are various aspects of temperature (e.g. yearly mean, lowest ever recorded), wind, oxygen (as anoxia) and toxic salts (pH is basically a proxy factor). Such a gradient will rarely be found as a sequence on the ground, it must be assembled, yet even so it will not be realistic because no factor occurs over its range with others held constant.

2. Abundance: For the Y-axis, abundance, Whittaker (1960, 1965) used stem counts or 'coverage estimates', given the vague term 'importance'. Biomass should be used (or ideally calorific value).

3. Biomass constancy: Biomass is often remarkably constant across species turnover because incoming radiation and energy→biomass conversion are both quite constant (Culmsee et al. 2010). Over wider environmental ranges, biomass will vary monotonically. Theoretical curves need to show similar behaviour.

Such diagrams need to be considered from three first principles, then applied to possible theory (Table 6.1).

Table 6.1 Relation of previous theories to the 12 possible combinations of distribution features described above (we have not included the genuine views of Clements or Gleason, because they were much more sophisticated than this analysis)

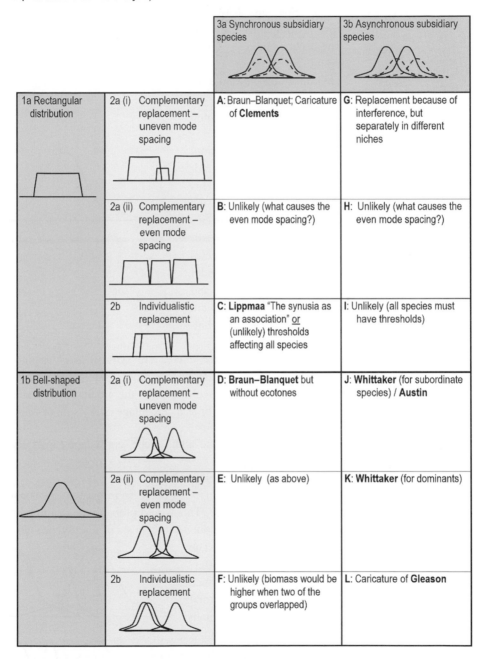

			3a Synchronous subsidiary species	3b Asynchronous subsidiary species
1a Rectangular distribution	2a (i)	Complementary replacement – uneven mode spacing	**A**: Braun–Blanquet; Caricature of **Clements**	**G**: Replacement because of interference, but separately in different niches
	2a (ii)	Complementary replacement – even mode spacing	**B**: Unlikely (what causes the even mode spacing?)	**H**: Unlikely (what causes the even mode spacing?)
	2b	Individualistic replacement	**C**: **Lippmaa** "The synusia as an association" or (unlikely) thresholds affecting all species	**I**: Unlikely (all species must have thresholds)
1b Bell-shaped distribution	2a (i)	Complementary replacement – uneven mode spacing	**D**: **Braun–Blanquet** but without ecotones	**J**: **Whittaker** (for subordinate species) / **Austin**
	2a (ii)	Complementary replacement – even mode spacing	**E**: Unlikely (as above)	**K**: **Whittaker** (for dominants)
	2b	Individualistic replacement	**F**: Unlikely (biomass would be higher when two of the groups overlapped)	**L**: Caricature of **Gleason**

1.   Shape of the distribution:

(a)   Rectangular   distribution: Boundaries are sharp and thus distributions are rect-angular/rhomboid.   This could be because of an environmental threshold or because of interference, though interference would not necessarily give this.

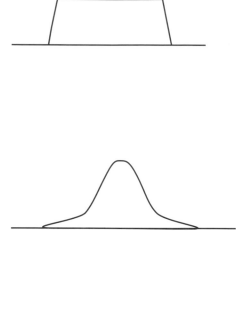

(b)   Bell-shaped   distribution: Abundance decreases grad-ually towards the limit. This could be because of the direct response of a species to the environmental gradi-ent or because two species can both tolerate a wider environmental range and the environmental gradient gradually changes the bal-ance of interference.

A subsidiary question is whether the curves are symmetric or asymmetric: in the extreme case rectangular on one side of the mode and bell-shaped on the other. However, symmetry depends on the scale used for the gradient (e.g. water availability expressed as a percentage of wet soil, or as a percentage of dry soil, or water potential?).

2.   Species replacement (within one alpha niche, e.g. canopy trees replacing each other):

(a)   Complementary replacement: A decrease in one species is complemented by an increase in another. This would probably be due to present interfer-ence, though it could be due to the ghosts of interference past.

(i) Uneven mode spacing: The species have to be differ-ent in modal abundance and/or niche width if total abundance is to remain approximately constant.

(ii) Even mode spacing: It is difficult to see how this neat situation would arise. Even coevolution would not necessarily give even spacing.

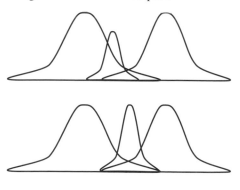

(b) Individualistic replacement: A decrease in one species is unrelated to an increase in another. With individualistic modes, constant biomass can be maintained only with skewed curves and/or different modal abundances.

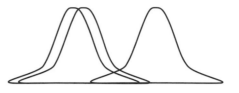

3. Stratification:

(a) Synchronous subsidiary species: When one species decreases, others (e.g. understorey species) follow. This pattern could be caused by an environmental threshold common to groups of species, or by coadaptation.

(b) Asynchronous subsidiary species: Distributions of subsidiary species are independent of those in other strata.

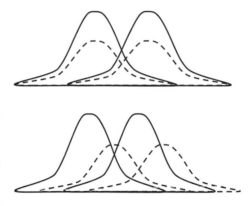

These three sets of alternatives give 12 possible combinations (Table 6.1). Those with even mode spacing (combinations **B, H, E, K**) seem unlikely, though **K** corresponds to Whittaker's concepts, at least for the dominant species in a community (Section 6.3). Combinations **I** and **F** seem unlikely. Combination **G** seems plausible, but we cannot match it with any existing theory. The other options can be matched with concepts in the literature, or at least with caricatures of them.

The various outcomes in Table 6.1 are depicted simplistically, in order to provide an overview of some of the key issues, and of some of the graphical models of species distribution that have been suggested. The potential for asymmetry in the response curves was noted above. An additional factor not explicitly included in the table is differences in species response when comparing fundamental and realised niches (Mueller-Dombois and Ellenberg 1974; Austin 2012). We return to these themes in Section 6.3.

6.2   The Clements/Gleason Synthesis

6.2.1   Clements

Frederick E. Clements saw a community as integrated: 'an organic entity exhibiting cooperation and division of labor' (Clements et al. 1929, p. 314; see also Clements

1905, p. 199) and thus 'greater than the sum of its constituent species' (Clements 1935, p. 35). He used the term 'complex organism' (never 'superorganism', a term inaccurately attributed to him by Whittaker and Woodwell 1972, and those who followed them). Phillips (1935), whom Clements and Shelford (1939, p. 24) cited with the greatest approval, elaborated on this: 'With properties definitely unpredictable from a knowledge of the individual organisms', i.e. with emergent properties (Wilson 2002). This implies deterministic structure: 'The bond of association is so strict...that the same seral stage may recur around the globe...with the same dominants and subdominants' (Clements et al. 1929, p. 315). Dice (1952) went further: 'All the species which are members of a given association...are adjusted more or less perfectly to one another'. For Clements, 'An association is similar throughout its extent in...general floristic compositions' (Weaver and Clements 1929, p. 46). These have been called the 'Integrated' (Goodall 1963) and 'Community-unit' concepts (Whittaker 1967). It is superficially that of the 1a/2a/2b boxes of Table 6.1a, but without any hint of even mode spacing, placing it in box **A**.

Clements was too good a field ecologist to take all this literally, writing that communities had 'more or less definite limits' forming a 'mosaic, in which the various pieces now stand out sharply, and are now obscure'. That is, '[A formation] can rarely have definite limits' (Pound and Clements 1900, pp. 313 and 315) and the 'ecotones are rarely sharply defined' (Clements 1905, p. 181) so that 'adjacent formations of the same general nature usually shade gradually into each other' (Clements 1907, p. 216). Describing variation within an association, Weaver and Clements (1929, p. 6) wrote: 'Practically all vegetation shows more or less striking differences every few feet'. Thus, Clements' formations and associations are simplifications – the same community rarely recurs (Wilson et al. 1996b). In fact, Clements' classifications were much more informal and broad-brush than those of the Braun–Blanquet SIGMA school in Continental Europe.

To those who criticise Clements for classifying vegetation, even as broadly as he did, we point out that classification is inevitable. Plant ecologists do it continually. Moreover, it is often required in order to identify conservation targets to the public and to government, as in the British National Vegetation Classification (Rodwell 1991–2000).

Tansley, another ecologist with great field experience, wrote: 'the complex of interactions between plants and their environment does lead to a certain degree of order...The same species are constantly present in the same kind of place and show the same groupings'. At equilibrium, he said, the association becomes 'the mature, integrated, self-maintaining quasi-organism' (Tansley 1920). Yet he criticised Clements' concepts, writing: 'the concept of the "complex organism"...is a false value, and can only mislead...because it is based either on illegitimate extension of the biological concept of organism. ...Climaxes may be considered as quasi-organisms'.

One might think that Braun-Blanquet (1932), who advocated formal hierarchical classification of vegetation into associations, alliances, orders and classes, would be close to Clements' views, yet he criticised it, writing: 'Clements has held that the plant community is an organic, closely organized, collectively reacting unit. ...Clements in a flight of imagination compares the climax community to an organism which has birth, growth, maturity, and death. . . .The facts show these opinions to be wholly incorrect' (Braun-Blanquet 1932, p. 315). There were certainly strong opinions on terms: – 'complex organism', 'quasi-organism', 'association' – and concepts such as 'more than the mere sum of its parts'. However, it is difficult to pin these down to testable features.

Whether communities are 'complex organisms' or not, the naming of them implies recurrence: that the same community will be found in several different locations. This has rarely been tested, but Wilson et al. (1996b) did so by quadrat sampling roadside communities in a range of environments across southern New Zealand.[1] The problem is defining 'the same' community. It would be unrealistic to expect exactly the same species complement, so a baseline is needed of how similar two remote quadrats should be in order to be regarded as the same. Wilson et al. answered this in two ways. The quadrats had been placed in adjacent pairs. One baseline was therefore the mean similarity between the two quadrats of a pair, making the question: 'does one ever come across another patch of vegetation as similar to this one as the patch next door is?'. Some next-door quadrats would happen to be quite different, e.g. in disturbance, so Wilson et al. omitted the 10 per cent of least similar adjacent pairs before taking the mean. The answer was basically 'no, the same community does not recur'; for only 19 per cent of sites was there another in the survey similar to it by this criterion (Figure 6.2). However, another comparison was available, since the pairs of quadrats at a site had themselves been placed in subsites 50 m apart. Using those subsites as the baseline, i.e. performing the test at a larger spatial extent, the percentage of sites with vegetation that occurred elsewhere in the survey increased to 83 per cent. We have to conclude that communities do recur, and in this Clements was right.

Deterministic, discrete structure could arise either by coevolution or by the assembly of preadapted species with only certain combinations of species being stable (Bazzaz 1987). Clements does not seem to have used the term 'coevolution', though he considered that the evolution of species occurred alongside the development of communities (Clements 1929, p. 202), and later workers made it more explicit, e.g. the 'interco-ordinated evolution' of Dice (1952). A consequence would be coevolutionary ecotones (Section 7.3.3) where one set of coevolved species gives way to another set, but it is hard to find an explicit statement of this concept. The fullest coevolutionary view is that of Dunbar (1960) who suggested that selection could operate at the level of the whole ecosystem: just as an individual can die and be replaced by one of genotype with higher fitness, so an ecosystem can be unstable, collapse to leave 'empty environmental space', and be replaced by a community from nearby with genetic differences in some of its species, giving it a higher stability (i.e. fitness). However, we note that collapse to empty space is not realistic. Otherwise, the concept is the group selection theory of Wilson (1992) applied to whole communities. Darnell (1970) had similar ideas, writing that 'the ecosystem. . .is. . .the basic selectional unit of evolution'. He suggested that species-level selection led to evolutionary adaptation, which led to

---

[1] Like the Botany Lawn (Section 5.13), the generation time of the ramets of all the species in roadside vegetation means that they will usually have had more ramet turnover generations to come to equilibrium than, say, most forests, and therefore more opportunity for interference and facilitation amongst the species.

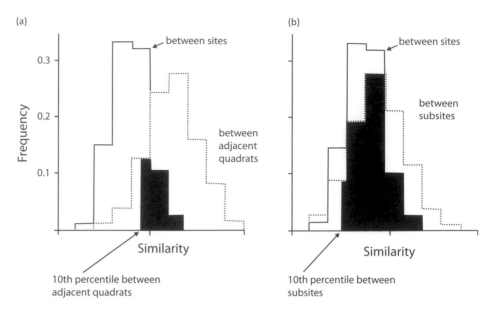

Figure 6.2 Does the same community recur? Comparison of between-site similarities in species composition along New Zealand roadsides with: (a) those between adjacent quadrats, and (b) those between subsites 50 m apart.
Reprinted from Wilson et al. (1996b), with permission from John Wiley and Sons Inc.

stability. This concept is not valid, for coevolution cannot lead directly to stability. Selection is for fitness of the individual, not for stability or any other property of the whole ecosystem.

Although many are ready to ridicule Clements' views, some contemporary ecologists have produced models in which the control of species composition is every bit as tight: mainly theoretical ecologists (e.g. Drake 1990), but also field ecologists such as Cody (1989).

6.2.2   Gleason, and the Clements/Gleason Consensus

Gleason's views are commonly seen as being of a random scatter of bell-shaped curves along a gradient (e.g. Whittaker 1975a; d). A literal application of this gives considerable variation in total abundance along the gradient, which is surely not intended. Gleason (1926, p. 18) did state that the physiological optima of the species of any association would cover a considerable range, and there are implications of random occurrence within ranges too ('accidents of seed dispersal', Gleason 1926, p. 19), but that is common to all ecological theories.

The great majority of present-day authors describe the theories of Clements and Gleason as opposites. For example, Keith et al. (2011) wrote: 'Clementsian communities are strongly delineated into specific community types because species are constrained by interspecific interactions (Clements 1916). In contrast, Gleasonian communities represent a situation whereby species respond individualistically to

Table 6.2 Comparison of the assumptions underlying the theories of Clements and Gleason

	Clements	Gleason
1. **Interference** is important	Yes	Yes
2. **The landscape** is a vegetational mosaic	Yes	Yes
3. **Repetition** of the stands of an association is never exact	Yes	Yes
4. **Priority** effects allow a species to preempt a site	Yes	Yes
5. **Associations** have fixed limits with ecotones between them	Yes	Yes
6. **Sharp ecotones** can occur along gradual environmental gradients due to switches	Yes	Yes
7. **Ecotonal species** occur	Yes	Yes
8. **Temporal fluctuation** occurs in species composition	Yes	Yes
9. **Succession**		
Successional pioneer species are the better dispersed ones	Yes	Yes
Climax species cannot establish early because of the environmental conditions	Yes	Yes
Pioneer species modify the environment, disfavouring themselves	Yes	Yes
Climax species modify the environment, disfavouring the pioneer species by shifting the competitive balance	Yes	Yes
10. **Classification** of plant communities is possible and necessary	Yes	Yes
11. **Assembly rules** exist	Probably	Probably not

See text for details of each entry.

environmental gradients and, as a result, community composition falls along a continuum (Gleason 1926)'. Very similar statements can be found, for example, in Presley et al. (2010), Chase and Myers (2011) and Willig et al. (2011). However, Gleason's concepts, like those of Clements, are widely misunderstood. True, his first theoretical paper (Gleason 1917) was explicitly presented as an alternative to the patterns and their causes espoused by Clements (1916) and he later declared, provocatively, that a stand of vegetation is a 'temporary and fluctuating[2] phenomenon' (Gleason 1939, p. 93). Clements et al. (1929, p. 315) wrote in return that Gleason's concept 'appears to involve a confusion of ideas as well as a contradiction of terms'.

Yet behind the invective, most of Gleason's views were identical to those of Clements (Table 6.2).

1.   **Interference**: The importance of negative interference was emphasised by Clements et al. (1929), but also by Gleason (1936) who wrote that when any two plants were growing together 'each interferes with the environment of the other' and that this interference 'may act either favourably or unfavourably' so that 'the vegetation...is the result of the interference'. The latter statement is as strong as any ecologist has ever made, and the very opposite of the no-interaction caricature of him often presented. In the process, he was among the first to suggest that facilitation ('favourable interference') is widespread.

---

[2] Though Clements' concept of 'annuation' seems close to Gleason's 'fluctuating'.

2. **The landscape**: Clements' understanding of the landscape was a mosaic of different formations/associations (Pound and Clements 1900), with 'the same species or formation in similar but separate situations' (Clements 1904, p. 163; 1907, p. 289), a situation he called alternation. Gleason's (1936, p. 447) concept was identical: 'a vegetational mosaic, composed of numerous types of vegetation, each repeated numberless times, but all united into a harmonious and extensive whole'.

3. **Repetition**: Gleason (1917) did state that, *contra* Clements, *exact* repetition of the same vegetation never occurs, but Clements did not expect this: 'No formation is uniform throughout its entire extent. …universal variation may be regarded as a law of formation structure' (Clements 1907, p. 221), echoed by Braun-Blanquet (1932, p. 22):'with few exceptions, no two bits of vegetation have precisely identical floristic composition'. It would be amazing for someone with Clements' depth of field experience to think otherwise.

4. **Priority**: Gleason (1917, pp. 472–473) envisaged that areas with similar preexisting environments could, because of accidents of migration, have differences in species composition that persisted some time, part of his 'individualistic' concept. Weaver and Clements (1929, p. 48) went further: 'It sometimes happens that one dominant occupies an area so completely as to exclude the others simply because it invaded first, although the habitat was equally suitable for all'. And, describing the situation when the climate returned to normal after years of drought causing prairie to replace forest, 'the hold exerted by the grasses through competition has delayed the return of forest' (Clements 1934, p. 65). This is one of the two interpretations of Egler's 'Initial Floristic Composition' theory (Egler 1954; see Wilson et al. 1992c), common to Gleason, Clements and Egler.

5. **Associations**: Gleason described associations as having 'limits. . .fixed by space and time' with 'tension zones' (i.e. ecotones) between them (Gleason 1927, pp. 325 and 313), every community necessarily having boundary and uniformity (Gleason 1936, p. 447). Clements was perhaps not quite so dogmatic: 'Ecotones are well marked between formations, particularly when the medium changes: they are less distinct within formations' (Clements 1904, p. 153), 'Each climax formation falls readily into two or more major subdivisions known as associations. Toward their edges these blend into each other more or less, making a transition area or ecotone.' (Clements 1920, p. 107), emphasising that the ecotone is often broad or confused.

6. **Sharp ecotones**: Gleason, sometimes associated with the term 'continuum', actually never used it. The closest to it would be in Clements' repeated discussion of zonation. Clements believed that narrow transition zones (ecotones) between associations could occur along gradual environmental gradients because of reaction (Clements 1904, p. 165). Similarly, Gleason (1917, p. 470) thought that at least in regions of 'genial environment and dense vegetation' there is reaction (a term that he used interchangeably with 'environmental control') with the result that: 'species of one association are then

excluded from the margin of the other by environmental control, when the nature of the physical factors alone would permit their immigration. The adjacent associations meet with a narrow transition zone, even though the variation in physical environment from one to the other is gradual'. Gleason's statement is a precise summary of Clements' view. Both are saying that often reaction causes a switch,[3] giving an ecotone between associations (switch ecotone, Section 7.3.3).

7. **Ecotonal species**: Both believed in these (Gleason 1926, p. 12; Clements and Shelford, 1939, p. 233).

8. **Temporal fluctuation** occurs in species' composition (Clements 1934, pp. 41–42, as 'annuation'; Gleason 1936, p. 448).

9. **Succession**: Their concepts of the mechanism of succession, and the role of reaction in it, were identical (Clements 1916; Gleason 1927).

10. **Classification**: Clements produced classifications of plant communities, though never again in the detail seen in Pound and Clements (1900), and Gleason (1939, p. 109) accepted the need: 'A classification of plant communities is certainly needed to enable us to talk about them conveniently'. He thought a 'precisely logical classification of communities' was not possible, with which Clements would doubtless have agreed (points 3 and 4 above).

11. **Assembly rules**: In terms of our four steps in community assembly (Section 1.5), both Clements and Gleason would have accepted A–C. Assembly rules (D) are apparent very occasionally when Clements writes on alternation: 'Owing to the accidents of migration and competition, similar areas within a habitat are not occupied by the same species, or group of species. A species found in one area will be replaced in another by a different one of the same, or a different genus. . . .Such genera and species may be termed *corresponding*' (Clements 1904, p. 173), and 'Such genera and species. . .must be essentially alike in. . .response to the habitat, though they may be entirely unrelated systematically' Clements (1907, p. 294). The concept here is of a niche in a community into which one species *or* another can fit, an assembly rule as strong as any. So far as we can tell, Gleason would not have accepted this.

Clements (1905) used the term 'complex organism' (never 'superorganism'), which Gleason (1917) disliked, but beyond terminology their concepts were almost identical (Table 6.2). There were probably personalities involved, at least in their approach to science.

Clements versus Gleason is a useful straw man in introductions to papers, e.g. Keith et al. (2011), Presley et al. (2010), Chase and Myers (2011), Willig et al. (2011) and Leibold and Mikkelson (2002), the latter describing 'the now well-known dispute between Clements (1916) and Gleason (1926). . .pitting the idea of 'discrete communities' against that of a 'continuum''), but it has also entered into serious discussion, e.g. 'the Gleasonian paradigm had overthrown the Clementsian one' (Simberloff 1980).

---

[3] Of course not in pre-Odum (1971) days using that term.

However, their concepts were almost identical, reflecting deep the understanding of plant communities that both had, and offer a strong basis for ecology today.[4]

6.3   Whittaker, Austin and Continuum Theory

6.3.1   The Continuum Concept

The *Continuum Concept* was proposed by Curtis and McIntosh (1951) and developed by Whittaker[5] (1967) and Austin (1985). It basically states that neither the modes nor the limits of species distributions are clustered along environmental gradients (1b/2a/3b in Table 6.1), in contrast to the naming of associations as 'discrete communities which are thought to be recognisable as distinct entities in the field' (Curtis and McIntosh 1951, p. 492).

Details differ between workers. Whittaker (1967) used proxy gradients – altitude or topography – but Austin and Smith (1989) apply *Continuum Theory* only to environmental gradients and explicitly not to proxy/complex gradients, resource gradients ('where the varying resource is consumed by plants in order to grow') or geographical position. Whittaker (1967, 1975a) envisaged that although minor species are positioned at random, character displacements (ghosts of interference past) result in regular spacing of dominants along the gradient (2a(ii) in Table 6.1). Austin (1985) knows nothing of such coevolution, for him the *Continuum Theory* is simply a description of the way vegetation is, not why, albeit with a strong focus on interpreting the patterns with respect to potential underlying mechanisms (Austin 2012). Extra hypotheses for Austin and Smith (1989) were that along *resource* gradients all species have the same limits, though not necessarily the same distribution, and that along *environmental* gradients species richness and evenness are both bimodal.

Austin and Smith (1989) suggest that the boundary will be sharp (1a in Table 6.1) when the limit is caused by extreme-environment filtering, but gradual (1b) towards the mesic position when it is caused by interference from other species, resulting in skewed distributions (Austin and Gaywood 1994). Indeed, all nine southeast Australia *Eucalyptus* species examined by Austin et al. (1994) showed significant skewing along a gradient of mean annual temperature, in the expected direction if 'mesic' was defined as 11.5°C. There are problems with this concept:

a.   'Mesic' is impossible to define.
b.   There are indeed environmental thresholds in some factors. For example, in many species the cut-off of water stress is −1,500 kPa for more than a day or two, though this is escaped by deciduous shrubs, by most succulents and by annuals. Yet, environmental filtering can also cause gradual reduction in abundance along a gradient. On the other hand, interference from an environmentally limited

---

[4] Jennifer Costanza (pers. comm.) asked: "Why do we tend to view the two as at opposite ends of the spectrum? Just because it's more fun that way?". Basically, yes.
[5] Who named it as the *Community Gradient/Ecocline* concept.

dominant species can cause a sharp cut-off in subordinates or in alternates, as at treeline.

c. A conclusion of skewness depends on the way the X-axis is expressed. For example a simple rainfall gradient assumes that the difference between 0 mm and 300 mm is equivalent to that between 2,000 and 2,300 mm. Who could believe that? Peppler-Lisbach and Kleyer (2009), working in deciduous forests in Germany, reported faster species turnover in the 3–4 range of a pH gradient than in pH 5–6, but this would be expected because it is in the pH 3–4 range that maximum turnover of pH-related factors such as exchangeable Al occurs; pH is largely a proxy gradient. It is usually impossible to know on what scale to express an environmental factor.

d. The conclusions will depend on the type of curve fitted and how skewed is 'skewed' (significance is not the best guide to effect size).

e. Skewness can be reliably determined only when the whole environmental range of the species has been sampled (M.P. Austin, pers. comm.), and in some parts of the range other factors may become limiting.

Continuum Theory has itself been misunderstood (Austin 2012). For example, Presley et al. (2010) stated as 'A fundamental principle in ecology' that species have Gaussian distributions along environmental gradients, citing Austin (1985), who in fact concluded that skewed, bimodal and other non-Gaussian curves predominate, at least in the literature. Presley et al. also cite Whittaker for this, but Whittaker (1967) stated that the curves could be bimodal. Bimodal curves would be interesting, being less likely to be an artefact of X-axis scaling, but do they occur? Austin (1985) commented: 'The occurrence of bimodal curves…seems well established', but he cites Whittaker whose declarations of bimodality have little support in his actual data (Wilson et al. 2004). We have not been able to find a good example of bimodality. Grime et al. (2007) show a bimodal curve in the British flora only for *Festuca ovina* along pH, and that may represent two subspecies (Wilkinson and Stace 1991).

6.3.2 Comparison with Other Theories

Whittaker (1967, 1975a) and Austin (1985) contrasted their *Continuum Theory* with what they termed the *Community-unit Theory* of Clements. Barbour et al.'s (1999) view that 'Whittaker was aggressive in his challenge of…Clementsian views on vegetation' would be common. However, the Whittaker/Austin theory differs from Clements' and Gleason's in wording, possibly in emphasis, but it is generally similar, or more extreme than their caricature of Clements:

Interference: Whittaker (1967, p. 228), Whittaker and Levin (1975, p. 30) and Austin and Smith (1989) saw interference as important in restricting species to their realised niche, though none stated it as strongly as Gleason (1936).

Sharp ecotones: Clements and Gleason attributed sharp boundaries along environmental gradients to switches. Whittaker attributed a sharp change in forest dominance with elevation in the Great Smoky Mountains to one species having

'a strong competitive advantage over other species in some range of the environmental gradient', and the beech forest/grassy bald ecotone as being due to
'special' or 'extreme' environments. The Clements/Gleason explanation seems
more developed, and matches the theme of this book.

Complex organism: Similarly to Clements' concept of the community as a complex
organism, Whittaker (1975a, p. 2) conceived the community as 'a distinctive
living system with its own composition, structure, ...development and function',
with (Whittaker and Woodwell 1972, p. 138) 'emergent characteristics'.

Coevolution: Whittaker saw coevolution as a cause of community structure: 'The
community is an assemblage of interacting and coevolving species' (Whittaker
and Woodwell 1972, p. 137); Clements did not.

Whittaker's theory was in some ways more of a 'community-unit theory' than Clements', even though he claimed the opposite.

6.3.3   Problems with the Continuum Concept

Every application of the Continuum Concept assumes that towards their limits species
become less abundant. Sometimes this is true: the plants at geographic margins can be
smaller, and/or less fecund (in seed number or quality, possibly because specialist
pollinators are sparse), and/or less fit (because of inbreeding depression). These features
can lead to populations that are scattered, smaller, sensitive to disturbance, restricted on
other niche axes (Section 1.4) and/or absent from some apparently suitable habitat
patches (Carey et al. 1995; Nantel and Gagnon 1999; Jump and Woodward 2003).
However, there are almost as many exceptions as examples for these trends – either no
difference or greater performance near the limit – except that population size is rarely
greater there (Abeli et al. 2014). There is a sudden cut-off at the northern limit of
Lactuca serriola (prickly lettuce) in Britain, with no reduction in vigour near the limit
(Carter and Prince 1985). This may not be exceptional. In an early and careful study at
Sugar Creek, Ohio, USA, of a 'tension zone' where over 120 species have geographical
boundaries, Griggs (1914) found that most species were abundant and flowered/fruited
successfully up to their geographical limit. Carter and Prince, and Griggs, could only
hypothesise that competition sharpened boundaries to make them abrupt. There can be
many, sometimes surprising, causes of a species' limit. Pigott and Huntley (1981) found
that the problem for *Tilia cordata* (linden) at its northern limit in England was that the
pollen tube could not grow fast enough in the low temperatures to reach the ovule,
leaving relictual populations now unable to reproduce. It seems that the behaviour of
species at the margins of their ranges as well as the causal factors for their limitation are
complex and unpredictable.

Whittaker (1975a, pp. 112–117) suggested that theories of community structure
could be distinguished by the shape of species/environment curves (Figure 6.1), but
his own attempts to do this are flawed (Wilson et al. 2004). Actually, the test cannot be
applied, for it is impossible to know the environmental gradient as seen by plants in
terms of: (a) what the limiting factor is (and it is very difficult to measure all possible

factors), (b) on what scale to express it (which will differ between species anyway) and (c) what interactions there are (it is unlikely that any environmental factor will be the limiting one throughout its range). In addition, the full range of a factor rarely exists. Questions such as whether distributions are nested would be answerable, but we know of no dataset able to address this. The character displacement of dominants that Whittaker proposed seems unlikely, if only because a species generally occurs with a range of others across space and time, and cannot evolve in sympathy with all of them (Goodall 1966; Jackson and Williams 2004).

Analysis of the shape of distributions along gradients seems to be a diversion from our search for the nature of the plant community. We accept the Clements/Gleason synthesis as above, which includes most of *Continuum Theory* anyway, but now with a clearer understanding that switches often produce boundaries along preexisting environmental gradients.

6.4   Hubbell and Chance

6.4.1   The Concept, and Unified Neutral Theory (UNT)

As we discussed in Chapter 3 (Section 3.10), a role has often been proposed for stochasticity/chance/random effects/disorderliness in the construction of plant communities, even a dominant role (Lippmaa 1939; Fowler 1990; Sykes et al. 1994; Richards 1996), in a scenario where many species are ecological equivalents[6] of each other, at least within guilds (e.g. equivalence between canopy tree species). In reality, although there can be probabilistic distributions for an individual case, chance does not exist. Everything happens according to the empirical laws of physics, and above the scale of the atom chance plays no part. Dispersal can be effectively unpredictable when caused by complicated processes such as eddy diffusion and animal movement, and the first species dispersed to a microsite might then hold it via cumulative competition for light (Section 2.2.1.2), inertia (Section 3.12) or a switch (Section 4.5), but this is unpredictability, not chance.

Hubbell and Foster (1986) made the concept of chance explicit, writing 'biotic interactions...are not very effective in stabilizing particular taxonomic assemblages, in causing competitive exclusion, or in preventing invasion of additional species' because there are 'ecologically equivalent species'. Therefore, 'chance and biological uncertainty may play a major rôle in shaping the population biology and community ecology of tropical tree communities'. Hubbell (2001) developed these concepts into a full 'Unified Neutral Theory' (UNT) in which species are equivalent in their demography and dispersal, i.e. in which niche differences play no role. He discovered, apparently almost as much to his surprise as to anyone else's, that many of the features of ecological communities that ecologists had long been discussing, such as relative

---

[6] This can be distinguished from the redundancy concept, where the species are equivalent in alpha niche but not in beta niche.

abundance distributions, species–area relations and island biogeography, could be predicted on this basis. UNT does not imply that even on one trophic level all species actually have the same niche: 'No ecologist in the world with even a modicum of field experience would seriously question the existence of niche differences among competing species' (Hubbell 2005). Rosindell et al. (2012) echoed this, emphasising UNT as a null model, useful for highlighting features of real-world communities that have to be explained by some non-neutral, non-UNT, deterministic mechanism that then must be added to the community model.

Hubbell's earlier work had described niche differences in the very tropical rainforest that he often takes as his example of chance: 'Some tree species are largely restricted to slopes, whereas others are predominant on flat ground or in the seasonal swamp. . . .Shade-tolerant shrubs and understorey trees are also recognizable guilds. Finally, there are gap-edge regeneration specialists' (Hubbell and Foster 1986). These effects would tend to cause aggregation within species, but the same workers demonstrated 'pervasive' negative effects of plants on neighbours that were of the same species, perhaps an adumbration of the zero-sum aspect of UNT but without its assumption of ecological equivalence. Such effects were confirmed when Uriarte et al. (2004) estimated the effect of neighbouring saplings on the diameter growth of other saplings on Barro Colorado Island, work in which Hubbell has been involved. For almost half of the species they could find species-specific effects, including larger effects (interpreted as greater interference) if the neighbours were conspecific, or confamilial, or in the same gap/shade-tolerant guild. All this emphasises that UNT is intended as a null model, not a best-fit model, and for this reason it does not feature in Table 6.1. Chase (2014) points out that communities are inevitably predictable at large spatial grain because of environmental filtering, and very liable to be determined by unpredictable birth and death at small grain.

6.4.2   Evidence

Various types of evidence have been used to suggest the operation of chance, and these are summarised below.

6.4.2.1   Relative Abundance Distributions (RAD, SAD)

Wootton (2005) parameterised a UNT model using a 12-year record of transitions in an intertidal community (sedentary animals and algae). Model predictions matched the observed relative abundance distribution (RAD, SAD, Section 5.10.2). However: (a) there was no alternative model (RAD curves tend to look all rather similar because they monotonically decrease) and many other curves could have fitted, (b) the confidence limits for the model prediction were wide, and (c) observed species abundance in mussel-removal plots bore no relation to the model's predictions. As Chave (2004) observed, many ecological models can result in the same RAD pattern, but that does not prove that any one of them is correct. However, the species will never be identical, and Fuentes (2004) showed the rather obvious effect that even slight differences in fitness (i.e. interference ability) resulted in quite different RAD and species–area curves, from those predicted by UNT, with the fitter species in high abundance ('overdominance'), generally giving concave-up curves quite similar to Zipf–Mandelbrot fits (Section 5.10.2). Jabot and Chave (2011) found the latter situation in a worldwide, plant-density-based dataset of 20–52-ha tropical rain forest plots: all significant deviations from a UNT-type RAD were due to overdominance, such as would result from cumulative competition for light (Section 2.2.1.2).

6.4.2.2   Temporal and Spatial Patterns

Campbell et al. (2010) analysed the occurrence of bird species in a forest in Maine, USA, over five years. After accounting for habitat (vegetation) preferences, the frequency distribution of the number of years a quadrat was occupied did not differ significantly from random for 12 out of the 20 species. However, lack of significant departure from the null model is no evidence that it is correct. Van der Maarel and Sykes (1997), proposing their 'Carousel model', envisaged random small-scale fluctuations in species composition (Section 4.2).

Spatial patterns have been seen as evidence of chance, especially for its prominence in in the establishment phase (e.g. Veblen and Stewart 1980; Merritt et al. 2010) and in lack of departure of co-occurrence patterns on real or virtual islands from null-model prediction (e.g. Wilson 1988d). These are weak arguments.

6.4.2.3   Lack of Environmental Correlations

Many studies have attempted to separate components of beta diversity due to environmental differences from those due to spatial autocorrelation, often equating the latter with the dispersal limitations envisaged by UNT (e.g. Gilbert and Lechowicz 2004; Dumbrell et al. 2010). For example, Myers et al. (2013) calculated vegetation dissimilarities between 36–38 0.12-ha forest plots in the Ozark Mountains, Missouri, USA, and Madidi National Park, Bolivia. To avoid overfitting, only 4–5 out of their 29 environmental factors were used. Environmental correlations were stronger than spatial ones in Missouri, spatial stronger than environmental in Bolivia. Still, less than half the vegetational differences were explained by the two together. Chase and Myers (2011) suggested a method based on beta dissimilarity along spatial and along environmental gradients for distinguishing between 'deterministic' (e.g. chequerboards arising from environment or interference) and chance processes. All of these studies use basically the same logic as early ones, attributing to chance any failure to find environmental correlations (e.g. McCune and Allen 1985; see Section 3.10), except that recent studies separate 'unexplained' into 'dispersal' and 'other unexplained'. It is not clear what processes are intended to be behind 'other unexplained', it can surely be only past disturbance, environmental effects not included in the model or dispersal effects not simply related to distance.

It is never possible to find a spatial gradient with no environmental change, nor to represent one as it exactly reflects the dispersal abilities of all species, even with sophisticated analyses such as principal coordinates of neighbour matrices. Likewise, it is never possible to fully characterise the environmental space (Section 6.3). Any environmental effects not included may be spatially autocorrelated, and therefore appear as spatial effects and be interpreted as dispersal limitation. Thus, dispersal and beta-niche effects cannot be reliably distinguished. Nor are the underlying environmental and spatial differences independent, since low dispersal can prevent ecological sorting, moderate dispersal can facilitate it, and high dispersal can reduce it by the spatial mass effect (Smith and Lundholm 2010).

6.4.2.4 Nugget in Spatial Autocorrelation

In a spatial autocorrelation dissimogram (i.e. dissimilarity vs distance plot) the intercept, i.e. the fitted dissimilarity/distance line extrapolated back to zero distance, has been interpreted as measurement errors and 'local random effects' (Fortin and Dale 2005). Brownstein et al. (2012) used this to estimate the degree of chance in 16 quite different plant communities in southern New Zealand. Randomness in species composition was higher where the species pool was larger, confirming theoretical speculation by previous workers.

6.4.3 Conclusion

If chance theory were correct, there would be no reason to expect community reassembly and hence no predictability. However, the reverse argument cannot be made: a failure to predict species composition well from the measured environmental factors is no evidence for chance (Section 3.10). What Hubbell has done is to remind us that any statement in community ecology must be made against the background of an appropriate null model.

6.5 Grime's C-S-R Theory

6.5.1 The Triangle

The C-S-R Triangle of Grime is a summary of habitats, of communities and of the species within them. The original description was of a triangle of habitats/communities, 'the spectrum of herbaceous vegetation types may be accommodated in...an equilateral triangle...[of] competition (**C**), stress (**S**) and disturbance (**D**)...' (Grime 1974). Traits of species were introduced as a means of placing communities within this triangle. Later, emphasis shifted to the triangle as primarily a summary of species' traits: a 'triangular model of primary plant strategies' (Grime 2001, p. xii), a 'functional classification of terrestrial plants' (Grime 2001, p. 3), with habitats/communities relegated to an explanation of how those plant strategies evolved (Grime 2001, p. 8). The triangle comprises (Figure 6.3):

- high-productivity, low-disturbance habitats/strongly competitive species (**C**),
- low-productivity habitats/stress-tolerant species (**S**) and
- high-disturbance habitats (**D**)/ruderal species (**R**).

This gives a **C-S-R** triangle of species 'strategies' and an equivalent **C-S-D** triangle of the sites they are in (Grime 1988).

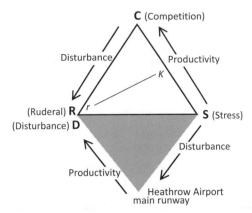

Figure 6.3 The C-S-R triangle of Grime (1979). In the 'Untenable triangle' the combination of high stress and disturbance makes it impossible for vegetation to exist.

In his original proposal of C-S-R theory, Grime (1974) characterised the **C-R** side of the triangle by RGR_{max} and the **S-R** side by plant size and litter accumulation. Elaborating on the theory (Grime 1977) correlated nine plant traits involving morphology, litter, growth, phenology and reproductive effort, and Hodgson et al. (1999) developed a method for placing a species within the triangle by weighting seven characters: plant height, width, flowering and leaf characters. Even a few simply obtained characters such as canopy height, flowering period and SLW can give good prediction of **C-S-R** category for most species (Bogaard et al. 1998; Hodgson et al. 1999), but a wider range of characters more related to the physiology of growth and stress resistance is desirable (Caccianiga et al. 2006). The C-S-R concept was supported by the analytical models of Bolker and Pacala (1999), showing that three, and only three, spatial strategies are possible. Their 'Exploitation' strategy can be matched with **C**, their 'Tolerance' strategy with **S** and their 'Colonisation' strategy with '**R**'. We note that Grime has also considered the ecosystem-level processes and the role of within-community, within-species genetic variation, but this is outside our scope. Finally, the disturbance axis (**C–R**) recalls the *r-K* selection of MacArthur and Wilson (1967) but puts it in a community context. However, C-S-R theory adds the distinction between **C** (*K* in productive sites, with competitor species) and **S** (*K* in unproductive sites, with stress-tolerating species) (Grime 2001; Figure 6.3). The model assumes that plants cannot grow where disturbance and stress are both high (the grey area in Figure 6.3), such as the middle of Heathrow Airport's main runway where the soil is too dry and low in nutrients (i.e. non-existent) and is disturbed every two minutes. The omission of the 'untenable triangle' omits half of the potential production/disturbance space and leaves the C-S-R triangle.

6.5.2   Strategy

The concept of strategy is old, and intuitive to every child. Put succinctly, the effort a plant or animal puts into one organ or activity is at the expense of another. Plant ecologists generally think of biomass, though calorific content might be more appropriate. Cody (1966) stated the concept eloquently: 'It is possible to think of organisms as having a limited amount of time and energy available for expenditure, and of natural selection as that force which operates in the allocation of this time and energy in the way that maximises the contribution of the genotype to following generations'. Harper and Ogden (1970) applied this concept to plants by examining the proportion of biomass and energy allocated to reproductive structures. Much consideration has been given to the selective advantages of particular reproductive strategies, formalised in terms of optimal strategy and later more correctly as evolutionarily stable strategy (Maynard Smith and Price 1973). The concept applies to a whole species, to ecotypes and to the plastic responses of individual ramets. It applies to vegetative allocation too, for example, shoot versus root allocation: 'the plant makes every endeavour to supply itself with adequate nutriment, and as if, when the food supply is low, it strives to make as much root growth as possible' (Brenchley 1916). This principle is implicit in C-S-R theory; it explains why no species can be a perfect competitor, a perfect stress tolerator and also a perfect ruderal. Indeed, Grime commonly refers to C-S-R as 'Strategy Theory'.

6.5.3   Stress

Stress is clearly defined in C-S-R theory as 'The external constraints which limit the rate of dry matter production of all or part of the vegetation', though this is complicated by the difficulty of measuring or even defining productivity (Section 1.2.2.2). 'Limit' also gives some problems. In relative usage it is clear that some habitats impose more limitation than others. But limitation compared to what? In some representations of the Triangle, e.g. fig. 40 of Grime (2001, p. 117), the relative importance of stress is on a scale 0–100. What is zero stress? The highest productivity found anywhere on earth? Or the highest productivity achieved by the same vegetation unit within its distributional limits? Or the highest productivity experienced over time at the same site, given annual climatic fluctuations and successional recovery from disturbance? The limit could also theoretically be defined relative to more favourable conditions than currently exist if species could have evolved to take the maximum productivity higher. Even if growth limitation is a relative scale, it still begs the question, relative to what?

There remains an additional problem that stress is defined in terms of vegetation production, i.e. by species. But which species? Take an alpine herbfield, where temperatures are low (Körner 2003b). Humans would consider this a stress[7] and so would most plants. Yet under climatic warming, heat-loving plants would be able to establish, and probably by interference exclude the alpines. How can we say the alpines were stressed before, when they were growing to their hearts' content, but that they are not under stress now that they are dead? And there may be more to their death than interference: some alpine species grow poorly in 'low-stress' sea level conditions,

---

[7] Except perhaps skiers.

probably because they lose carbohydrate in the warmer winter temperatures there (Stewart and Bannister 1973). We would think that stress would ideally be measured by the phytometer approach of Clements and Goldsmith (1924): planting the same species into a range of communities and measuring its growth. However, Grime has chosen to define stress on a whole-community basis and on the basis of the plants presently occurring, and is clear and consistent in that.

Perhaps the most difficult habitat for C-S-R theory is forests. The dominant trees of tropical rainforests might be seen from their characters as the ultimate competitors (Grime 2001, his table 6), but Hubbell (2005) described them as the 'competitive (stress tolerator) functional group', with characteristics typical of **S** species: tolerance of low light levels (as juveniles), long life spans, high resistance to pests and herbivores. This rather depends on how the dominants regenerate. If they grow fast from seed or from suppressed seedlings after disturbance they could be **C** species, almost **R**. Others have seen the dominants as species that are shade tolerant, either growing slowly up through the canopy, or sitting 'conservatively' and making bursts of growth during temporary gaps (Kubota et al. 1994), in either case making them **S** species as juveniles but not as adults. The understorey plants of evergreen forests tend to be slow-growing evergreens, **S** species, tolerant of low light (Grime, pers. comm.). Characterising a whole forest site as low/high light stress is very difficult, since in a productive environment there will always be some species low in the canopy that have to tolerate the stress of shade from taller plants, yet the canopy species are little shaded (Pigott 1980).

But is light the limiting factor to herbs and seedlings on the forest floor? Trenching experiments have shown that competition for nutrients is often more limiting to them (Coomes and Grubb 2000). Grime had envisaged that any community would comprise a mixture of species with different C-S-R status because of the overlap in species' ecological ranges, or of microhabitat variation, but in a forest the **C** (canopy) and **S** (understorey) species 'grow together in vegetation...*because* they possess different strategies'[8] (Pigott 1980). Moreover, shade is a stress, but not one 'external' to the plant community. So what does 'external' mean? In fact, Grime's definition of competition, with 'the same quantum of light' seems to make shade (by other plants) into competition.

Since the *r-K* spectrum is widely accepted, the crux of C-S-R theory is that different types of stress favour similar species traits (Grime 1988), resulting in consistent features in **S** species (Grime 1977, table 2; 2001, table 6). Such species grow slowly, at least in their natural habitat. Leaves can therefore be produced only infrequently, so they must function for more than a year. This results in a whole suite of leaf characters, e.g. small, evergreen, often stiff and tough, needle-like, high SLW, low percentage nitrogen, abundant defence compounds and low maximum photosynthetic rate, one end of the 'leaf economics spectrum' (Wright et al. 2004). Reich et al. (2003) found a compelling negative correlation between leaf lifespan and net photosynthetic capacity, albeit with scatter, and a slightly weaker one via leaf N, a pattern now expanded into a 'leaf

---

[8] Italics ours.

economics spectrum' with additional traits borrowed from the 'fast-slow' 'plant economics spectrum' (Reich 2014). There are obvious similarities here to Grime's **C-S** spectrum, and the leaf costs/amortisation theory of Orians and Solbrig (1977), thus: C-S-R = r/K theory + Leaf Amortization theory (Wilson and Lee 2000). Some physiologists have proposed a unifying stress mechanism, perhaps with a stress hormone (Lichtenthaler 1996), and rather in parallel Grime (1988; 2001, p. 71) has concluded since the original formulation of C-S-R that the common underlying stress in all **S** habitats is a deficit of major mineral nutrients, either directly or as a result of other stresses. This is unproven (Craine 2005).

It has sometimes been suggested that low RGR_{max} is directly adaptive in stress environments (e.g. Hunt and Hope-Simpson 1990). However, adaptation to stress environments is by relatively high RGR in those environments, not by low RGR in a hypothetical optimal environment. Low RGR_{max} is adaptive to stress environments only via a strategic tradeoff: 'It is possible that genetic characteristics conducive to rapid growth in productive conditions become disadvantageous when the same plants are subjected to environmental extremes' (Grime and Hunt 1975).

There are considerable limitations to similarity. Thus, saltmarsh, alpine, waterlogging-site and arid species are very different. Even within one stress site not all species are adapted in the same way. A dramatic example of this is seen in the wide range of life forms found in deserts with unpredictable rainfall, such as those in North America. Stem and leaf succulents (e.g. in Cactaceae, Crassulaceae and Euphorbiaceae) avoid drought by water storage, but annuals/ephemerals avoid water stress as adults by dying and surviving as dormant seeds. Other species are tolerators, notably shrubs with very low water potentials in their tissues in dry periods, shedding leaves and even branches, but remaining alive.

6.5.4   Disturbance

Grime's (2001) definition of disturbance is unambiguous: 'The mechanisms which limit the plant biomass by causing its partial or total destruction'. This refers to the whole community, but brings the problem that what is a disturbance for one species might not be for another (paralleling one of the criticisms relating to stress). For example, the mowing disturbance of Burke and Grime (1996) will have disturbed only the tall species. Selective grazing is another example; short or unpalatable species might be described as 'disturbance avoiders' in contrast to 'disturbance tolerators', but it is not clear how this distinction fits into C-S-R theory. How does C-S-R theory incorporate autogenic disturbance (Section 2.5)? One answer to this is that C-S-R theory is largely about the characters of species and that species of different C-S-R status can co-occur. Where does the removed plant material go, in terms of loss of CO_2, nutrient recycling, etc.? However, productivity, stress and disturbance are all defined per site, and there is also a C-S-R triangle of sites, for example Grime (2001, his figure 40) shows 'habitats experiencing intermediate intensities of competition, stress, and disturbance'. The destruction of biomass includes dead biomass (Grime 2001, p. 80), which is logical since it is an important community effector (Section 2.4). Fire is a disturbance that usually removes dead as well as live biomass. Decomposition removes biomass and is not thought of as a disturbance. Perhaps 'partial or total destruction' means more or less instantaneous removal. Part of the concept of disturbance is that resources are released (Davis et al. 2000). This certainly applies to light, and perhaps to soil nutrients, but severe burning can lower nutrients, notably through nitrogen volatilisation (Certini 2005).

6.5.5 Competition

The C-S-R concept of competitiveness as an overall plant attribute, i.e. that a species that is a superior competitor for one resource is also a superior competitor for all other resources, is controversial (e.g. Grubb 1985), but is eminently testable. Contrasting shoot competitive ability (for light) with root competitive ability (for water and the major nutrients) for the same species in the same conditions, the data assembled by Wilson (1988c) indicate 13 (22 per cent) cases where the relative competitive abilities of two species were in a different direction between shoot and root competition, and 46 (78 per cent) where they were in the same direction, a significant difference. This supports the idea of a general competitive ability,[9] but not that it is invariably true. Perhaps other forms of interference were operating in the 13 cases. Non-transitivity of interference would make a nonsense of the idea of overall competitive ability, but it seems to be rare or non-existent anyway (Section 3.5). Another prediction of C-S-R theory is that competitive intensity will be lower in stress sites (Grime 2001, pp. 35–37). Grime (*op. cit.*) writes: 'Some ecologists are extremely reluctant to recognise the declining importance of competition for resources in unproductive habitats'. We count ourselves among them (Section 6.7).

6.5.6 Species/Character Tests

The basic assumption of C-S-R theory, as of the leaf economics spectrum concept, is the concept of strategy; that there are 'design constraints' that limit viable character combinations (Grime 1988; Grime et al. 2007). Grime et al. (1987) made a general test by using a range of characters to classify species with cluster analysis and then looking for correlation between the resulting groups and the three C-S-R 'strategies'. They found, in one analysis, a group of low-stature, evergreen species with 'tough' foliage comparable to the **S** group. Grime et al. (1997) used 67 characters, including experimental responses, to ordinate 43 species. They could informally overlay a C-S-R triangle on the ordination diagram, though the fit of a scatter to a triangle could be questioned: a triangle can be placed across any swarm of points, and in fact a trapezoid fits better. There was also a good fit between this ordination and that derived in Grime et al. (2007) from field distributions: e.g. the three species in the **C** corner of the character ordination are in that corner in the distribution ordination, with comparable fits for the **S** and **R** corners, giving some support to **C-S-R** theory. A more direct test of

---

[9] Assuming that competition was the dominant type of interference.

these tradeoffs would be to demonstrate that there were no species outside the triangle, but no **C-S-R** triangle has been constructed that would allow this test, and we cannot see how one could be. There are species just outside the triangles in Grime (1974) and Grime et al. (1997), but that depends on just where the origin of the axes are drawn.

Other tests can be made by determining whether species of the right type occur in the right habitats. For example, Madon and Médail (1997) examined the distribution of species in a Mediterranean grassland. Sites with a high cover of **S** species (but how they were designated as **S** species is unclear) also contained a higher cover of annuals, but annuals do occur in **S** sites as well as **R** ones and indeed are frequent in most such semiarid grasslands and in some deserts. This emphasises that C-S-R theory is a generalisation, not a law of the type that physicists can have. Caccianiga et al. (2006) attempted to test C-S-R theory on a chronosequence of glacial moraines in Italy. The concept is valid and it was a brave attempt. They inferred a succession[10] from communities dominated by **R** species to those dominated by **S** ones. Such a change is predicted by C-S-R theory, but there are problems. The prediction of C-S-R theory is for an intermediate trend towards the **C** corner, and surely the **S** corner would not be reached within the <200 years of their dataset, though both of these may have been because the environment at 2,400–3,000 m a.s.l. is a high-stress one. Most problematic is that under C-S-R theory the **R**→**S** change that Caccianiga et al. found is more characteristic of secondary succession; in a stress habitat and involving a primary succession, which theirs certainly was, the trend should be **S**→**CSR**→**S** (Grime 2001, p. 242).

An experimental approach is perhaps better, since one can be sure what the habitat differences are. Moog et al. (2005) at 14 sites in southwest Germany applied four basic treatments: sheep grazing, mulching with hay, burning in winter and control ('succession'). The vegetation resulting 25 years later was classified in terms of **C-S-R** composition, using visually estimated cover and calculating species' **C-S-R** rankings by the method of Hodgson et al. (1999). There were some changes in community C-S-R status that agreed with the theory. For example, grazing and twice-yearly mulching, both presumably disturbances, each reduced **C**-ness by c. 0.35 (compared to the control) and increased **R**-ness by 0.4–0.5, both as predicted by C-S-R theory. Grazing increased **S**-ness by c. 0.25, explained by the authors as via the herbivory defence of **S**-strategists, or by nutrient removal (though it is not clear whether grazing will reduce nutrient availability through removal of animal products or increase it through nutrient recycling). Burning increased **S**-ness by c. 0.2, explained as an indirect effect, that burning favoured species with rhizomes that happened to be **S**-strategists. However, large differences in C-S-R status were found between the same treatment at different sites, up to 1.0 difference.

6.5.7   Do Succession Pathways Provide a Test of C-S-R?

The clearest successional tests of C-S-R seem to be with secondary succession. Grime's (1979) interpretation was that for sites of differing productivity there would be separate

---

[10] By space-for-time substitution.

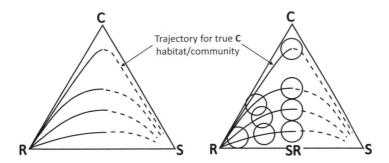

Figure 6.4 (a) C-S-R theory and specialist pioneers. Grime suggested that sites with differing degrees of stress would follow different pathways. (- - -) indicates the part of the succession that will probably be slow. (b) The closeness near the R corner of the lines heading towards C and SR indicates that there is no room nor need for a range of strategies, making it possible for the same species to occur along pathways in habitats of different productivity/stress. We have added a trajectory that goes via a true C habitat/community, such as a *Phragmites australis* reedswamp (Grime et al. 2007).

secondary successional pathways, all starting from the **R** corner, and all ending (eventually) in the **S** corner (Figure 4.3 and Figure 6.4a). At the start in the **R** corner, the **C** and **S** succession trajectories are very close, i.e. with very similar environments, leading to the same C-S-R *traits* and the opportunity for the same ruderal *species* to occur (Figure 6.4b; Wilson and Lee 2000).

Considering the first of the three **S** habitat types discussed by Wilson and Lee (2000), i.e. arid lands, the main pioneer of semiarid degraded land in Naiman Banner County, Inner Mongolia, is *Agriophyllum squarrosum*, a specialist pioneer of dunes in semiarid areas (Zhang et al. 2005). In Sonoran Desert oldfields, pioneers include the very widespread weed *Taraxacum officinale* (dandelion), but also species such as *Salsola kali* (tumbleweed), a ruderal annual of dry, often alkaline areas (Castellanos et al. 2005). These examples add to those documented by Wilson and Lee (2000), indicating that in arid habitats the majority of secondary pioneers do conform to the prediction of Figure 6.4 in not being restricted to deserts. The secondary pioneers of saltmarsh gaps, the second habitat type considered, are generally species of the lower saltmarsh, as would be expected, since all species that occur on salt marshes, ruderal or not, have to be quite salt tolerant: therefore there is no agreement with the Figure 6.4 prediction (though it is not certain that Grime would categorise saltmarsh as an **S** habitat). Moving to the third type, alpine stress, the species present in mid-succession in southern New Zealand alpine grassland include the dicotyledons *Anisotome aromatica*, *Plantago novae-zelandiae*, *Colobanthus strictus* and *Epilobium alsinoides* (Lloyd et al. 2003); the last extends down to the lowlands, but the others are basically montane/subalpine in range. Wilson and Lee (2000) gave an example from alpine Scotland where the first colonists were not specialist pioneers, but species of several **S** habitats. However, pioneers in the Andean alpine oldfields include the very widespread ruderal *Erodium cicutarium* (stork's bill; **R/CSR** in Grime et al. 2007), *Poa annua* (**R**) and *Rumex acetosella* (sheep's sorrel; **SR/CSR**) (Sarmiento et al. 2003). There seems to be no consistency for alpine pioneers.

Grime (2001) shows primary succession in **S** environments as occurring within a corner of the triangle near **S**. Wilson and Lee (2000) argued that only specialised species would be able to tolerate the environment of an **S** site. This gives the likelihood of autosuccession, with no specialised secondary pioneer species. Again, C-S-R theory strictly predicts the same type of species, but if there is space for fewer different niches there will tend to be fewer different species. Wilson and Lee (2000) tested this prediction in relation to four types of stress. For desert, Allen's (1991) suggestion that autosuccession is common is not supported by older literature nor by Castellanos et al. (2005) in the Sonoran Desert, though the evidence of Zhang et al. (2005) from China is mixed. Autosuccession is often seen on saltmarsh, especially on the more **S** lower saltmarsh. In alpine environments Wilson and Lee cited two examples, one where autosuccession was occurring and one where it was not. Sarmiento et al. (2003) found that in high-Andean oldfield succession, of the eight most abundant species in the undisturbed community, four were absent the first year after abandonment, three others were present only in traces and the remaining one made up less than 1 per cent of the cover – no autosuccession here. In arctic tundra, another habitat cited by Grime (1979) as high-stress, there are usually pioneers, but autosuccession is occasionally seen (Wilson and Lee 2000). Overall, there is some tendency, no more, for autosuccession to occur in the most extreme **S** habitats. It occasionally occurs in mesic habitats too (Wilson and Lee 2000). Support for C-S-R via predictions for succession is mixed.

6.5.8   Species Richness

Grime (1973) proposed that there would be a humped-back relation between productivity and species richness, i.e. with maximal richness at intermediate productivity. He ties this in with C-S-R theory: only a few **S** and **R** species are able to tolerate the stress or disturbance of low-productivity sites, and only a few **C** species are able to withstand competitive exclusion at high productivity, but there are many intermediate species, mixing with **C**, **S** and **R** species along the productivity gradient (Grime 2001, p. 263). The evidence for the hump is mixed, especially because many investigators have tested only for quadratic curvature, not for a downturn at both low and especially high productivity (a notable exception being Mittelbach et al. 2001). Although the humped-back and C-S-R theories specify only 'competition', high productivity will in practice be achieved only with adequate mineral nutrients and water, so that competition will be predominantly for light, its cumulative effects suppressing some species and reducing richness (Section 2.2; Hautier et al. 2009).

6.5.9   Conclusions

There have been many more criticisms of C-S-R theory, but most of them have missed its point (Wilson and Lee 2000). It is easy to categorise communities and perhaps species in C-S–R terms in a general way, though several of the detailed predictions of C-S-R theory are very difficult to test, reducing the value of the theory as an explanatory model for the structure of plant communities. Even for predictions that are more easily tested, there has been little quality evidence. However, the evidence so far is that predictions from C-S-R succeed more often than they fail. It is a useful generalisation.

6.6   Tilman's Theories

Tilman (e.g. Titman 1976; Tilman 1982, 1988) has produced a number of ideas. Here we emphasise those with a community focus. The concepts have been described as having 'a hard centre but woolly edges': that is, there is a solid core of irrefutable mathematics, but it is not always clear how to apply it to the real world of higher plants. The issue of competition is considered in comparison with Grime's theory (Section 6.7.1).

Tilman [Titman] (1976) concluded from his first experiments: 'long-term coexistence of competing species was observed only when the growth rate of each species was limited by a different nutrient'. This is standard Gaussian competitive exclusion, but his R* theory suggests how it happens. He later incorporated spatial heterogeneity (Tilman 1988), and then embraced the interference/dispersal tradeoff model (Tilman 1994; this volume Section 3.7). Craine (2005) has documented the developments and retreats of Tilman's theories.

6.6.1   Strategy

Like Grime, Tilman has also moved to an emphasis on strategy with his ALLOCATE model of plant growth and competition (Tilman 1988), emphasising shoot versus root strategy. The latter has a long history of theory and observation (Wilson 1988a), but Tilman has usefully put it into the context of the community.

6.6.2   The Competitive Process: R*

Tilman's (1982) R^* theory is that a particular species in a particular set of environmental conditions has, for a particular resource R, a value R*, which is the lowest [R] (i.e. concentration of R) at which it can grow in monoculture. Above its R* the species can grow, absorb R, and will therefore lower [R] towards R*. In mixtures of species growing where R is limiting, as [R] becomes lower each species will drop out when [R] drops below its R* and it can no longer grow. Apparently no species will enter during this process. Eventually, one species will be left: that which can tolerate the lowest [R], and soon [R] will be at its R*. To summarise, the species with the lowest R* will be the superior competitor. The model is simple, but application to the real-world terrestrial communities is not. Moreover, its predictions assume that only competition is operating, not other forms of species–species interference.

6.6.2.1   Major Soil Nutrients (NPK)

Tilman (1981) found that the R* model explained which species of alga dominated in a microcosm experiment with inorganic nutrient limitation, in constant temperature, a constant and uniform light regime and with the solution well mixed by flow and shaking. Miller et al. (2005) surveyed the literature and found 11 similar plankton microcosm experiments, and when analysed by Wilson et al. (2007b), all tended to support the theory (*contra* Miller et al.'s own conclusion). Moreover, of another 32 such studies in the literature, 30 supported it, with the other two equivocal.

The situation is not so clear-cut in soil where the environment is variable in time and space. Tilman and Wedin (1991a, 1991b) found that in field plots at Cedar Creek the outcome of interference in soil with low N was predicted by R* in some cases. Comparing the grasses *Agrostis scabra* (bent) with *Schizachyrium scoparium* (little bluestem) the two performed approximately equally in monoculture, but in mixture *S. scoparium* was the clear winner. By R* theory, it should have reduced [N] in the soil (both nitrate and ammonium) to a lower level than *A. scabra* and indeed available [N] in monoculture as measured by KCl extraction was lower. It should also have been able to grow at a lower [N], but the experimental results do not tell us one way or the other: *A. scabra* suffered in mixture at the low [N] that *S. scoparium* produced, but not necessarily because of the low [N] since it suffered almost as much in mixture in the two higher N levels. Very similar effects were seen in mixture between the *A. scabra* and *Andropogon gerardii* (big bluestem) and in a less clear-cut way in mixtures between *A. scabra* and *Agropyron repens* (quackgrass). The results from these two species pairs are ambiguous: perhaps *A. scabra* is more efficient at N uptake, but suffers in light competition, and indeed it grew shorter than other species, including *S. scoparium*, in the experiments of Tilman (1986). Fargione and Tilman (2006) found clear support for R* in overall patterns in a Cedar Creek experiment. Among 13 species, those species with a low R* for soil N were less depressed[11] in mixture, as judged by their above-ground yield, with only one exception.

The best test of R* theory is possibly that by Dybzinski and Tilman (2007), so we analyse their results in more detail. They grew six grass species in monocultures, in eight two-species combinations and in a four-species combination, on a range of sand/ soil mixtures giving various N levels in soil. Nutrients P, K, Ca, Mg, S, minor-nutrients and summer-time water were added, to ensure N was the major soil factor limiting growth. Strips of biomass were harvested at intervals until the tenth year. Light at the soil surface was measured just before the final harvest (R* for light = I*), and soil N (KCl extract) for three further years. They report N* (i.e. R* for nitrogen) at a range of soil N values (which is odd, surely the concept is that R* is the value to which a species will reduce N until it can no longer grow, whatever the starting [N]?). We use N* and I* values at the lowest N (N* at low N, I* at low N) and those at the highest N (N* at high N, I* at high N) from their figures 1b and 1d plus biomass and vegetative height in monoculture from their figures 1a and 1c, as predictors of the outcome of interference in two-species plots (their figures 2–4; d–f). In no case was the outcome affected by soil N treatment. According to R* theory the best predictor of the interference outcome should be N* at low N, but in fact this is the worst predictor (Table 6.3,

---

[11] Compared to monoculture.

Table 6.3 Predictors of interaction outcome in the experiment of Dybzinski and Tilman (2007)

Comparison	Predictor	Agreement with R* theory	Disagreement with R* theory	Four-species mixture
1	N* at low N	5 / 8	3 / 8	*Panicum* & *Bouteloua* major discrepancy
2	I* at low N	6 / 8	2 / 8	Two minor discrepancies
3	N* at high N	7 / 8	1 / 8	One minor discrepancy
4	I* at high N	7 / 8	1 / 8	*Schizachyrium* and *Panicum* minor discrepancy
5	Biomass in monoculture	7 / 8	1 / 8	*Schizachyrium* and *Panicum* minor discrepancy
6	Height in monoculture	6 / 8	2 / 8	*Schizachyrium* and *Panicum* minor discrepancy

Comparison 1); N* at high soil N (Comparison 3) is much better. Nitrogen would be expected to give more shoot growth, resulting in lower light at the soil surface, which it did for all species, so that I* [light] at high N might be an effective predictor, and it is (Comparison 4), though biomass in monoculture was equally effective (Comparison 5). In none of the six comparisons was the prediction of outcome in the four-species mixture perfect, but it was generally good. This is hardly a triumph for R* theory: biomass in monoculture predicted the interference outcome as well as any of the R* comparisons.

It is difficult to apply R* to real-world soil nutrients. Plants will lower [N]/[P]/[K] to some extent. However, these elements are always being added in mineralisation from organic matter via decomposers, in normal and occult precipitation (i.e. from fog), in the settling of atmospheric particulates and in leaf leachates. In the case of N, the mineralisation is to ammonium, taken up thus by some plants but for others first nitrified to nitrate and taken up in that form. Nitrogen can also be fixed from the air by free-living soil microorganisms and in larger quantities by symbioses with legumes (Fabaceae) and several other species, perhaps thus added to the soil or perhaps not. P and K can be made available by hydrolysis of minerals, especially of feldspars and apatite, a process that can be speeded by exudates from plant roots. Against all this, nutrients can become unavailable, immobilised from solution via exchangeable form to unavailable forms (organic or mineral), or taken up by bacteria and other microorganisms. They can be lost to the system through leaching, runoff and soil erosion, and in the case of N by denitrification/volatilisation. Fire can increase loss by runoff, erosion and volatilisation, and animals can cause local loss of nutrients through redistribution. Vitousek (2004) describes many components of nutrient cycles in Hawai'i. These addition/loss processes are all dependent on water; they are affected by various environmental factors (soil pH) as well as the microflora, animals and the plants themselves. As a result, all are patchily available in two-dimensional space, and since most nutrient processes start at the soil surface, there is usually also considerable variation with depth. Availability also varies markedly in time, stochastically and with season. Nitrogen will normally be at its most abundant in spring, when maximum growth occurs (Tilman and Wedin [1991a] measured soil N in summer). Plant roots will hardly lower total [P], since most is insoluble. Unlike N, with its fast-diffusing NH_3 and NO_3 ions, most available P is immobile so plants cannot rely on diffusion to acquire the element from the soil as a whole, but have to forage for it by growth. The cylinder around the root from which P is scavenged can be as narrow as 1 mm in radius (Kraus et al. 1987). Such localised depletion makes an overall [P] value pointless (Craine 2005). This basic difference between competition for mobile N and that for immobile P was pointed out by Bray (1954). It is perhaps unsurprising that there are difficulties in the application of R* theory to soil nutrients, given all these complications.

6.6.2.2   Water

Water is intermittently available across depths and times: a complication for R* theory. Most water lands on the surface and perhaps moves down, but water can also be available from deep aquifers and by hydraulic lift. It can be lost by evaporation. Thus, plants are not all using the same resource base of water in space (deep-rooted shrubs and perennial grasses versus shallow-rooted cacti) or time (ephemeral annuals versus perennials). Thus, there are also difficulties when applying R* to water.

6.6.2.3   Light

In competition for light, canopy species reduce the resource availability below them, but not above them. R* theory knows nothing about directions. By R* theory, the climax canopy species would reduce lower-stratum light to low levels, and be able to tolerate these low levels and hence regenerate. But this depends on whether there is continuous regeneration, large-gap regeneration or single-tree replacement, and in the latter two cases whether the next generation is from dispersal, the seed bank, from suppressed seedlings or from advanced regeneration. All these modes of regeneration occur, often within the same community, different species regenerating by different modes (e.g. Lusk and Ogden 1992; Thomas and Bazzaz 1999), but there seems to have been no review of their relative importance.

Assuming continuous below-the-canopy regeneration, R* predicts that shade-tolerant plants will dominate by having a lower light compensation point. Kitajima (1994) compared 13 tree species of Barro Colorado Island rainforest. Shade tolerance was determined as the survival rate of seedlings under a shade cloth that gave light intensity very similar to that of the forest understorey, with supporting evidence from field observations of mortality in the understorey and from the requirement or preference of the species for light gaps. Survival in shade was not well correlated with the light compensation point ($r = +0.27$) nor with dark respiration ($r = +0.25$ on a mass basis). Eschenbach et al. (1998) examined tree species of North Borneo lowland dipterocarp forests in the field. Light compensation points were attained mainly between 6 and 9 μmol photons $m^{-2} s^{-1}$ but were higher for pioneering species. This supports an R* interpretation in continuous regeneration, but the presence of pioneer species reminds us that gap regeneration is occurring. The truth is that regeneration in forests, and probably in some other communities, is too complex to fit R* theory.

6.6.2.4   Conclusion on R*

None of these complications occur in environmentally constant, homogeneous, nutrient-limited habitats such as the laboratory tank with planktonic algae that Tilman had in mind when he formed R* theory, and it usually gives correct predictions for them (Wilson et al. 2007b). In real habitats, R* theory is not only very complex to test, but it is often impossible to see how to apply it or test it, though it has given some correct predictions at Cedar Creek. There are many, obvious simplifications in the model. It would have been very useful had R* been able to predict competitive ability, for previous attempts to find empirical correlations between interference abilities and plant characters have generally failed (e.g. Jokinen 1991). An exception is the obvious correlation of height when competition for light is important (e.g. Balyan et al. 1991), a correlation expected under C-S-R theory, but not under R* theory. There is a problem that under R* theory exclusion by interference would lead to only one species remaining, yet communities almost always comprise many species, not least in Tilman's own site at Cedar Creek where a 49-year oldfield (the second-oldest field) contained 12 species per 0.5 m^2 quadrat (Inouye et al. 1987). There was not even a downward trend at Cedar Creek: among the 22 fields the highest richness was in the 49-year field and the overall trend, although non-significant, was for richness to *increase* with age.

6.6.3   Succession

Tilman (1982) generated a resource-ratio theory of succession, starting from the observation that at his Cedar Creek experimental site soil nitrogen increased during secondary succession (this is often the case, though it is difficult to know what fraction of soil N is available to plants). From this he theorised that the early-successional species would have low N* (R* for N) and therefore be better competitors at low N; late-successional species would require high N, but it would be available then and they would be better competitors for other factors in those conditions, probably light. He performed experiments with coworkers and concluded that later successional species at Cedar Creek do not necessarily have a higher N requirement or response (Tilman 1986, 1987; Tilman and Cowan 1989). Comparing results from different Cedar Creek studies, the statement about N *requirement* is true: the modal nitrogen content of the soil in which various species grow there (Tilman and Wedin 1991a) is not significantly related[12] to their RGR at low N (Tilman and Cowan 1989) nor[13] to their growth at high N. The species' *response* to N (RGR at high N/RGR at low N, data as above) is clearly related[14] to modal soil N (Figure 6.5), but unfortunately for Tilman's theory not consistently to their successional position. The low-responding *Agrostis scabra* does appear early on and peak at c. five years (Tilman and Wedin 1991a), but the high-requiring and high-responding *Poa pratensis* (meadow grass) peaks at c. 15 years, whereas *Schizachyrium scoparium* is hardly present then, and peaks at c. 45 years.

---

[12] Spearman's rank correlation $r_s = -0.45$, with RGR taken from the graphs of Tilman and Cowan (1989) at 150 mg N/kg of soil.

[13] $r_s = -0.24$, RGR at 1,500 mg N.   [14] $r_s = +0.84$, $p < 0.05$.

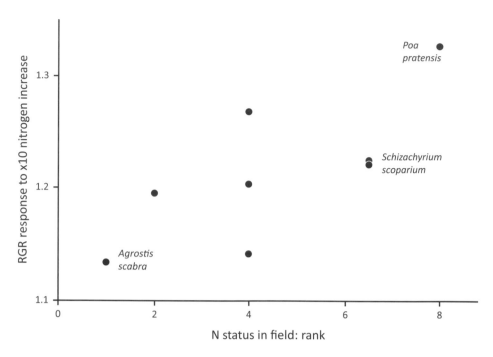

Figure 6.5 The experimental response to N compared to the rank of species in a successional/N field gradient.
See Tilman (1986, 1987), Tilman and Cowan (1989), Tilman and Wedin (1991a).

Harpole and Tilman (2006) and Fargione and Tilman (2006) produced similar partial support by correlating previously determined N* values with relative abundance in three seminatural or experimental areas at Cedar Creek. This assumes that interference ability and abundance in a mixture will be correlated and this is not necessarily so (Section 3.5). The correlations were highly significant but reflect only that the three abundant species have low N*, whilst that for other species covers the range from low to high.

It is also difficult to generalise Tilman's ideas on succession. He emphasised the increase in soil nutrient status, especially of nitrogen, during both primary and secondary succession (Tilman 1988). Soil nitrogen indeed normally increases through succession. However, there is evidence of a decrease in soil N over some hundreds to thousands of years (Richardson et al. 2004; Crews et al. 1995,[15] examining a 4,100,000-year 'chronosequence' in Hawai'i; Parfitt et al. 2005 examining trends in three New Zealand 'chronosequences' from 1,000 years to 14,000–291,000 years). Moreover, phosphate is often a major limitation to plant growth during succession and it decreases with time almost from the beginning (Chapin et al. 1994; Richardson et al. 2004).

---

[15] The timing is difficult to establish since the evidence is naturally from sites of different successional age that differ also in factors such as initial substratum and altitude.

6.6.4   Species Richness

Tilman (1982) reached a conclusion on species richness similar to Grime's humped-back theory. He similarly explained low richness under low productivity partly by there being few species capable of tolerating high stress, though he also suggested that environmental heterogeneity would be low there, with all microsites equally stressful (others have suggested that temporal variability leads to species' extinctions, and that smaller populations there do not reach viable size). Tilman agreed with Grime in explaining a decrease in richness at high productivity by greater competitive exclusion. Later, finding that nitrogen application led to a reduction in species richness in the Cedar Creek oldfields, he converged with Grime's conclusion that this effect was due to shading suppression by live plant material and litter (Tilman 1993).

For Grime the issue is connected to his humped-back model of species richness: at low productivity/standing-crop R and C species will be absent. Wardle (2002) uses this argument to comment that Tilman's R* model is 'difficult to reconcile with the frequently observed humped-back relationship between diversity and productivity', because according to Tilman, competition, and hence competitive exclusion, will be no greater at high biomass. Wardle's statement is misleading for several reasons: (a) valid demonstration of the humped-back relation is not usual (Section 6.5.8); (b) the usual relation has been with richness, not diversity; (c) productivity is difficult to measure (Section 1.2.2.2), and has hardly ever been measured in such studies – only above-ground standing crop; and (d) the logic is based on the downturn in richness at high standing crop being due to competitive exclusion, which even Wardle admits is only 'a likely reason'.

With respect to the question of the 'Paradox of the Plankton' (Chapter 3), it would at first appear that R* theory could not generate enough niches to support the richnesses generally observed in real communities. However central to the theory is the idea that two species competing for two resources can coexist within particular ranges of the ratios of those two resources; thus any spatial heterogeneity in those ratios could provide a mechanism for the coexistence of many more than two species. Whilst theoretically possible (Tilman 1994), this hypothesis is extremely difficult to test empirically, especially in terrestrial communities with all of their additional complexities (Section 6.6.2).

6.6.5   Conclusions

Tilman followed the example of MacArthur in producing formal models of how communities might work. He and others have been able to test his models (Miller et al. 2005). R* theory has proved highly robust for microorganisms in experimental conditions, though it is hard to see how it will extend to embryophytes. Tilman has published many results with frank admission of their conflict with R* theory, which has earned him considerable respect, though this leaves the theory hanging (Craine 2005). When Tilman attempted to put successional processes on a more formal basis, reality proved to be more complicated. We discuss below his ideas on the intensity of competition along gradients of productivity.

Tilman tried to find neat patterns in ecology. It was brave of him, but these theories are not always effective at encapsulating the complexities of the real world.

6.7 Where Will Competition Be Strongest?

6.7.1 Grime's and Tilman's Hypotheses

Tilman's (1988) view is that competition will be equally important in unproductive (high stress) and productive (low-stress) environments. The logic seems to be that if resource availability is too high for there to be competition, the plants will grow until availability has been reduced so that there *is* competition for it, or until another resource becomes limiting. If so, we agree. Grime's (1974, 2001) contrary theory is that in the very **S** corner of the C-S-R Triangle there is no competition because neighbours are too limited by the physical environment to interact. *Near* the **S** corner, competition is low. This conflicts with almost universal observation. Almost everywhere plants fill the available area. Having grown, how does the biomass reach its environmental maximum ('constant final yield'), unless through competition or a similar density-dependent process? [The discussion has been in terms of competition, but similar arguments apply to other forms of interference.]

An explanation for the maintenance of populations without competition in high-stress environments, such as deserts, has been a low probability of plant establishment and/or high mortality (this assumes individuals, not biomass, but this simplification is always made in this argument, and we can ignore it for the moment). However, an establishment rate that happens to exactly balance mortality is infinitely unlikely. Even a slight deficit of establishment over mortality would give a long-term RGR (e.g. population growth rate) less than 0.0 and a population that declines to zero, and even a slight excess of establishment would give RGR greater than 0.0 and a population that climbs towards infinity. Neither can be seen in extant populations. The logical conclusion is that in all persistent populations of a species, establishment and/or mortality must be abundance-dependent, and the most likely explanation is interference. We conclude that in all environments plants will increase in abundance until they are limited by interference, probably competition. Eventually competition becomes *complete* even if ecologists, as outsiders, cannot easily see what the critical resource is (by *complete* competition we mean that the community has reached constant final yield, i.e. environmental carrying capacity). There are three riders to this. (1) Seral communities: competition will be absent in the very earliest stages of succession (Clements et al. 1929). (2) A spatial mass effect may be supporting a population that is not self-sustaining (Section 3.11), though it seems unlikely this is a major explanation. (3) Another process with negative abundance-dependence, e.g. specific pest pressure (Section 3.4), or other forms of interference, might hold abundance too low for competition to occur.

Grace (1991) made another important point: 'both Grime and Tilman discuss gradients in habitat productivity as if it makes no difference whether they are gradients in [resources] or non-resource factors'. The problem goes further: many studies have examined the effect of a mineral nutrient such as N in a system where competition is probably for light, or even necessarily for light by the experimental design. For example, Stern and Donald (1962) added N to a grass and a clover growing with their roots separated. In this example the gradient is in the resource N, but the competition for light. The true distinction is between a gradient in a factor for which there is competition and in a factor for which there is no competition. But the factor for which there is competition may change along the gradient. If soil nutrients or water are limiting initially, and they are added, the limitation will be removed and competition will shift to being for light (Tilman 1988). Moreover, the environmental conditions will then be so different that it will be difficult to say whether competition is less, the same or greater (see Section 6.7.2). Perhaps an even greater problem is that once competition is for light, it will probably be cumulative because of its asymmetry. Indeed, using a simple mathematical model DeMalach et al. (2016) have demonstrated the importance of asymmetric light competition, with model outcomes that support either Grime or Tilman, depending on the degree of competitive asymmetry specified. Other complications are the change in species composition that will occur along gradients and the lack of a generally agreed index of competition. However, most of the confusion that has grown about this topic seems to come from ignoring what resource competition is *for*.

6.7.2   The Growth-Rate Artefact

A huge complication in experiments testing between the Grime and Tilman ideas on competition is that if plants are put in a pot (or planted in the field a certain distance apart) in a higher-productivity environment, the plants will by definition grow faster, and thus come into competition sooner. Therefore, if competition is measured at a fixed time after planting, it will appear to be greater in more productive conditions. The same situation occurs after natural disturbance. Eventually, competition will be *complete* right along the productivity gradient because the plants will grow until carrying capacity is reached. Therefore competitive intensity cannot be measured as the final outcome either. Competition is like death: it is not a question of if or how much, but of when. This problem of the growth-rate artefact is removed when the experiment indicates *lower* competition at higher-productivity conditions, but it is difficult to accept results in only one direction. Other complications are: (a) resources will be exhausted sooner when they are in shorter supply and (b) resource availability can be influenced by the plants themselves.

6.7.3   Wilson, Agnew and Roxburgh Hypothesis

Firstly, we reject the concept that has been put forward of the 'importance of competition' relative to other constraints on growth, contrasting with the 'intensity of competition'. Freckleton et al. (2009) argue that 'importance' cannot be measured in less than one generation, though that actually applies to competition/interference ability too (Section 2.2). Interference can be measured as reduction in growth compared to an isolated plant, though there will be self-interference (e.g. shading) from the seedling stage onwards. But physical constraints on growth must mean compared to optimal values of all physical conditions. This cannot be achieved in a growth cabinet, in a greenhouse (against which the comparison has sometimes been made) or still less in the field, and if it could be achieved in a growth cabinet what would the relevance be? Freckleton et al. (2009) also ask: importance for what population/community process? However, in our view the whole question is flawed. Stresses such as cold temperatures or toxic soil conditions are basically constant, community-independent, whereas competition and other forms of interference must ultimately limit population growth.

We give our own view, from first principles.

Along a beta-niche gradient (i.e. a gradient of conditions, of non-resources) or of resources for which competition is not occurring, competition (for other resources) will be of constant intensity, i.e. complete, but it will appear to be greater in more productive conditions because of the growth-rate artefact. In fact, how can competition be measured? If plants are removed from a community, others will grow or establish to fill the gap. This will happen faster in some communities than others, but this is the growth-rate artefact again.

Along a gradient of a resource for which there is competition, competition for that resource will be strongest when the resource is in shortest supply. There can be exceptions, e.g. the mobility of some soil nutrients can be higher when they are present at higher concentrations, and this can result in greater below-ground competition (Vaidyanathan et al. 1968; Wilson and Newman 1987), and the same could apply to water. An additional complication is that as the availability of resource X increases along a gradient, the plants may change from competing for X to competing for resource Y, in fact by our argument above, this is almost certain to happen.

We therefore believe in the ubiquity of competition. But could an exception be the widely spaced plants in deserts?

6.7.3.1   Deserts

Many ecologists, from Shreve (1942) through Went (1955) to Mirkin (1994), have denied that desert plants compete, an exception to the universality of competition. The desert habitat is indeed stressful and the plants are typically spaced above-ground. This concept was fuelled by studies that failed to find a regular spatial pattern of individual plants in deserts, and sometimes found clumping instead (e.g. Gulmon et al. 1979). The idea was often that plant populations in deserts were kept below 100 per cent occupancy by unfavourable probabilities of colonisation and death. We discussed the flaws in this logic in Section 6.7.3, showing that there must be *complete* interference, so if it is manifestly not above-ground, it must be below-ground. Ecologists have often under-estimated the intensity of competition where there seem to be unvegetated gaps between plants; in fact the cacti and shrubs may be spaced-out, but their root systems are not (Woodell et al. 1969). The existence of intense competition for water has been demonstrated by finding negative correlations between plant sizes and distance apart (Yeaton and Cody 1976) and by relief of plant water stress and increase in plant growth upon removing neighbours (Fonteyn and Mahall 1978; Robberecht et al. 1983; Fowler 1986a; Kadmon and Shmida 1990). In fact, the effect of competition on plant spatial pattern has been best demonstrated in desert communities. Clements knew all this, of course: 'The open spacing of desert shrubs in particular suggests some indirect influence in explanation, but studies of the root systems demonstrate that this is a result of competition for water where the deficit is great' (Clements et al. 1929, p. 317). Remember the riders above. (1) Seral communities: clearly many desert plants are old (notably cacti and yuccas), so the whole system is clearly not in a continually seral state.[16] (2) Spatial mass effect: most deserts are too extensive for the vegetation to depend on diaspores from other habitats. Neither '1' nor '2' would give the regular distribution or pattern often seen in arid areas. (3) Other abundance-dependent processes: one possibility is a community-wide version of pest pressure (Section 3.4). Plant density might be held so low by herbivory that the plants never come close enough to compete or otherwise interfere. If the density starts to increase, the population of herbivores will increase to reduce it again, to the point where the herbivores will move to somewhere else, or the local herbivore population will die out. Herbivore pest pressure is an unlikely cause because the plants are close enough for vertebrate herbivores, and even invertebrate ones, to move readily between them. Diseases are not likely to be seriously inhibited by the distances, so they could not mediate such a process. Moreover, plants of unproductive sites usually have more defences (Grime 2001). We do not believe that this community-wide pest pressure can operate, but we have no logic nor evidence to refute it. Fire would be more likely or more severe at higher abundance, but fire typically disturbs patches, not all individuals within the population of a species, and in the deserts of many parts of the world fire is rare (and even where fire does occur it has the potential to promote diversity via spatially mediated mechanisms such as the 'Intermediate Disturbance Hypothesis' (Section 3.6)). This leaves some form of interference as the only possibility. Allelopathy may be important in some cases, but the very consistent finding of spacing in dry environments with quite different species present excludes it as the general explanation. We conclude that the intensity of competition is likely to be close to *complete* in all deserts, and moreover it will be for water (perhaps secondarily for NPK).

We should also consider the possibility of plant–environment feedback, specifically switches. A desert surface cannot be uniform since each plant reacts on its environment, notably trapping soil nutrients and water runoff. This can give a pattern, as in 'tiger bush' (Section 4.5.4.2), which could be regular, without even the plants' root systems coming into contact (between patches), though with the potential for intense competition for the spatially concentrated resources within-patches.

6.7.3.2   Tests

The complete absence of competition would be testable. However, we argue above that this is not tenable, and moreover no habitat can fall exactly in the **S** corner so the question does not arise. We have to test degrees of competition along an **S–C** gradient, which is possible but difficult. The literature is unclear on how to measure the intensity

---

[16] According to C-S-R theory stress sites cannot have too much disturbance, or no plants can grow (Grime 2001).

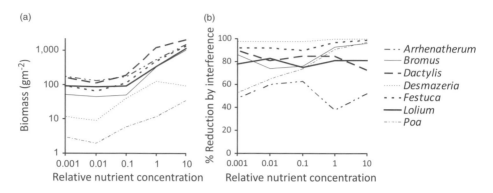

Figure 6.6 The effect of nutrient concentration on (a) biomass and (b) competitive ability, from the experimental results of Campbell and Grime (1992).

of competition; we shall use a species' percentage reduction from monoculture, ideally in RGR but sometimes in biomass.

Considering a gradient of environment conditions, La Peyre et al. (2001) grew three species of salt/freshwater marshes in monoculture and competition along a salinity gradient. A measure of the overall intensity of competition was almost constant along the gradient, and once allowance is made for dead material, the competitive response of the individual species varied remarkably little. Similarly, Cahill (1999) found no consistent change difference in above-ground competition in his oldfield experiment between the two NPK levels. Although N, P and K are resources, in the latter comparison there was competition only for light, so here N, P and K are conditions. These two pieces of work support our thesis that along a gradient of conditions, competition will be of constant intensity.

For a gradient in a resource for which competition is occurring, a relevant experiment is that of Campbell and Grime (1992), growing seven species in outdoor plots with a range of nutrient levels and disturbance regimes. Nutrients promoted growth considerably (Figure 6.6a; note the log scales). Campbell and Grime declare that *Arrhenatherum elatius* (oat grass) is a plant of fertile soils, and *Festuca ovina*, *Bromus erectus* and *Desmazeria rigida* (fern grass) are plants of infertile soils, but actually the nutrient response does not differ between species ($P = 0.989$[17]). The species differ in the effect of competition on them (Figure 6.6b; $P < 0.001$[18]), but there is no significant overall effect of nutrient supply on the intensity of competition ($P = 0.072$[19]), failing to support Grime's assumption. Goldberg et al. (1999) found in a meta-analysis that there was a tendency for competitive intensity to *decrease* more often than increase with productivity, in general conformity with our theory, and Peltzer and Wilson (2001) found no significant trend with standing crop, used as a proxy for productivity. However, the experiment of Campbell and Grime, as many of those surveyed by Goldberg, has the restriction that it is not possible to tell whether competition was for the resource (NPK)

[17] Test for heterogeneity of slopes on a log–log basis.
[18] By analysis of covariance, with log of nutrient concentration as the independent variate.
[19] For a joint residual regression.

that varied along the gradient. This restricts very considerably the range of investigations available for critical tests.

The experiments of Peltzer et al. (1998) in Saskatchewan, Canada, and Cahill (1999)
in Pennsylvania, USA, were both in oldfields, planting seeds or seedlings into plots
where shoot competition was prevented by tying back the vegetation, root competition
was either prevented or allowed by using plastic tubes, and fertiliser was added or not
(N in the case of Pelzer et al. and NPK in the case of Cahill). Both studies showed
somewhat greater below-ground competitive effects when soil resources were in shorter
supply. This confirms our theoretical argument above, the conclusion of Wilson (1988c)
surveying experiments on root competition, and the still limited evidence now available,
that competitive intensity is highest when resources are in shortest supply. In all these
experiments the gradient is one of soil nutrients and competition must be for either
mineral nutrients or water, but generally it is not possible to tell which. However, Cahill
recorded soil moisture with gypsum blocks, and found no significant difference between
treatments, implying so far as one can from non-significance that the competition was
for mineral nutrients. The study that comes closest to answering the question is that of
Wilson and Tilman (1991) at Cedar Creek, an experiment similar in most respects to
those of Peltzer et al. and Cahill. It is known from other work that nitrogen is the
limiting mineral nutrient in the oldfield at Cedar Creek, and other nutrients were applied
to all treatments to make absolutely certain of this (right down to Cu, Co, Mn and Mo).
Only N (ammonium nitrate) differed between treatments. It therefore seems likely that
this is competition for N along a gradient of N supply. Moreover, RGR is available to
judge the result. In all three species used the below-ground competitive effect was
greater at low soil N supply. An experiment with one of the same species confirmed this
(Wilson and Tilman 1993).

The overall evidence overwhelmingly supports our contention that competition for
resource X will generally be most severe when X is in shorter supply. It is surprising
that anyone thought otherwise.

6.7.3.3 Conclusion

Our model contrasts with Grime's and Tilman's hypotheses and proposes that along an
environmental gradient of an unlimiting or unlimited resource, competition will be
uniformly intense. Along the gradient of a limiting resource competition will be most
severe when it is in shortest supply. If the resource for which competition is occurring
changes, the question is too difficult to answer.

6.8   Where Will Facilitation Be Strongest?

The existence of facilitation has long been recognised, that one species can have a
positive effect on another due to reaction (Section 2.3). For example, Warming (1909)
noted 'There are also obvious cases in which different species are of service to each
other. The carpet of moss in a pine-forest, for example, protects the soil from desiccation and is thus useful to the pine, yet, on the other hand, it profits from the shade cast by

the latter'. Its importance was recognised by Gleason (1936; Section 6.2). Bertness and Callaway (1994) proposed a 'general principle' that facilitation would occur mainly in habitats with high physical stress or consumer pressure, presumably along a stress/pressure gradient of the environmental factor mediating the facilitation, and termed this the 'Stress–gradient hypothesis' (SGH). It is logical: reaction caused by facilitation cannot ameliorate an environment that is already optimal. More recently, a Humped SGH has been proposed: that at the highest stress, facilitation would decrease again (Maestre and Cortina 2004). Michalet et al. (2014) suggested this could be because at high stress: (a) facilitation collapses or (b) it is overwhelmed by more intense competition. However, both the SGH and Humped SGH are controversial, as theories and as to evidence for them (Soliveres et al. 2014).

Several recent meta-analyses of the SGH (e.g. He et al. 2013) have been contradictory. However, we believe it is necessary to consider the mechanisms behind each type of stress before making general SGH conclusions. Field studies based on positive co-occurrences between species are insufficient evidence of facilitation. The association may reflect overlapping beta niches for particular resources rather than interspecific interdependence: community ecology needs to be a rigorous science, especially when experiments are possible. The same problem besets planting of a species into one type of patch (e.g. beneath a nurse shrub) versus into another (the open matrix): the patches may have been different in the first place. We therefore consider here only experiments on the response to species' removals, particularly since observational studies might contradict results from experiments; positive versus negative effects of neighbours (Maestre et al. 2005).

Associational defence (Section 2.7.3.2) is a type of facilitation, perhaps effected by a spiny shrub. It will be inoperative when grazers are absent and effective when grazing pressure is higher, but at the highest pressures grazers may utilise the unpalatable species itself or navigate through its defences. Brooker et al. (2006) demonstrated this by clipping (or not) *Calluna vulgaris* (heather) as a defender of *Pinus sylvestris* (Scots pine) saplings against different intensities of *Cervus elaphus* (red deer) browsing. Is this a Humped SGH?

Water facilitation has been suggested in arid areas. There can be more species or greater abundance of them under nurse shrubs/trees than in the open. Improved water availability is possible there through shelter from evapotranspiration, faster water infiltration, higher soil organic matter and thus water-holding capacity, or hydraulic lift (e.g. Howard et al. 2012, in an arid area of Australia, see Section 2.3.1). What could be the mechanisms for a Humped SGH then? Shelter from evaporation by neighbours may collapse if the plants are further apart in drier conditions, thus there may be fewer species to act as facilitators, and facilitation might well be outweighed by competition for sparse water. Indeed, Maalouf et al. (2012) reported, from responses to neighbour removal in a French limestone grassland under severe experimental drought, that both interference and facilitation, demonstrated in the wettest conditions, 'collapsed' in the driest. We argued above that at least interference must exist, but it is possible that slow growth in dry conditions prevented effects being demonstrable within the 18 months of the experiment. Likewise, nurse shrubs/trees could provide nutrient facilitation via nutrient lift and through accumulation of litter, dust and perhaps dead animals (Howard et al. 2012). This would have more effect where nutrients are more limiting, but competition for nutrients would be greater then, giving a Humped SGH. We know of no test.

High altitude is normally seen as high-stress (though not necessarily for the species growing there: Section 6.5.3). Callaway et al. (2002) report an 11-site experiment in which small target plants of a number of species were released from plant interactions by removing above-ground parts of other plants growing within a 10-cm radius. At lower altitudes, interference was revealed to have been operating at most sites (with non-significant facilitation at three), but at altitudes 340–1,600 m higher facilitation was operating (significant for 9 out of 11 sites). There was a significant tendency for facilitation to be greater at sites with a lower summer temperature, suggesting the facilitation was due to shelter from cooling winds. Kikvidze et al. (2006) reported a similar experiment from subalpine Georgia (Caucasus), showing interference in the wetter months May–July, but facilitation in the drier August, suggesting water facilitation. Investigations that have shown correlative evidence of facilitation in the alpine zone – species tending to occur within cushions rather than in the open ground between – have generally shown higher soil moisture and nutrients within cushions, but ameliorated temperatures only in some studies (e.g. Cavieres et al. 2008). All three factors could be operating together.

Salt and anoxia are stresses for most species. On an American saltmarsh dense *Juncus gerardii* ameliorates soil anoxia and salinity, allowing associated *Iva frutescens* (marsh elder) to survive (Hacker and Bertness 1999). Soliveres et al. (2014) give this as their example of greater facilitation at intermediate levels of stress. Indeed, in the high-elevation marsh (low stress) there is no flooding and negligible salinity so facilitation by *J. gerardii* is not needed, and the species is sparse anyway. In the low-elevation marsh (high stress via demonstrated high salinity and low aeration) facilitation by *J. gerardii* is not possible because it is absent. Here the mechanism of the Humped SGH is clear, but whether it is a deep concept, that *J. gerardii* cannot facilitate if it is not present, is doubtful.

All the generalisations of the effect of stress on the interference/facilitation balance are probably simplistic: Grant et al. (2014) found that drought stress resulted in facilitation for some species but interference for others, and that even these effects could be opposite depending on the associated species, the latter result perhaps indicating little more than the effect of an external factor on the competitive balance.

6.9   Conclusion

Community ecologists have tended to produce models and then try to make the facts fit them. Anna Bio (2000) neatly criticised this in her thesis title, 'Does vegetation suit our models?'. This tendency has led to a plethora of concepts and theories for understanding communities, and it has also been suggested that the discipline of community ecology largely comprises a collection of special cases, each with their own contingencies, thus making meaningful generalisations untenable (Lawton 1999). Vellend (2016) has, however, done an admirable job of trying to bring some order to this chaos, through his four-axis classification of community ecology theories into 'selection', 'drift', 'speciation' and 'dispersal'. The schema appears to work well, probably because it is based on fundamental processes, not phenomenological outcomes. But even with a consistent organising framework, the community ecology theoretical landscape remains complex.

Frederick E. Clements' concepts were based on vast field knowledge, acute observation and pioneering experimental work. His generalisations remain true and we dedicate this book to him for his insights. Both Henry Gleason and Clements pointed out that vegetation is sometimes discontinuous along environmental gradients due to the operation of a switch.[20] This is true even of many boundaries that ecologists categorise as 'environmental' such as a riverbank, or that between a saltmeadow and a saltpan. The methods of Mike Austin generally indicate continuous variation along environmental gradients (Section 6.3), but that is because they have been used at larger spatial extents than the ones on which most switches occur. Nevertheless, there is increasing interest in geographic-scale switches, especially those involved with climatic change (Section 4.5.4.1). We note that Hubbell's Unified Neutral Theory is (probably) not intended to be a model of how communities are, but a null model that has been useful for highlighting the processes that are occurring.

Intrinsic guilds (Section 5.7.3) and Philip Grime's C-S-R theory both address the functional role of species. Intrinsic guilds are based in the structure found in the community itself. C-S-R uses an *a priori* triangle originally inspired by habitats in the field, but with underpinnings from *r-K* and leaf amortisation theories, as a template for all vegetation, including succession, stability and the relation of species richness to productivity (the Humped-back theory). It led to the excellent Integrated Screening Programme functional-trait database (Grime et al. 1997). Other datasets should aspire to this quality. David Tilman has produced theories on many aspects of vegetation, including the mechanism of competition, where competition will be most intense, how resources will change during succession, whether and how species will coexist, species diversity, biomass allocation, etc. Some of these theories have been effectively disproved, even by Tilman himself, but perhaps that is because the theories were put in a more testable form than Grime's. Tilman's concept that competition will be equal along a productivity gradient is close to the truth, but his R* approach seems to be too simplistic for embryophytes. However, C-S-R theory seems closer to reality, in that the ecologist in the field will often interpret communities in terms of C-S-R, rarely in terms of Tilman's ideas.

So 'Does vegetation suit our models?'. None of the models of plant communities yet produced have high synthetic or predictive value. None are well enough established to be applied to practical problems, and the dangers of doing so are exemplified by attempts to apply species–area curves and 'Island Biogeography' theory to reserve design. Why should this be the case? We believe one major impediment to progress has been the focus of many theories on making predictions on the *outcomes* of community assembly, without due recognition of, or integration with, the underlying *mechanisms*. We will return to this theme in the Chapter 7.

---

[20] Of course, in pre-Odum (1971) days not using that term.

7 Synthesis

7.1 Introduction

We have attempted to examine the construction and structure, not the natural history, of terrestrial plant communities by comprehensively describing the nature of plants and their interactions. We have repeatedly dwelt on the complexity introduced by variation: in biota, in phenotypes and in the environment, in both time and space. We have examined the current models of plant distribution, Grime's C-S-R theory and the various concepts of Tilman, and find that although they contain useful and realistic concepts, they do not help greatly with our main enquiry. We persist in examining the natural world. It is so much altered by human disturbance that many might expect that few natural situations of stable vegetation remain, yet there are measurable properties of coexistence between species in such an unprepossessing habitat as a mown lawn. Therefore we take heart and expect there to be real generalisations to be made about the functioning of communities.

Over the previous chapters we have established three clear aspects of plant communities: (1) almost all comprise many species (**The 'Paradox of the Plankton'** [Section 7.2]), (2) they are spatially and temporally **heterogeneous** (Section 7.3) and (3) ecologists must hope there are some rules governing the assembly of species in them – **assembly rules** – or there is no science in plant community science (Section 7.4). The processes that we emphasise in generating community patterns are **reaction** and **switches** (Section 7.5). These three aspects and the key processes of reaction and switches are summarised below.

7.2 The 'Paradox of the Plankton'

The 12 mechanisms that can allow species to coexist are clear, but their relative importance is not. We concluded in Chapter 3 that the major reason for species coexistence is *alpha-niche differentiation*. The ecosystem processes enumerated by Reichle et al. (1975; see also Section 1.1.2) must change both stochastically and predictably on all timescales, and also spatially, especially during the year in seasonal climates – the energy base (affected by irradiance), the reservoir of energy, nutrient cycling (through mineralisation rates) and rate regulation (temperature, water availability, herbivory) – and this variation should easily be sufficient for coexistence mechanisms based on *environmental fluctuation* and associated environmental heterogeneity (Chesson 2000a). *Pest pressure* may be important; Gillett (1962) and Petermann et al. (2008) suggested that it is a major mechanism, but that remains to be proved. The *spatial mass effect* must be very common, but is difficult to quantify. *Allogenic disturbance* is clearly common and its potential to promote coexistence is well established, at least theoretically (Shea et al. 2004). Allogenic disturbance can be reinforced by the lesser-known *autogenic disturbance* (Section 2.5), which at small spatial grain can also allow coexistence. Autogenic disturbance could also be a component of *cyclic succession*, and recent understanding of negative feedback via the soil microbiota increases the likelihood that it occurs (Section 4.4). *Inertia*, both temporal and spatial, is clearly widespread, though it is equalising, not stabilising. We suggest that other mechanisms are of minor importance.

Several mechanisms of coexistence are therefore well established (though we do not discount the possibility of additional mechanisms being uncovered), and these provide the building blocks for our community assembly rules. Given the multiplicity of mechanisms and the potential for them to allow the coexistence of a very large number of plant species, perhaps the 'Paradox of the Plankton' question should be inverted, and rather than asking why there are so many species in a given habitat, perhaps we should be asking why there are so few? We have uncovered snippets of how these processes might be operating across a range of communities, and that gives us great hope. The challenge now is to evaluate how they could be contributing, separately and in interaction with one another, to the promotion of species coexistence more generally, as well as for particular communities.

7.3   Heterogeneity

Environments and communities are always heterogeneous in space and time (Robertson et al. 1988; Farley and Fitter 1999), but the cause can be either allogenic or autogenic.

7.3.1   Allogenic Heterogeneity

Preexisting, underlying environmental heterogeneity is inescapable on any land surface. Deserts appear uniform, but there is always topographic variation that affects plant growth, for example damper depressions. Alluvial plains seem homogeneous but receive non-uniform deposits as rivers flood and meander. Similar processes occur on saltmarshes. Since species necessarily differ in their environmental tolerances (Section 1.4), the result must be allogenic community heterogeneity. Therefore any investigation of community processes has to use areas where allogenic environmental heterogeneity is minimal, and/or allow for it (e.g. using patch models: Section 5.2.4).

Allogenic disturbance is another cause of patchiness at almost all spatial grains (Section 3.6), but we must ask: how much of a landscape has its species composition still affected by past disturbance, e.g. in forests? This is difficult, because: (a) no studies have been able to achieve sufficient timespan and (b) identifying past impact is almost impossible. The sporadic data suggest that 20–40 per cent of typical forests consist of gaps, though this surely differs between habitats and forest types, and 'gap' is not easy to define. It is even more difficult to define how much of a forest still bears the impact of past disturbance in species composition different from the matrix, i.e. different from the climax composition. For example, in the seasonal dry evergreen forest of western Thailand, large-scale gaps (up to several square kilometres) caused by cyclones, fires or drought, plus gaps in the order of 350 m × 350 m created every 20 years, plus gaps due to the death of single trees, may mean that almost none of the area has reached climax, and there is no matrix of undisturbed forest (Baker et al. 2005, pers. comm.; Middendorp et al. 2013). We need more knowledge of openings in forests and all natural plant communities, for gaps occur in all vegetation types.

7.3.2   Autogenic Heterogeneity

Vegetation heterogeneity can occur without preexisting environmental differences if switches are operating. Switches can magnify patchiness, not cause it, so in order to give initial patchiness 'random' dispersal of species has to be assumed. The arrival and establishment of propagules must be limited, since too many propagules would give a uniform cover with no patchiness. Subsequent invasion must be by infiltration invasion, or the patchiness will rapidly be extinguished, but infiltration invasion may be pervasive (Section 1.5). Alternatively, the initial patchiness may be caused by patchy disturbance, allogenic or autogenic, with different species establishing in different patches because of either a different propagule rain or different prevailing environmental conditions. The initial heterogeneity can also be caused by minor environmental differences.

The colonists must then react on their environment, but all plants do (Section 2.2; Chapter 2), and the species must differ in their reaction, since by definition, they are differentiated (Section 1.4). The reaction can be abiotic or biotic, the latter notably via the soil biota. If the reaction of the species is in the direction that favours themselves, a switch operates (Section 4.5) leading to a mosaic of communities, in its fullest form a mosaic of ASSs (Section 4.6). The spatial grain of these processes can vary from geographic, e.g. closed forest versus savannah in northern Australia (Section 4.5.4.1), to a single shrub in arid environments that accumulates soil beneath itself, with shade, and regenerates there. At smaller grains, the mosaic may be a shifting one, e.g. the very visible mosaic of tree islands in the low alpine, permanent but perhaps moving (Wilson and Agnew 1992, p. 296).

The requirements of this process are thus: (a) infiltration invasion, disturbance patches or minor environmental variation leading to differing initial species composition; and (b) reinforcement by switches. We believe heterogeneity due to ASSs is more common than has been realised, though the evidence for them is rarer than has been realised (Mason et al. 2007).

Cyclic succession (Section 4.4), if it occurs, could also give autogenic heterogeneity, and such a process could involve a heterotroph. A dramatic example is 'fairy rings', where fungi promote plant growth in some of the concentric zones and depress it in others, apparently via nutrient depletion and/or the release of plant growth-promoting substances (Edwards 1988; Choi et al. 2010).

Autogenic heterogeneity is an understudied area of plant community ecology, and there are a number of aspects that deserve greater attention:

1. **Direction**: Reaction *favouring* the present species as opposed to invaders will produce autogenic heterogeneity if it magnifies small initial differences via a switch. Reaction *disfavouring* the present species will lead in all patches towards Clementsian climax, even if rarely reaching it (Section 4.3); a mosaic will result only if the succession is cyclic.

2. **Magnitude**: Major reactions clearly cause major heterogeneity. Many reactions are small, but they can produce heterogeneity if all species are reacting on the environment in the same direction.

3. **Transience**: The light regime differs beneath different species, though much more is known about differences in total light transmittance/reflectance than of effects on spectral composition. Reaction on light is transient, but can be maintained by a switch (Section 4.5.4.3). Soil reactions occur much more slowly, but can be much more persistent. Notably, podsolisation is reversed only very slowly. However, there is little information on the degree to which the species within one geographic pool differ in their reaction, and soil reaction is known rigorously (from replicated, randomised experiments, as opposed to sampling beneath trees that may be growing in different microsites anyway) only from plantations or communities that are close to monocultures.

4. **Type of switch**: A Type 1 (one-sided) switch, which seems to be the most common, cannot give rise to a permanent ASS mosaic. For this, a switch of Types 2–4 is required (Section 4.5.2).

All this can occur, but our current understanding comes from multiple fragments of information. Full details of any one situation are unknown.

Small-spatial-grain heterogeneity can also be caused by the morphology of a species – a single ramet for many trees and shrubs or a clone for many herbaceous plants – if within-species interactions such as competition, autoallelopathy or autogenic disturbance such as in crown shyness keep them apart. This is most obvious in the gaps between plants in the desert, though whether the 'gaps' can be regarded as gaps, and therefore whether heterogeneity exists, can depend on whether we look above-ground where there is release from cumulative competition for light (Section 2.2.1.2) or below-ground where competition may be intense (Section 6.7.4).

7.3.3   Ecotones

Heterogeneity is most obvious when the edges between patches are sharp, i.e. ecotones. Switches of any type (1–4; Section 4.5.2) can sharpen edges in a mosaic, or on a gradient, forming an ecotone. However, there are six types of ecotone (Lloyd et al. 2000; Walker et al. 2003c; Figure 7.1): (1) *Environmental (geomorphological) ecotone*, caused by a sharp change in the preexisting environment (e.g. a riverbank),

Environmental ecotone
The ecotone occurs where there is a sudden change in the underlying physical environment.

Anthropogenic ecotone
There is no difference in the underlying physical environment. Logging has removed trees from one area, and forbs and grasses have invaded.

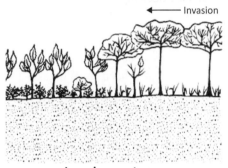

◄───── Invasion

Invasion ecotone
One species is invading from the right, displacing many of the species.

Threshold ecotone
The environment changes gradually, but at a certain point on the gradient the environment (here, soil) becomes unfavourable for one set of species.

Community coadapted ecotone
The environment does not change, but one set of coadapted species is replaced by another such set. The two sets cannot mingle because mixtures are unstable.

Switch ecotone
A small initial difference in vegetation or environment (here soil) has been magnified, and the boundary sharpened by a positive-feedback switch.

Figure 7.1 Six types of ecotone.

(2) *Anthropogenic ecotone*, due to human intervention (e.g. a forest-logging edge), (3) *Invasion ecotone*, where there is phalanx invasion (Section 1.5.2), (4) *Threshold ecotone*, where some value of an environmental factor is critical for the survival of species, (5) *Switch ecotone*, where a switch of any type (types 1–4) sharpens a gradual,

preexisting environmental gradient or (types 2–4) creates a mosaic (Section 4.5.3), and (6) *Community-coadaptation ecotone*, where one set of coadapted species gives way to another set. *Environmental, anthropogenic* and *invasion ecotones* are trivial in terms of community structure, and their cause is usually obvious. Sharp timberlines, an apparent example, are sometimes, perhaps always, caused by a temperature switch (Section 4.5.4.3). We know of no examples of *threshold ecotones*, where several species reach their physiological limits together, and suggest they may not exist. A *community-coadaptation ecotone*, where one set of coadapted species is replaced by another set, seems to be based on a caricature of Clements' views that he would not recognise; it is difficult to believe in it.

7.3.4   Heterogeneity and Community Structure

Spatial analysis on its own is rather uninformative for community structure. Goodall (1954) argued that if a community has real existence it should show homogeneity of composition within its boundaries, an argument echoed by Whittaker (1975a). However, integrated communities could well show small-scale heterogeneity, especially since switches can cause autogenic heterogeneity. F. E. Clements who wrote 'the community is a complex organism...greater than the sum of its constituent species' (Clements 1935) also wrote 'Practically all vegetation shows more or less striking differences every few feet' (Weaver and Clements 1929, p. 6). 'Individualistic' communities along an environmental gradient, with no coevolution, can show sharp boundaries when the gradient is sharpened by a switch (Section 4.5), as envisaged by both Clements and Gleason (Section 6.2.2). In fact, we suggest that all persistent boundaries that are neither anthropogenic nor correlated with a sharp environmental change are caused by a switch. One of the key roles for heterogeneity in determining community structure is via the potential for spatial and temporal niche differentiation.

7.3.5   Conclusion

Almost certainly both allogenic and autogenic community heterogeneity occurs simultaneously in all communities, and interact. However, it is rarely clear how much of the small-scale patchiness that ecologists see around them is due to disturbance, how much to environmental variation, and how much to reaction giving autogenic heterogeneity via switches or cyclic succession. We call for further investigation, especially of the role of autogenic heterogeneity.

7.4   Assembly Rules

At any point within vegetation heterogeneity there can be: (a) a stable community (but of course with continual allogenic change, tracking the environment), (b) an ASS, or (c) a seral stage of directional or cyclic succession (Chapter 4). However, our discussion would be little more than natural history were there not some regularities, or rules, in how the states are assembled, and assembly rules are the third aspect of plant communities that we emphasise. They are less well established than the mere coexistence of species (Section 7.2) or the existence of heterogeneity (Section 7.3), but more crucial for community structure.

7.4.1   Assembly Rules and Alternative Stable States

Assembly rules and ASSs are related concepts, in that from a pool of species only some combinations are stable. The differences are of degree:

(a)   Alternative stable states could conceptually be maintained by autogenic disturbance or parasitism, but we cannot think of any actual mechanisms for either. In practice, they are always produced by switches. Petraitis and Hoffman (2010) declared that positive feedbacks (i.e. switches) were not necessary for ASSs, but their mathematical models contain no alternative ecological mechanism. A switch generally depends on a considerable degree of physical reaction, sufficient to make the state stable in the face of all but the more extreme environmental fluctuations, though it could depend on interactions via heterotrophs (Section 2.7). Assembly rules are also caused by reaction, be it more subtle, and again possibly via heterotrophs on a small-scale, and again autogenic disturbance or parasitism cannot be ruled out.

(b)   Alternative stable states are usually envisaged to exist either at different times or in different places over scales of hundreds of metres or more. Assembly rules are envisaged at a small, within-community scale, but this difference cannot be absolute since desert patchiness in the order of 0.1 m could be ASSs, and Diamond's (1975) original assembly rules operated among islands up to 1,000 km apart.

(c)   Because of the difference in spatial scale, transient reactions, such as via light, are among the likely causes of assembly rules, whereas gross changes in soil composition are likely effectors for ASSs.

(d)   The scope for assembly rules is wider, for example specifying a relative abundance distribution (RAD) or specifying guild proportions whilst leaving open which species are involved, whereas there will be a limited number of specified ASSs, often only two, each with its own set of environmental conditions and species.

7.4.2   Coevolution versus Preadaptation

Any assembly rules could arise through either coevolution or preadaptation. Dice (1952) and Whittaker (1975b) were convinced about coevolution. One of the strongest advocates of this was Goodall (1963), who argued that a group of species that grew together in common types of site would adapt evolutionarily to those site conditions and to each other by '...positive feedback'. He still retained this view in 2009 (D.W. Goodall, pers. comm.): 'In this sense the plant community may sometimes be

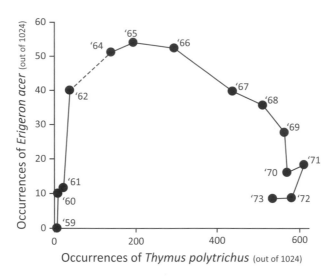

Figure 7.2 Trends in the shoot frequency of *Erigeron acer* and *Thymus polytrichus* (= *T. drucei*, thyme) in the 10×160-cm plot in the Breckland, eastern England, plotted from the data of Watt (1981).

said with justice to have evolved as a whole'. We believe that coevolution between plant species is unlikely, even at the ecotypic level.

One problem for coevolution is that equilibrium in a plant community is rare: there is constant reassortment of species due to environmental change or autogenic processes. Coevolution would still be viable if groups of species moved around the landscape together, but the pollen record tells us they do not (Section 5.9). Neither do species stay associated on much finer timescales. For example, in Watt's (1981) records from Breckland, *Erigeron acer* (fleabane) first increased as *Thymus polytrichus* (= *T. drucei*, thyme) stayed constant, then stayed essentially constant as *T. polytrichus* increased, then decreased as *T. polytrichus* remained constant (Figure 7.2). Much more analysis of local time/space relations like this is needed.

Another problem for a coevolutionary explanation is spatial variation. Since environmental heterogeneity exists right down to the smallest scales, it is not predictable even within a community which species a plant will have as a neighbour. Part of this heterogeneity is caused by the reaction of one plant affecting another: effects between different species, between ramets of the same species and between modules of the same ramet. If coevolution comprises the adjustment of one species characters to match those of another species, a species cannot coevolve to match all these different assemblages. An even greater spatial problem is that species normally occur in several communities, and the characters of a whole species cannot coevolve to be optimal in each (Gleason 1926; Goodall 1966).

Furthermore, we argued in Chapter 1 that evolutionary change in plants is often slow. All this makes coevolution of species traits impossible in heterogeneous communities, and hardly likely even within homogeneous ones if they existed. Therefore, when assembly rules are found, they are likely to be due to the assembly of preadapted species, that happen from their evolutionary histories in a variety of contexts to have the right characters for the job. Preadaptation is the key to community ecology.

That is not to deny coevolution. It likely occurs between some plants and their obligate pollinators and dispersers, though even there it is possible that the characters of plant and/or pollinator/disperser originally arose from coevolution with another partner species, or evolution for some quite other reason, i.e. preadaptation is the cause. The relationship between host and parasite (here including partial parasites) is close, but most angiosperm root parasites are generalists. A rare specialist is *Epifagus virginiana* (beech drops) occurring only on *Fagus grandifolia* (American beech; Musselman and Press 1995). Species of *Rafflesia* are parasitic (mainly below-ground) only on lianas of *Tetrastigma* spp. Some species of *Rafflesia* appear to be specific to particular *Tetrastigma* species and vice versa, though unclear taxonomy in *Tetrastigma* and geographic restriction make this uncertain. Many aerial parasites such as 'mistletoes' and *Cuscuta* spp. (dodder) are generalists, though some *Cuscuta* species show preferences (Kelly et al. 1988). New Zealand genera of Loranthaceae have species that usually parasitise *Nothofagus* trees, but not exclusively, for example, *Alepis flavida* occurs mainly on *N. solandri*, but with c. 20 per cent of its occurrences on nine other species (Norton 1997). Partial preference does not suggest coevolution. Epiphytes generally have little host specificity (Benzing 2004), but the epiphytic filmy fern *Hymenophyllum malingii* (≡ *Sphaerocionium applanatum*) is found in New Zealand mainly on the tree *Libocedrus bidwillii* and in Tasmania perhaps exclusively on the trunks of *Athrotaxis selaginoides* (P.J. Dalton, pers. comm.); the two host trees being in different subfamilies of Cupressaceae. It is difficult to tell whether species of lower strata, e.g. in a forest, are associated with particular canopy species and might therefore have coevolved with them, since both may be responding to the same environmental variation. Wilson and Allen (1990) investigated a site in southern New Zealand where *Nothofagus menziesii* was invading, slowly enough for the uninvaded and invaded areas being close but with time for the understorey to have changed in response to the canopy. There were no detectable differences between the understories of the two very different canopies: the multilayer, rather open canopy of the podocarp/angiosperms versus the more closed, level *N. menziesii* canopy.

7.4.3   Assembly Rules Conclusion

Ecologists cannot claim to understand plant communities without knowing what restrictions there are on species coexistence, when they occur and where. However, it is difficult to test for the presence of assembly rules: (a) without knowing what rules are likely, and (b) when it is so easy to obtain negative or invalid results, for example by not using a patch model. As we noted earlier, there are many traps for the unwary null modeller.

Nevertheless, there is overwhelming evidence that assembly rules do exist (Chapter 5), refuting claims to the contrary by Hubbell (2005) and Grime (2006). The best evidence is from character-based rules, and the future probably lies therein. The characters must be carefully selected as those likely to reflect the alpha niche. They will often not be easy to measure. Analyses of plant communities by their characters rather than by the names of their species can be traced to Raunkiaer's (1909) 'biological spectra' of formations and facies based on his earlier 'life forms', expanded by Jan Barkman (1979) into his concept of vegetation texture using a range of characters such as leaf size, leaf consistency, leaf orientation and, ideally, root traits. Among character-based assembly rules, some distributional evidence supports guild proportionality, as does the successional study of Fukami et al. (2005). There is little support from removal experiments, probably because of high experimental error. The use of a priori guilds has limitations, and we strongly advocate seeking intrinsic guilds (as in Wilson and Roxburgh 1994, 2001; Wilson and Whittaker 1995; Wilson and Gitay 1999). The use of texture (i.e. the range of traits) instead of discrete guilds avoids classification, but does not avoid the problem of character choice. Methods for the determination of intrinsic guild structure on a continuous scale rather than in categories remain to be developed, as do those for intrinsic texture, i.e. determining the characters of the species to use from the properties of the communities.

The evidence for assembly rules so far comes mainly from herbaceous communities, with the only comprehensive body of evidence being from the University of Otago Botany Lawn (Section 5.13). We would be cautious about the demonstration of an assembly rule in any single study, but the coherent conclusions from this site are compelling. The evidence from this and other sites suggests that canopy relations are important, even in the shortest communities such as lawns, saltmarshes and sand dunes. This may be partly because of the types of communities that have been examined so far, and it may reflect the use of easily measured characters. When Stubbs and Wilson (2004) utilised a wider range of characters in a sand dune community, the results implied community structuring by mode of foraging for water and soil nutrients. The evidence on temporal even–spacing of flowering and fruiting implies that phenological niche differentiation is important in restricting species assembly too, though great care is needed in examining that evidence. However, the failure of British roadside communities to reassemble when the species involved have been introduced to New Zealand (Wilson et al. 2000b) indicates that the restrictions on community assembly rules are often weak, and probably that alternative stable combinations exist (Section 5.12.2).

The other urgent need is to understand how each particular assembly-rule operates, i.e. what reaction in what environmental factor or resource, or other type of effect, is caused by each species that limits the ways others can associate with it. There is little evidence as to how frequent the various types of autogenic disturbance are, and what assembly rules they might cause. Interactions via heterotrophs are very likely to cause limitations to community assembly – assembly rules – but the evidence is frustratingly sparse. Such effects are often mentioned to advocate companion planting in horticulture, and the apprentices' gardens at Kew all contain the obligatory marigolds, but we found searches of the scientific literature for evidence almost fruitless. There are a few studies on the effects of applying insecticide or fungicide at the community level, but many simply report the effect on species diversity. Careful examination is needed of the cascade of effects that are caused and their role in assembly rules.

Assembly rules are based on mechanisms, but the net result is efficiently summarised in the concept of stability, and represented mathematically in the community matrix (Section 4.8). Introduced to ecology by May (1972), it was not until c. 30 years later that matrix values were determined for a real plant community, with predictions from the matrix compared with recovery of the community from experimental perturbation (Roxburgh and Wilson 2000a, 2000b). A community matrix summarises the dynamics of the community, but only one at an equilibrium and perturbed a very small amount (Figure 4.11a). 'Invasibility analysis', based on quantifying species' long-term low-density growth rates (Figure 4.11b), more generally provides insights into the presence of stabilising mechanisms (but not necessarily assembly rules, nor stability *sensu* Section 4.8), but without the restrictive assumptions of equilibrium and small perturbation. Regardless of the theoretical framework used, making the link between species coexistence mechanisms and how they are embedded within assembly rules that, for example, provide regularity in co-occurrence patterns, maintain relative abundances, and allow communities to persist stably over time is a challenging task. In general, more realistic models are needed, parameterised from real communities, to provide insight into the actual biological mechanisms at work. For assembly rules more generally, as Zhou Enlai said when asked of the effect of the French Revolution on subsequent history: '[it is] too early to say'.[1]

7.5   Processes: Reaction and Switches

The nature of the plant community depends on the characteristics of plants (Section 1.2). They do not consistently have 'individuals' but are colonies of modules. Dead plants persist as ghosts via the effects of their litter, part of their extended phenotype, indefinitely if a switch is operating. In these ways, the species plays the part of an individual in the community, reacting on the environment and thus on associated biota. These characteristics of plants produce a range of interactions within and between species (Box 7.1), some of which need to be given more consideration in a community context. All this will make it difficult for vegetation to suit our simple models.

A few of the interactions between plants are direct, for example parasites, strangling lianes and the fascinating and understudied effects of shaking and physical abrasion (Section 2.5.1). Interactions via herbivores, fungi and microbes are more important. However, the most important types of interaction are via the environment, i.e. physical reaction. Reaction is the plant community, for without reaction there would be little more than a collection of plants (Gleason 1936). We do not believe there are occasional species that are 'ecosystem engineers' or ones that perform 'niche construction'; rather all plants modify their environment, i.e. all produce reaction. Strong reaction can result in community change: climax or cyclic succession, or a switch (Box 4.1). The resulting pattern is potentially complex.

---

[1] It was later claimed that he misunderstood the question, and was referring to unrest in France in 1968.

Box 7.1 Types of interaction between plants

At the species (or within-species) level
    Negative effects, via:
        Reaction (i.e. interference): competition, allelopathy, spectral interference, etc.
        Litter
        Parasitism
        Autogenic disturbance
        Heterotrophs
    Positive effects (facilitation): commensalism, mutualism, altruistic facilitation, via heterotrophs[2]
At the community level
    Guild/community X gives a relative disadvantage to itself:
        The effect is abundance-independent: altruistic facilitation and/or
            autointerference = **succession-to-climax** or **cyclic succession**
        The effect disappears at low abundance of X (negative feedback) = **stability**
    Guild/community X gives a relative advantage to itself = **switch**

Recent work has emphasised biotic reaction, notably on the soil biota. Some studies have found effects positive for the species causing them, which would drive a switch (e.g. Viketoft 2008 for nematodes; Bezemer et al. 2010 for several groups). However, negative soil feedbacks have been shown in many experiments, which would result in succession-towards-climax or, perhaps more likely in the pseudoclimax communities studied, cyclic succession. However, soil feedbacks do not always seem to be via the soil fauna/microflora, and when demonstrably by the latter, it is usually unclear by which specific group of organisms. The effects are clearly contingent on the fauna/microflora present. For example, Gundale et al. (2014) found that greenhouse growth of *Pinus contorta* (lodgepole pine) was **depressed** 21 per cent by soil inoculum from Canada (where it is native), but **increased** 27 per cent by inoculum from Sweden (where it is exotic, and even though the Swedish soil came from several *P. contorta* plantations). Although there have been many recent studies on soil feedbacks, much remains to be known before it will be possible to know the effects on community structure.

The role of switches in generating spatial and temporal heterogeneity has been considerably underestimated: reinforcing the current state, sharpening edges (Section 7.3.3) or accelerating/delaying succession. Switches of types 2–4 (zero-sum, symmetric and two-factor) can produce ASSs. Erwin Adema produced evidence of this in Dutch dune slacks (Section 4.6.4), perhaps the best of all terrestrial examples, but he has remained modest about it. Many workers have seen ASSs as being common. They may be, but unless we are credulous, the hard evidence for them is vanishingly small (Section 4.6; Mason et al. 2007). Elsewhere in the literature there has been hand-waving

---

[2] It seemed necessary to divide facilitation by whether the effects were reciprocal. However, we could also divide facilitation by its cause, notably reaction versus interactions via heterotrophs.

about ASSs and diagrams of hysteresis, with minimal consideration of the mechanisms involved, still less evidence for them. Sometimes (e.g. Lortie et al. 2004) it has not even been realised that ASSs will always be caused by a switch. This is especially important because superficially observed ASSs and hysteresis could be due simply to temporal inertia, e.g. to plants taking their time about dying (Section 3.12). Thus, whilst we suggest that the importance of switches has been considerably underestimated, we must also point out that for only the lake turbidity system (Section 4.5.4.5) have all the steps of any one switch been demonstrated.

7.6   Conclusion

We have examined the available evidence on the nature and functioning of terrestrial plant communities in an attempt to come to a new understanding. Too often, theories formulated for animal communities are applied unthinkingly to plants, ignoring their particular characteristics as colonies of modules, often genetic mosaics, their reaction on the environment, and their particular types of alpha-niche differentiation (Chapters 1 and 3). The formation of a plant assemblage, and our account, start with the arrival of species from the available pool. A wide variety of plant–plant interactions then occur (Chapter 2). Parasitism is occasional and autogenic disturbance more common, but the majority of plant–plant interactions are based on reaction, including those via litter and heterotrophs. Almost all these interactions are different from those between animals, notably autogenic disturbance and litter effects. All species react on their physical and biotic environment, but our knowledge of the complexity of reaction is fragmentary. An obvious example is the reaction of tree species on underlying soil, which is known largely from *post hoc* observation below trees in mixed communities or unreplicated plantations (Section 7.3.2). Reliable evidence from replicated, randomised experiments is sparse. Information on biotic reaction is accumulating but even when demonstrated, the mechanism is usually unclear (Chapters 1 and 4). This work is not easy, but is essential if we are to understand plant communities.

The surviving species from these interactions comprise a plant community, with several coexisting species save under exceptional circumstances. There are many gaps in our knowledge of the 12 mechanisms involved in coexistence (Chapter 3). Of the ones we believe the most significant, evidence is now being obtained on *environmental fluctuation* by parameterising models, and there is widespread but fragmentary knowledge on *pest pressure*. The *spatial mass effect* is extraordinarily difficult to quantify, as is the significance of *allogenic disturbance*. *Niche differentiation* is easily quantified (relatively), but the extent to which it enables coexistence is rarely known. However, the most pressing need is for integrated knowledge of the 12 mechanisms in particular plant communities.

Assembly rules must apply if the interactions between species results in repeated patterns. They should exist even in such an unprepossessing habitat as a mown lawn, as we have demonstrated (Section 5.13). However, great care is needed, for inappropriate test statistics and null models can lead to spurious reports. The species combinations produced by assembly-rule processes might be persistent, but we can regard them as a real community only if there is more than temporal inertia, i.e. if there is true, Lyapunov stability (Chapter 4). Amazingly, this is known only for that same mown lawn, which turned out to be on the cusp of stability. Plant community ecology is poorly developed when it cannot explain the coexistence of species in any one community, and knows the stability of only one.

Plant ecology has moved beyond Warming's 'easy task' of describing those plant distributions caused by preexisting environmental heterogeneity, which used proxies such as successions in time, species–area curves and ordinations of species in community space. Switches appear to us to be the key concept, for reaction must always be present, and if it favours the plants causing it, then a switch will operate. We see switches as the only known process of community discrimination. They are the necessary cause of persistent sharp edges (ecotones) where no other discontinuity can be found, such as change in the underlying physical environment or human activity. Switches are also the necessary cause of persistent autogenic community heterogeneity, i.e. ASSs. Whilst reaction in the opposite direction, i.e. disfavouring the species causing it,[3] is the primary driver of succession-towards-climax, switches are the cause of the more interesting aspects, *viz.*, delay, acceleration and perhaps alternative pathways. Cyclic succession as envisaged by Watt (1947) may not occur, but hints are accumulating of small-grain cyclic succession, with the mechanism often unclear but probably mediated by biotic reaction via the microflora.

Biological communities are too complex for overarching, simplistic theories to be testable. Many, such as changes in competition and facilitation along gradients, have been proposed with little consideration of the mechanisms that would cause them, as our reductionist analysis above demonstrates. Many ecological theories, such as Whittaker's (1967) suggested coevolution of species along environmental gradients, are made unlikely by the fact that most species occur with different associates, in several different communities and at different times on scales of years (van der Maarel and Sykes 1997), decades (Watt 1981) and millennia (Section 5.9). Preadaptation will be much more common than coevolution. Organismic theories of community cohesion, such as those attributed to Clements, are unrealistic, but Clements' actual writings, with the almost identical statements of Gleason, are largely summaries of field reality.

Consideration of the Clements/Gleason synthesis provides a strong framework for explaining the heterogeneity that we always see in plant communities. Ecologists need to obtain hard evidence for (or against) their concepts of succession-to-climax, switches, priority effects and (perhaps for Clements) assembly rules, as well as Watt's cyclic succession. Information is also critically needed on the reaction of plants on the environment. It is the mechanism behind almost all those processes, but its role in them needs to be known.

The process of the accession of species into communities reviewed above was summarised graphically in Figure 1.2. Figure 7.3 builds upon this framework, adding

---

[3] That is, facilitation, *sensu* Connell and Slatyer (1977).

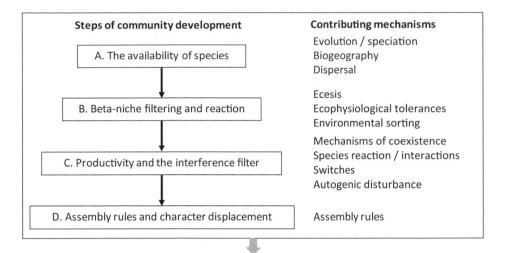

Figure 7.3 Relationship between the steps of community development, the underlying contributing mechanisms and their ultimate expression as community-level outcomes. The grey arrow denotes the missing link between many current community ecology theories that focus on the predictions of outcomes, and the mechanistic basis of those outcomes.

additional details on the underlying mechanisms, and their possible outcomes at the community level.

The list of mechanisms in Figure 7.3 (A–D) and their possible outcomes (E) is non-exhaustive, and this simple representation ignores the complexity of the interactions among the components (such as the 12 coexistence mechanisms, or the myriad steps underlying ecesis); nevertheless we believe it provides a useful summary of many of the key concepts. Our consideration of these relationships in this book, and our current knowledge of them, has led us to three main conclusions regarding the future of plant community ecology:

1.   Over recent years ecologists have sought to improve the link between theory and practice. We believe John Lawton's 1991 observation that 'Too many ecologists still believe that theory is cheap and largely irrelevant, and that salvation lies in having both feet firmly embedded in the local swamp. Likewise, too many theoreticians wouldn't recognize a swamp if they fell in it' (Lawton 1991) is no longer generally true; a good example being the investigation of fluctuation-dependent coexistence mechanisms by Peter Chesson and coworkers (Section 3.3), where theory and empiricism were developed hand-in-hand. Theoretical ecology has become mainstream, but to ensure the further advancement of community ecology it is imperative that a close relationship between theory and practice continues to be forged.

2.   There is empirical evidence to support many of the mechanisms that underlie community assembly (Figure 7.3); in some cases strong evidence (e.g. the evolutionary process; modes of dispersal). There are, however, many knowledge gaps. They include the potential importance of autogenic disturbance, the need for a better understanding of reaction, and more focussed attention on the contribution of plant interactions and coexistence mechanisms to specific assembly rules. A key to improving our understanding of community structure is the identification, across a range of community types, of the mechanisms that operate, and how they contribute to the patterns of assembly and community dynamics that we observe today.

3.   Our conclusion in Chapter 6 was that many theories of plant community ecology have sought to seek predictions for patterns in the outcomes of community development, without necessarily giving due recognition to, or even consideration of, the underlying mechanistic basis. If we are truly seeking to develop a predictive capacity for community ecology, then there needs to be a greater focus on integrating our understanding of the community patterns we can observe (Figure 7.3, E), with the underlying mechanisms that gave rise to those patterns.

The world's ecosystems are undergoing rapid change, for example the direct and indirect impacts of anthropogenic climate change, ongoing clearing of native vegetation, and the human-assisted transportation of species into new environments. As terrestrial plants provide the energy source on which much of life on earth depends, understanding and predicting how plant communities will respond to these drivers is a key management challenge. Central to this is the task of understanding how plant communities are structured – that is, of how species respond both plastically and genetically to change, of how dispersal and environmental filtering act to constrain local species pools, and of how species interactions act to further place constraints on community organization – the assembly rules and associated community assembly mechanisms that are the major topics of this book.

From our review of the evidence for community assembly mechanisms and their outcomes, our overall conclusion is that, whilst there exists empirical evidence in support of particular mechanisms, and for particular repeated patterns of assembly, as a discipline we have been generally deficient in bridging the gap between the two. What is needed is a greater effort to bring a mechanistic understanding to our theories of community ecology. Only by seeking hard evidence underpinned with mechanistic explanations, and by further bridging the gap that exists between theory and field reality, can the science of plant community ecology hope to advance.

References

Aarssen, L.W. (1983) Ecological combining ability and competitive combining ability in plants: towards a general evolutionary theory of coexistence in systems of competition. *American Naturalist*, **122**, 707–731.

Aarssen, L.W. (1985) Interpretation of the evolutionary consequences of competition in plants: an experimental approach. *Oikos*, **45**, 99–109.

Aarssen, L.W. (1988) 'Pecking order' of four plant species from pastures of different ages. *Oikos*, **51**, 3–12.

Aarssen, L.W. (1989) Competitive ability and species coexistence: a 'plant's-eye' view. *Oikos*, **56**, 386–401.

Aarssen, L.W. & Turkington, R. (1985) Vegetation dynamics and neighbour associations in pasture-community evolution. *Journal of Ecology*, **73**, 585–603.

Abades, S.R., Gaxiola, A. & Marquet, P.A. (2014) Fire, percolation thresholds and the savanna forest transition: a neutral model approach. *Journal of Ecology*, **102**, 1386–1393.

Abeli, T., Gentili, R., Mondoni, A., Orsenigo, S. & Rossi, G. (2014) Effects of marginality on plant population performance. *Journal of Biogeography*, **41**, 239–249.

Abrams, M.D. & Scott, M.L. (1989) Disturbance-mediated accelerated succession in two Michigan forest types. *Forest Science*, **35**, 42–49.

Abrams, P.A. (1990) Ecological vs evolutionary consequences of competition. *Oikos*, **57**, 147–151.

Abul-Fatih, H.A. & Bazzaz, F.A. (1979) The biology of *Ambrosia trifida* L. I. Influence of species removal on the organization of the plant community. *New Phytologist*, **83**, 813–816.

Adams, J.M., Fang, W., Callaway, R.M., Cipollini, D. & Newell, E. (2009) A cross-continental test of the Enemy Release Hypothesis: leaf herbivory on *Acer platanoides* (L.) is three times lower in North America than in its native Europe. *Biological Invasions*, **11**, 1005–1016.

Adema, E.B. & Grootjans, A.P. (2003) Possible positive-feedback mechanisms: plants change abiotic soil parameters in wet calcareous dune slacks. *Plant Ecology*, **167**, 141–149.

Adema, E.B., Grootjans, A.P., Petersen, J. & Grijpstra, J. (2002) Alternative stable states in a wet calcareous dune slack in the Netherlands. *Journal of Vegetation Science*, **13**, 107–114.

Adema, E.B., Van de Koppel, J., Meijer, H.A.J. & Grootjans, A.P. (2005) Enhanced nitrogen loss may explain alternative stable states in dune slack succession. *Oikos*, **109**, 374–386.

Adler, P.B., Lambers, J.H.R., Kyriakidis, P.C., Guan, Q. & Levine, J.M. (2006) Climate variability has a stabilizing effect on the coexistence of prairie grasses. *Proceedings of the National Academy of Sciences of the U.S.A.*, **103**, 12793–12798.

Adler, P.B., Lambers, J.H.R. & Levine, J.M. (2009) Weak effect of climate variability on coexistence in a sagebrush steppe community. *Ecology*, **90**, 3303–3312.

Agnew, A.D.Q., Rapson, G., Sykes, M.T. & Wilson, J.B. (1993b) The functional ecology of *Empodisma minus* (Hook.f.) Johnson & Cutler in New Zealand ombrotrophic mires. *New Phytologist*, **124**, 703–710.

Agnew, A.D.Q., Wilson, J.B. & Sykes, M.T. (1993a) A vegetation switch as the cause of a forest/mire ecotone in New Zealand. *Journal of Vegetation Science*, **4**, 273–278.

Agrawal, A.A. (2004) Resistance and susceptibility of milkweed: competition, root herbivory, and plant genetic variation. *Ecology*, **85**, 2118–2133.

Agrawal, A.A., Kotanen, P.M., Mitchell, C.E., Power, A.G., Godsoe, W. & Klironomos, J. (2005) Enemy release? An experiment with congeneric plant pairs and diverse above- and below-ground enemies. *Ecology*, **86**, 2979–2989.

Agrawal, A.A., Lau, J.A. & Hambäck, P.A. (2006) Community heterogeneity and the evolution of interactions between plants and insect herbivores. *Quarterly Review of Biology*, **81**, 349–376.

Aide, T.M. (1987) Limbfalls: a major cause of sapling mortality for tropical forest plants. *Biotropica*, **19**, 284–285.

Aizen, M.A. & Vazquez, D.P. (2006) Flowering phenologies of hummingbird plants from the temperate forest of southern South America: is there evidence of competitive displacement? *Ecography*, **29**, 357–366.

Akashi, N., Kohyama, T. & Matsui, K. (2003) Lateral and vertical crown associations in mixed forests. *Ecological Research*, **18**, 455–461.

Alexander, H.D. & Dunton, K.H. (2002) Freshwater inundation effects on emergent vegetation of a hypersaline salt marsh. *Estuaries*, **25**, 1426–1435.

Allan, E., van Ruijven, J. & Crawley, M.J. (2010) Foliar fungal pathogens and grassland biodiversity. *Ecology*, **91**, 2572–2582.

Allan, E., Weisser, W., Weigelt, A., Roscher, C., Fischer, M. & Hillebrand, H. (2011) More diverse plant communities have higher functioning over time due to turnover in complementary dominant species. *Proceedings of the National Academy of Sciences of the U.S.A.*, **108**, 17034–17039.

Allen, B.P., Pauley, E.F. & Sharitz, R.R. (1997) Hurricane impacts on liana populations in an old growth southeastern bottomland forest. *Journal of the Torrey Botanical Society*, **124**, 34–42.

Allen, E.A. (1991) Temporal and spatial organization of desert plant communities. In *Semiarid Lands and Deserts: Soil Resource and Reclamation* (ed. J. Skujins), pp. 193–208. Dekker, New York.

Allen, E.B. (1988) Some trajectories of succession in Wyoming sagebrush grassland: implications for restoration. In *The Reconstruction of Disturbed Arid Lands: An Ecological Approach* (ed. E.B. Allen), pp. 89–112. American Association for the Advancement of Science, Washington.

Allen, E.B. & Forman, R.T.T. (1976) Plant species removals and old-field community structure and stability. *Ecology*, **57**, 1233–1243.

Allen, R.B. & Peet, R.K. (1990) Gradient analysis of forests of the Sangre de Cristo Range, Colorado. *Canadian Journal of Botany*, **68**, 193–201.

Allen, R.B., Wilson, J.B. & Mason, C.R. (1995) Vegetation change following exclusion of grazing animals in depleted grassland, Central Otago, New Zealand. *Journal of Vegetation Science*, **6**, 615–626.

Alriksson, A. & Eriksson, H.M. (1998) Variations in mineral nutrient and C distribution in the soil and vegetation compartments of five temperate tree species in NE Sweden. *Forest Ecology and Management*, **108**, 261–273.

Andersen, D.C. (1987) Below-ground herbivory in natural communities: a review emphasising fossorial animals. *Quarterly Review of Biology*, **62**, 261–286.

Andersen, U.V. (1995) Resistance of Danish coastal vegetation types to human trampling. *Biological Conservation*, **71**, 223–230.

Anderson, R.L., Foster, D.R. & Motzkin, G. (2003) Integrating lateral expansion into models of peatland development in temperate New England. *Journal of Ecology*, **91**, 68–76.

Angers, D.A. & Caron, J. (1998) Plant-induced changes in soil structure: processes and feedbacks. *Biogeochemistry*, **42**, 55–72.

Angert, A.L., Huxman, T.E., Chesson, P. & Venable, L. (2009) Functional tradeoffs determine species coexistence via the storage effect. *Proceedings of the National Academy of Sciences of the U.S.A.*, **106**, 11641–11645.

Appanah, S. & Putz, F.E. (1984) Climber abundance in virgin dipterocarp forest and the effect of pre-felling climber cutting on logging damage. *Malaysian Forester*, **47**, 335–342.

Arceo-Gómez, G. & Ashman, T.L. (2014) Heterospecific pollen receipt affects self pollen more than outcross pollen: implications for mixed-mating plants. *Ecology*, **95**, 2946–2952.

Archer, S., Scifres, C., Bassham, C.R. & Maggio, R. (1988) Autogenic succession in a subtropical savanna: conversion of grassland to thorn woodland. *Ecological Monographs*, **58**, 111–127.

Arianoutsou, M. (1989) Timing of litter production in a maquis ecosystem of North-East Greece. *Acta Oecologica Oecologia Plantarum*, **10**, 371–378.

Arizaga, S. & Ezcurra, E. (2002) Propagation mechanisms in *Agave macroacantha* (Agavaceae), a tropical arid-land succulent rosette. *American Journal of Botany*, **89**, 632–641.

Armbruster, W.S. (1986) Reproductive interactions between sympatric Dalechampia species: are natural assemblages "random" or organized? *Ecology*, **67**, 522–533.

Armbruster, W.S. (1995) The origins and detection of plant community structure: reproductive versus vegetative processes. *Folia Geobotanica et Phytotaxonomica*, **30**, 483–497.

Armbruster, W.S., Edwards, M.E. & Debevec, E.M. (1994) Floral character displacement generates assemblage structure of Western Australian triggerplants (Stylidium). *Ecology*, **75**, 315–329.

Armentrout, S.M. & Pieper, R.D. (1988) Plant distribution surrounding Rocky Mountain pinyon pine and one-seed juniper in south-central New Mexico. *Journal of Range Management*, **41**, 139–143.

Ashton, P.S., Givnish, T.J. & Appanah, S. (1988) Staggered flowering in the Dipterocarpaceae: new insights into floral induction and the evolution of mast fruiting in the aseasonal tropics. *American Naturalist*, **132**, 44–66.

Atsatt, P.R. & O'Dowd, D.J. (1976) Plant defense guilds. *Science*, **193**, 24–29.

Attiwill, P.M. (1994) The disturbance of forest ecosystems: the ecological basis for conservative management. *Forest Ecology and Management*, **63**, 247–300.

Attiwill, P.M. & Wilson, B. (2003) *Ecology: An Australian Perspective*. Oxford University Press, Melbourne.

Auerbach, M. & Shmida, A. (1993) Vegetation change along an altitudinal gradient on Mt Hermon, Israel – no evidence for discrete communities. *Journal of Ecology*, **81**, 25–33.

Augustine, D.J., Frelich, L.E. & Jordan, P.A. (1998) Evidence for two alternate stable states in an ungulate grazing system. *Ecological Applications*, **8**, 1260–1269.

Augusto, L., Ranger, J., Binkley, D. & Rothe, A. (2002) Impact of several common tree species of European temperate forests on soil fertility. *Annals of Forest Science*, **59**, 233–253.

Austin, M.P. (1982) The use of a relative physiological performance value in the prediction of performance in multispecies mixtures from monoculture performance. *Journal of Ecology*, **70**, 559–570.

Austin, M.P. (1985) Continuum concept, ordination methods, and niche theory. *Annual Review of Ecology and Systematics*, **16**, 39–61.

Austin, M.P. (2012) Vegetation and environment: discontinuities and continuities. In *Vegetation Ecology* (eds E. van der Maarel & J. Franklin), pp. 71–106. Wiley-Blackwell, Malden, MA.

Austin, M.P. & Gaywood, M.J. (1994) Current problems of environmental gradients and species response curves in relation to continuum theory. *Journal of Vegetation Science*, **5**, 473–482.

Austin, M.P., Nicholls, A.O., Doherty, M.D. & Meyers, J.A. (1994) Determining species response functions to an environmental gradient by means of a beta-function. *Journal of Vegetation Science*, **5**, 215–228.

Austin, M.P. & Smith, T.M. (1989) A new model for the continuum concept. *Vegetatio*, **83**, 35–47.

Awasthi, O.P., Sharma, E. & Palni, L.M.S. (1995) Stemflow: a source of nutrients in some naturally growing epiphytic orchids of the Sikkim Himalaya. *Annals of Botany*, **75**, 5–11.

Babikova, Z., Gilbert, L., Bruce, T.J.A., Birkett, M., Caulfield, J.C., Woodcock, C., Pickett, J.A. & Johnson, D. (2013) Underground signals carried through common mycelial networks warn neighbouring plants of aphid attack. *Ecology Letters*, **16**, 835–843.

Badano, E.I., Bustamante, R.O., Villarroel, E., Marquet, P.A. & Cavieres, L.A. (2015) Facilitation by nurse plants regulates community invasibility in harsh environments. *Journal of Vegetation Science*, **26**, 756–767.

Bais, H.P., Weir, T.L., Perry, L.G., Gilroy, S. & Vivanco, J.M. (2006) The role of root exudates in rhizosphere interactions with plants and other organisms. *Annual Review of Plant Biology*, **57**, 233–266.

Baker, P.J., Bunyavejchewin, S., Oliver, C.D. & Ashton, P.S. (2005) Disturbance history and historical stand dynamics of a seasonal tropical forest in western Thailand. *Ecological Monographs*, **75**, 317–343.

Bakker, E.S., Olff, H., Vandenberghe, C., de Maeyer, K., Smit, R., Gleichman, J.M. & Vera, F.W.M. (2004) Ecological anachronisms in the recruitment of temperate light-demanding tree species in wooded pastures. *Journal of Applied Ecology*, **41**, 571–582.

Ball, M.C., Egerton, J., Lutze, J.L., Gutschick, V.P. & Cunningham, R.B. (2002) Mechanisms of competition: thermal inhibition of tree seedling growth by grass. *Oecologia*, **133**, 120–130.

Ballaré, C.L., Scopel, A.L., Roush, M.L. & Radosevich, S.R. (1995) How plants find light in patchy canopies. A comparison between wild-type and phytochrome-B-deficient mutant plants of cucumber. *Functional Ecology*, **9**, 859–868.

Balslev, H., Valencia, R., Paz y Mino, G., Christensen, H. & Nielsen, I. (1998) Species count of vascular plants in one hectare of humid lowland forest in Amazonian Ecuador In *Forest Biodiversity in North, Central and South America, and the Caribbean* (eds F. Dallmeier & J.A. Comiskey), pp. 585–594. UNESCO, Paris.

Balyan, R.S., Malik, R.K., Panwar, R.S. & Singh, S. (1991) Competitive ability of winter wheat cultivars with wild oat (*Avena ludoviciana*). *Weed Science*, **39**, 154–158.

Baraza, E., Zamora, R. & Hodar, J.A. (2006) Conditional outcomes in plant-herbivore interactions: neighbours matter. *Oikos*, **113**, 148–156.

Barbour, M.G., Burk, J.H., Pitts, W.D., Gilliam, F.S. & Schwartz, M.W. (1999) *Terrestrial Plant Ecology*, 3rd edn. Benjamin/Cummings, Menlo Park CA, USA.

Bardgett, R.D., Smith, R.S., Shiel, R.S., Peacock, S., Simkin, J.M., Quirk, H. & Hobbs, P.J. (2006) Parasitic plants indirectly regulate below-ground properties in grassland ecosystems. *Nature*, **439**, 969–972.

Barkman, J.J. (1979) The investigation of vegetation texture and structure. In *The Study of Vegetation* (ed. M.J.A. Werger), pp. 123–160. Junk, The Hague.

Barkman, J.J., Masselink, A.K. & de Vries, B.W.L. (1977) Uber das microclimat in wacholderfluren. In *Vegetation und Klima* (ed. H. Dieschke), pp. 123–160. Junk, The Hague.

Barnes, B.V. (1966) The clonal growth habit of American aspens. *Ecology*, **47**, 439–447.

Barot, S. & Gignoux, J. (2004) Mechanisms promoting plant coexistence: can all the proposed processes be reconciled? *Oikos*, **106**, 185–192.

Barreiro, R., Guiamét, J.J., Beltrano, J. & Montaldi, E.R. (1992) Regulation of the photosynthetic capacity of primary bean leaves by the red-far-red ratio and photosynthetic photon flux density of incident light. *Physiologia Plantarum*, **85**, 97–101.

Barrington, D.S. & Paris, C.A. (2007) Refugia and migration in the quaternary history of the New England flora. *Rhodora*, **109**, 369–386.

Barros, C., Thuiller, W., Georges, D., Boulangeat, I. & Münkemüller, T. (2016) N-dimensional hypervolumes to study stability of complex ecosystems. *Ecology Letters*, **19**, 729–742.

Bartha, S., Czárán, T. & Oborny, B. (1995) Spatial constraints masking community assembly rules: a simulation study. *Folia Geobotanica et Phytotaxonomica*, **30**, 471–482.

Bartha, S., Meiners, S.J., Pickett, S.T.A. & Cadenasso, M.L. (2003) Plant colonization windows in a mesic old field succession. *Applied Vegetation Science*, **6**, 205–212.

Barto, E.K., Powell, J.R. & Cipollini, D. (2010) How novel are the chemical weapons of garlic mustard in North American forest understories? *Biological Invasions*, **12**, 3465–3471.

Batanouny, K.H. (1981) *Ecology and Flora of Qatar*. Alden, Oxford.

Batlla, D., Kruk, C. & Benech-Arnold, R.L. (2000) Very early detection of canopy presence by seeds through perception of subtle modifications in red:far red signals. *Functional Ecology*, **14**, 195–202.

Baustian, J.J., Mendelssohn, I.A. & Hester, M.W. (2012) Vegetation's importance in regulating surface elevation in a coastal salt marsh facing elevated rates of sea level rise. *Global Change Biology*, **18**, 3377–3382.

Bazzaz, F.A. (1983) Characteristics of populations in natural and man-modified ecosystems. In *Disturbance and Ecosystems: Components of Response* (eds H.A. Mooney & M. Godron), pp. 259–275. Springer, Berlin.

Bazzaz, F.A. (1987) Experimental studies on the evolution of niche in successional plant populations. In *Colonization, Succession and Stability* (ed. A.J. Gray), pp. 245–272. Blackwell, Oxford.

Bazzaz, F.A. & McConnaughay, K.D.M. (1992) Plant-plant interactions in elevated CO_2 environments. *Australian Journal of Botany*, **40**, 547–563.

Beans, C.M. (2014) The case for character displacement in plants. *Ecology and Evolution*, **4**, 862–875.

Beaton, L.L., VanZandt, P.A., Esselman, E.J. & Knight, T.M. (2011) Comparison of the herbivore defense and competitive ability of ancestral and modern genotypes of an invasive plant, *Lespedeza cuneata*. *Oikos*, **120**, 1413–1419.

Beckage, B., Kloeppel, B.D., Yeakley, J., Taylor, S.F. & Coleman, D.C. (2008) Differential effects of understory and overstory gaps on tree regeneration. *Journal of the Torrey Botanical Society*, **135**, 1–11.

Becks, L., Ellner, S.P., Jones, L.E. & Hairston, N.G. (2010) Reduction of adaptive genetic diversity radically alters eco-evolutionary community dynamics. *Ecology Letters*, **13**, 989–997.

Begon, M., Harper, J.L. & Townsend, J.R. (1996) *Ecology: Individuals, Populations and Communities*, 3rd edn. Blackwell, Oxford.

Belleau, A., Leduc, A., Lecomte, N. & Bergeron, Y. (2011) Forest succession rate and pathways on different surface deposit types in the boreal forest of northwestern Quebec. *Ecoscience*, **18**, 329–340.

Belsky, A.J. (1994) Influences of trees on savanna productivity: tests of shade, nutrients, and tree-grass competition. *Ecology*, **75**, 922–932.

Belsky, A.J., Mwonga, S.M., Amundson, R.G., Duxbury, J.M. & Ali, A.R. (1993) Comparative effects of isolated trees on their undercanopy environments in high-rainfall and low-rainfall savannas. *Journal of Applied Ecology*, **30**, 143–155.

Belyea, L.R. & Clymo, R.S. (1998) Do hollows control the rate of peat bog growth? In *Patterned Mires and Mire Pools* (eds V. Standen, J.H. Tallis & R. Meade), pp. 55–65. Mires Research Group, BES, Durham, UK.

Bengtsson, J., Fagerstrom, T. & Rydin, H. (1994) Competition and coexistence in plant communities. *Trends in Ecology and Evolution*, **9**, 246–250.

Bennett, J.A. & Cahill, J.F. (2013) Conservatism of responses to environmental change is rare under natural conditions in a native grassland. *Perspectives in Plant Ecology Evolution and Systematics*, **15**, 328–337.

Bennett, J.A. & Pärtel, M. (2017) Predicting species establishment using absent species and functional neighborhoods. *Ecology and Evolution*, **7**, 2223–2237.

Bennington, J.B. & Bambach, R.K. (1996) Statistical testing for paleocommunity recurrence: are similar fossil assemblages ever the same? *Palaeoecology*, **127**, 107–133.

Benzing, D.H. (2004) Vascular epiphytes. In *Forest Canopies* (eds M. Lowman & H.B. Rinker), pp. 175–211. Elsevier Academic, Amsterdam, NL.

Berendse, F. (1994) Litter decomposability – a neglected component of plant fitness. *Journal of Ecology*, **82**, 187–190.

Berendse, F. (1999) Implications of increased litter production for plant biodiversity. *Trends in Ecology and Evolution*, **14**, 4–5.

Berendse, F. & Aerts, R. (1984) Competition between *Erica tetralix* L. and *Molinia caerulea* (L.) Moench as affected by the availability of nutrients. *Acta Oecologica Oecologia Plantarum*, **5**, 3–14.

Berendse, F., Oudhof, H. & Bol, J. (1987) A comparative study on nutrient cycling in wet heathland ecosystems. I. Litter production and nutrient losses from the plant. *Oecologia*, **74**, 174–184.

Berendse, F., Schmitz, M. & de Visser, W. (1994) Experimental manipulation of succession in heathland ecosystems. *Oecologia*, **100**, 38–44.

Berg, S.S. & Dunkerley, D.L. (2004) Patterned mulga near Alice Springs, Central Australia, and the potential threat of firewood collection in this vegetation. *Journal of Arid Environments*, **59**, 313–350.

Berg-Binder, M.C. & Suarez, A.V. (2012) Testing the directed dispersal hypothesis: are native ant mounds (Formica sp.) favorable microhabitats for an invasive plant? *Oecologia*, **169**, 763–772.

Bergelson, J. (1990) Life after death: site pre-emption by the remains of Poa annua. *Ecology*, **71**, 2157–2165.

Bergengren, J.C., Thompson, S.L., Pollard, D. & DeConto, R.M. (2001) Modeling global climate-vegetation interactions in a doubled CO_2 world. *Climatic Change*, **50**, 31–75.

Berkley, H.A., Kendall, B.E., Mitarai, S. & Siegel, D.A. (2010) Turbulent dispersal promotes species co-existence. *Ecology Letters*, **13**, 360–371.

Bernard-Verdier, M., Navas, M.L., Vellend, M., Violle, C., Fayolle, A. & Garnier, E. (2012) Community assembly along a soil depth gradient: contrasting patterns of plant trait convergence and divergence in a Mediterranean rangeland. *Journal of Ecology*, **100**, 1422–1433.

Berntson, G.M. & Wayne, P.M. (2000) Characterizing the size dependence of resource acquisition within crowded plant populations. *Ecology*, **81**, 1072–1085.

Bertness, M.D. & Callaway, R. (1994) Positive interactions in communities. *Trends in Ecology and Evolution*, **9**, 191–193.

Bertness, M.D. & Hacker, S.D. (1994) Physical stress and positive associations among marsh plants. *American Naturalist*, **144**, 363–372.

Bertness, M.D. & Yeh, S.M. (1994) Cooperative and competitive interactions in the recruitment of marsh elders. *Ecology*, **75**, 2416–2429.

Bever, J.D. (2003) Soil community feedback and the coexistence of competitors: conceptual frameworks and empirical tests. *New Phytologist*, **157**, 465–473.

Bezemer, T.M., Fountain, M.T., Barea, J.M., Christensen, S., Dekker, S.C., Duyts, H., van Hal, R., Harvey, J.A., Hedlund, K., Maraun, M., Mikola, J., Mladenov, A.G., Robin, C., de Ruiter, P.C., Scheu, S., Setälä, H., Šmilauer, P. & van der Putten, W.H. (2010) Divergent composition but similar function of soil food webs of individual plants: plant species and community effects. *Ecology*, **91**, 3027–3036.

Bhatt, M.V., Khandelwal, A. & Dudley, S.A. (2011) Kin recognition, not competitive interactions, predicts root allocation in young *Cakile edentula* seedling pairs. *New Phytologist*, **189**, 1135–1142.

Biddington, N.L. & Dearman, A.S. (1985) The effects of mechanically-induced stress on the growth of cauliflower, lettuce and celery seedlings. *Annals of Botany*, **55**, 109–119.

Biedrzycki, M.L. & Bais, H.P. (2010) Kin recognition in plants: a mysterious behaviour unsolved. *Journal of Experimental Botany*, **61**, 4123–4128.

Binkley, D. & Valentine, D. (1991) 50-year biogeochemical effects of green ash, white-pine, and Norway spruce in a replicated experiment. *Forest Ecology and Management*, **40**, 13–25.

Bio, A.M.F. (2000) Does vegetation suit our models? Data and model assumptions and the assessment of species distribution in space. PhD thesis, Utrecht University.

Birkett, M.A., Campbell, C.A., Chamberlain, K., Guerrieri, E., Hick, A.J., Martin, J.L., Matthes, M., Napier, J.A., Pettersson, J., Pickett, J.A., Poppy, G.M., Pow, E.M., Pye, B.J., Smart, L.E., Wadhams, G.H., Wadhams, L.J. & Woodcock, C.M. (2000) New roles for cis-jasmone as an insect semiochemical and in plant defense. *Proceedings of the National Academy of Sciences of the U.S.A.*, **97**, 9329–9334.

Birks, H.J.B. (1993) Quaternary palaeoecology and vegetation science: current contributions and possible future developments. *Review of Palaeobotany and Palynology*, **79**, 153–177.

Birouste, M., Kazakou, E., Blanchard, A. & Roumet, C. (2012) Plant traits and decomposition: are the relationships for roots comparable to those for leaves? *Annals of Botany*, **109**, 463–472.

Bittebiere, A.K., Renaud, N., Clement, B. & Mony, C. (2012) Morphological response to competition for light in the clonal *Trifolium repens* (Fabaceae). *American Journal of Botany*, **99**, 646–654.

Black, J.N. (1958) Competition between plants of different initial seed sizes in swards of subterranean clover (*Trifolium subterraneum* L.) with particular reference to leaf area and the light microclimate. *Australian Journal of Agricultural Research*, **9**, 299–318.

Blair, A.C. & Wolfe, L.M. (2004) The evolution of an invasive plant: an experimental study with *Silene latifolia*. *Ecology*, **85**, 3035–3042.

Blair, B. (2001) Effect of soil nutrient heterogeneity on the symmetry of belowground competition. *Plant Ecology*, **156**, 199–203.

Blindow, I., Andersson, G., Hargeby, A. & Johansson, S. (1993) Long-term pattern of alternative stable states in two shallow eutrophic lakes. *Freshwater Biology*, **30**, 159–167.

Blindow, I., Hargeby, A. & Andersson, G. (2002) Seasonal changes of mechanisms maintaining clear water in a shallow lake with abundant Chara vegetation. *Aquatic Botany*, **72**, 315–334.

Blondel, J. (2003) Guilds or functional groups: does it matter? *Oikos*, **100**, 223–231.

Bobiwash, K., Schultz, S.T. & Schoen, D.J. (2013) Somatic deleterious mutation rate in a woody plant: estimation from phenotypic data. *Heredity*, **111**, 338–344.

Bode, M., Bode, L. & Armsworth, P.R. (2011) Different dispersal abilities allow reef fish to co-exist. *Proceedings of the National Academy of Sciences of the U.S.A.*, **108**, 16317–16321.

Boege, K. (2004) Induced responses in three tropical dry forest plant species – direct and indirect effects on herbivory. *Oikos*, **107**, 541–548.

Bogaard, A., Hodgson, J.G., Wilson, P.J. & Band, S.R. (1998) An index of weed size for assessing the soil productivity of ancient crop fields. *Vegetation History and Archaeobotany*, **7**, 17–22.

Bolker, B.M. & Pacala, S.W. (1999) Spatial moment equations for plant competition: understanding spatial strategies and the advantages of short dispersal. *American Naturalist*, **153**, 575–602.

Bonan, G.B., Pollard, D. & Thompson, S.L. (1992) Effects of boreal forest vegetation on global climate. *Nature*, **359**, 716–718.

Bond, W.J. (1993) Keystone species. In *Ecosystem Functioning and Biodiversity* (eds E.-D. Schulze & H.A. Mooney), pp. 237–253. Springer, Berlin.

Booth, B.D. & Larson, D.W. (1999) Impact of language, history and choice of system on the study of assembly rules. In *Ecological Assembly Rules: Perspectives, Advances, Retreats* (eds E. Weiher & P.A. Keddy), pp. 206–232. Cambridge University Press, Cambridge, UK.

Booth, M.G. (2004) Mycorrhizal networks mediate overstorey-understorey competition in a temperate forest. *Ecology Letters*, **7**, 538–546.

Borrelli, J.J. (2015) Selection against instability: stable subgraphs are most frequent in empirical food webs. *Oikos*, **124**, 1583–1588.

Bossdorf, O., Auge, H., Lafuma, L., Rogers, W.E., Siemann, E. & Prati, D. (2005) Phenotypic and genetic differentiation between native and introduced plant populations. *Oecologia*, **144**, 1–11.

Bossuyt, B., Honnay, O. & Hermy, M. (2005) Evidence for community assembly constraints during succession in dune slack plant communities. *Plant Ecology*, **178**, 201–209.

Bosy, J.L. & Reader, R.J. (1995) Mechanisms underlying the suppression of forb seedling emergence by grass (*Poa pratensis*) litter. *Functional Ecology*, **9**, 635–639.

Boyce, C.K., Lee, J.E., Feild, T.S., Brodribb, T.J. & Zwieniecki, M.A. (2010) Angiosperms helped put the rain in the rainforests: the impact of plant physiological evolution on tropical biodiversity. *Annals of the Missouri Botanical Garden*, **97**, 527–540.

Boyd, R. & Jaffré, T. (2001) Phytoenrichment of soil Ni content by *Sebertia acuminata* in New Caledonia and the concept of elemental allelopathy. *South African Journal of Science*, **97**, 535–538.

Braam, J. & Davis, R. (1990) Rain-, wind-, and touch-induced expression of calmodulin and calmodulin-related genes in Arabidopsis. *Cell*, **60**, 357–364.

Bracken, M.E.S. & Low, N.H.N. (2012) Realistic losses of rare species disproportionately impact higher trophic levels. *Ecology Letters*, **15**, 461–467.

Bradshaw, A.D. (1965) Evolutionary significance of phenotypic plasticity in plants. *Advances in Genetics*, **13**, 115–155.

Braun-Blanquet, J. (1932) *Plant Sociology: The Study of Plant Communities*. McGraw-Hill, New York.

Bray, R.H. (1954) A nutrient mobility concept of soil-plant relationships. *Soil Science*, **78**, 9–22.

Brenchley, W.E. (1916) The effect of the concentration of the nutrient solution on the growth of barley and wheat in water cultures. *Annals of Botany*, **30**, 77–90.

Brittingham, S. & Walker, L.R. (2000) Facilitation of *Yucca brevifolia* recruitment by Mojave Desert shrubs. *Western North American Naturalist*, **60**, 374–383.

Brokaw, N. & Busing, R.T. (2000) Niche versus chance and tree diversity in forest gaps. *Trends in Ecology and Evolution*, **15**, 183–188.

Brooker, R.W., Scott, D., Palmer, S.C.F. & Swaine, E. (2006) Transient facilitative effects of heather on Scots pine along a grazing disturbance gradient in Scottish moorland. *Journal of Ecology*, **94**, 637–645.

Brooks, M.L., D'Antonio, C.M., Richardson, D.M., Grace, J.B., Keeley, J.E., DiTomaso, J.M., Hobbs, R.J., Pellant, M. & Pyke, D. (2004) Effects of invasive alien plants on fire regimes. *BioScience*, **54**, 677–688.

Brown, C., Law, R., Illian, J.B. & Burslem, D.F.R.P. (2011) Linking ecological processes with spatial and non-spatial patterns in plant communities. *Journal of Ecology*, **99**, 1402–1414.

Brown, P.M. & Sieg, C.H. (1999) Historical variability in fire at the ponderosa pine – Northern Great Plains prairie ecotone, southeastern Black Hills, South Dakota. *Ecoscience*, **6**, 539–547.

Brown, V.K. (1993) Herbivory: a structuring force in plant communities. In *Individuals, Populations and Patterns in Ecology* (eds S.R. Leather, K. Walters, N. Mills & A. Watt), pp. 299–308. Intercept, Andover.

Brown, V.K. & Gange, A.C. (1989a) Herbivory by soil-dwelling insects depresses plant species richness. *Functional Ecology*, **3**, 667–671.

Brown, V.K. & Gange, A.C. (1989b) Differential effects of above- and below-ground insect herbivory during early plant succession. *Oikos*, **54**, 67–76.

Brownstein, G., Steel, J.B., Porter, S., Gray, A., Wilson, C., Wilson, P.G. & Wilson, J.B. (2012) Chance in plant communities: a new approach to its measurement using the nugget from spatial autocorrelation. *Journal of Ecology*, **100**, 987–996.

Buchmann, N., Kao, W.-Y. & Ehleringer, J.R. (1996) Carbon dioxide concentrations within forest canopies – variation with time, stand structure, and vegetation type. *Global Change Biology*, **2**, 421–432.

Buck-Sorlin, G.H. & Bell, A.D. (1998) A quantification of shoot shedding in the pedunculate oak (*Quercus robur* L.). *Botanical Journal of the Linnean Society*, **127**, 371–391.

Buck-Sorlin, G.H. & Bell, A.D. (2000) Crown architecture in *Quercus petraea* and *Q. robur*: the fate of buds and shoots in relation to age, position and environmental perturbation. *Forestry*, **73**, 331–349.

Bullock, J.M., Joe Franklin, J., Stevenson, M.J., Silvertown, J., Coulson, S.J., Gregory, S.J. & Richard Tofts, R. (2001) A plant trait analysis of responses to grazing in a long-term experiment. *Journal of Applied Ecology*, **38**, 253–267.

Burdon, J.J., Wennström, A., Ericson, L., Müller, W.J. & Morton, R. (1992) Density-dependent mortality in *Pinus sylvestris* caused by the snow blight pathogen *Phacidium infestans*. *Oecologia*, **90**, 74–79.

Burke, D.J. (2012) Shared mycorrhizal networks of forest herbs: does the presence of conspecific and heterospecific adult plants affect seedling growth and nutrient acquisition? *Botany*, **90**, 1048–1057.

Burke, M.J.W. & Grime, J.P. (1996) An experimental study of plant community invasibility. *Ecology*, **77**, 776–790.

Burns, K.C. (2005) Is there limiting similarity in the phenology of fleshy fruits? *Journal of Vegetation Science*, **16**, 617–624.

Busing, R.T. (1996) Estimation of tree replacement patterns in an Appalachian Picea-Abies forest. *Journal of Vegetation Science*, **7**, 685–694.

Buss, L.W. & Jackson, J.B.C. (1979) Competitive networks: nontransitive competitive relationships in cryptic coral reef environments. *American Naturalist*, **113**, 223–234.

Bycroft, C.M., Nicolaou, N., Smith, B. & Wilson, J.B. (1993) Community structure (niche limitation and guild proportionality) in relation to the effect of spatial scale, in a Nothofagus forest sampled with a circular transect. *New Zealand Journal of Ecology*, **17**, 95–101.

Caccianiga, M., Luzzaro, A., Pierce, S., Ceriani, R.M. & Cerabolini, B. (2006) The functional basis of a primary succession resolved by C-S-R classification. *Oikos*, **112**, 10–20.

Cadisch, G. & Giller, K.E. (1997) *Driven by Nature: Plant Litter Quality and Decomposition.* CAB International, Cambridge.

Cadotte, M.W., Dinnage, R. & Tilman, D. (2012) Phylogenetic diversity promotes ecosystem stability. *Ecology*, **93**, 223–233.

Cahill, J.F. (1999) Fertilization effects on interactions between above- and belowground competition in an old field. *Ecology*, **80**, 466–480.

Cairney, J.W.G. & Ashford, A.E. (2002) Biology of mycorrhizal associations of epacrids (Ericaceae). *New Phytologist*, **154**, 305–326.

Calcagno, V., Sun, C., Schmitz, O.J. & Loreau, M. (2011) Keystone predation and plant species coexistence: the role of carnivore hunting mode. *American Naturalist*, **177**, 1–13.

Caldwell, M.M., Dawson, T.E. & Richards, J.H. (1998) Hydraulic lift: consequences of water efflux from the roots of plants. *Oecologia*, **113**, 151–161.

Callaway, R.M. (1995) Positive interactions among plants. *Botanical Review*, **61**, 306–349.

Callaway, R.M. (1998) Are positive interactions species-specific? *Oikos*, **82**, 202–207.

Callaway, R.M. & Aschehoug, E.T. (2000) Invasive plants versus their new and old neighbors: a mechanism for exotic invasion. *Science*, **290**, 521–523.

Callaway, R.M., Brooker, R.W., Choler, P., Kikvidze, Z., Lortie, C.J., Michalet, R., Paolini, L., Pugnaire, F.I., Newingham, B., Aschehoug, E.T., Armas, C., Kikodze, D. & Cook, B.J. (2002) Positive interactions among alpine plants increase with stress. *Nature*, **417**, 844–848.

Callaway, R.M., Ridenour, W.M., Laboski, T., Weir, T. & Vivanco, J.M. (2005) Natural selection for resistance to the allelopathic effects of invasive plants. *Journal of Ecology*, **93**, 576–583.

Callaway, R.M., Thelen, G.C., Rodriguez, A. & Holben, W.E. (2004) Soil biota and exotic plant invasion. *Nature*, **427**, 731–733.

Campbell, B.D. & Grime, J.P. (1992) An experimental test of plant strategy theory. *Ecology*, **73**, 15–29.

Campbell, B.D., Grime, J.P. & Mackey, J.M.L. (1992) Shoot thrust and its role in plant competition. *Journal of Ecology*, **80**, 633–641.

Campbell, S.P., Witham, J.W. & Hunter, M.L. (2010) Stochasticity as an alternative to deterministic explanations for patterns of habitat use by birds. *Ecological Monographs*, **80**, 287–302.

Cape, J.N., Brown, A.H.F., Robertson, S.M.C., Howson, G. & Paterson, I.S. (1991) Interspecies comparisons of throughfall and stemflow at three sites in northern Britain. *Forest Ecology and Management*, **46**, 165–177.

Cappellato, R. & Peters, N.E. (1995) Dry deposition and canopy leaching rates in deciduous and coniferous forests of the Georgia Piedmont: an assessment of a regression model. *Journal of Hydrology*, **169**, 131–150.

Cappuccino, N. & Carpenter, D. (2005) Invasive exotic plants suffer less herbivory than non-invasive exotic plants. *Biology Letters*, **1**, 435–438.

Cardinale, B.J., Gross, K., Fritschie, K., Flombaum, P., Fox, J.W., Rixen, C., van Ruijven, J., Reich, P.B., Scherer-Lorenzen, M. & Wilsey, B.J. (2013) Biodiversity simultaneously enhances the production and stability of community biomass, but the effects are independent. *Ecology*, **94**, 1697–1707.

Cardinale, B.J., Wright, J.P., Cadotte, M.W., Carroll, I.T., Hector, A., Srivastava, D.S., Loreau, M. & Weis, J.J. (2007) Impacts of plant diversity on biomass production increase through time because of species complementarity. *Proceedings of the National Academy of Sciences of the U.S.A.*, **104**, 18123–18128.

Carey, P.D. & Watkinson, A.R. (1993) The dispersal and fates of seeds of the winter annual grass *Vulpia ciliata*. *Journal of Ecology*, **81**, 759–767.

Carey, P.D., Watkinson, A.R. & Gerard, F.F.O. (1995) The determinants of the distribution and abundance of the winter annual grass *Vulpia ciliata* ssp. *ambigua*. *Journal of Ecology*, **83**, 177–187.

Carlucci, M.B., Debastiani, V.J., Pillar, V.D. & Duarte, L.D.S. (2015) Between- and within-species trait variability and the assembly of sapling communities in forest patches. *Journal of Vegetation Science*, **26**, 21–31.

Caron, M.N., Kneeshaw, D.D., Degrandpre, L., Kauhanen, H. & Kuuluvainen, T. (2009) Canopy gap characteristics and disturbance dynamics in old- growth *Picea abies* stands in northern Fennoscandia: is the forest in quasi-equilibrium? *Annales Botanici Fennici*, **46**, 251–262.

Carroll, I.T., Cardinale, B.J. & Nisbet, R.M. (2011) Niche and fitness differences relate the maintenance of diversity to ecosystem function. *Ecology*, **92**, 1157–1165.

Carson, W.P. & Root, R.B. (2000) Herbivory and plant species coexistence: community regulation by an outbreaking phytophagous insect. *Ecological Monographs*, **70**, 73–99.

Carter, R.N. & Prince, S.D. (1985) The geographical distribution of prickly lettuce (*Lactuca serriola*). I. A general survey of its habitats and performance in Britain. *Journal of Ecology*, **73**, 27–38.

Casas, G., Scrosati, R. & Piriz, M.L. (2004) The invasive kelp *Undaria pinnatifida* (Phaeophyceae, Laminariales) reduces native seaweed diversity in Nuevo Gulf (Patagonia, Argentina). *Biological Invasions*, **6**, 411–416.

Case, T.J. (1991) Invasion resistance, species build-up and community collapse in metapopulation models with interspecific competition. *Biological Journal of the Linnean Society*, **42**, 239–266.

Cash, F.B.J., Conn, A., Coutts, S., Stephen, M., Mason, N.W.H. & Wilson, J.B. (2012) Assembly rules operate only in equilibrium communities: Is it true? *Austral Ecology*, **37**, 903–914.

Castellanos, A.E., Martinez, M.J., Llano, J.M., Halvorson, W.L., Espiricueta, M. & Espejel, I. (2005) Successional trends in Sonoran Desert abandoned agricultural fields in northern Mexico. *Journal of Arid Environments*, **60**, 437–455.

Caswell, H. (1976) Community structure: a neutral model analysis. *Ecological Monographs*, **46**, 327–354.

Cavender-Bares, J., Ackerly, D.D., Baum, D.A. & Bazzaz, F.A. (2004) Phylogenetic overdispersion in Floridian oak communities. *American Naturalist*, **163**, 823–843.

Cavieres, L.A., Badano, E.I., Sierra-Almeida, A. & Molina-Montenegro, M.A. (2007) Microclimatic modifications of cushion plants and their consequences for seedling survival of native and non-native herbaceous species in the high Andes of central Chile. *Arctic Antarctic and Alpine Research*, **39**, 229–236.

Cavieres, L.A., Quiroz, C.L. & Molina-Montenegro, M.A. (2008) Facilitation of the non-native Taraxacum officinale by native nurse cushion species in the high Andes of central Chile: are there differences between nurses? *Functional Ecology*, **22**, 148–156.

Certini, G. (2005) Effects of fire on properties of forest soils: a review. *Oecologia*, **143**, 1–10.

Challinor, D. (1968) Alteration of surface soil characteristics by four tree species. *Ecology*, **49**, 286–290.

Chambers, J.Q., Higuchi, N., Schimel, J.P., Ferreira, L.V. & Melack, J.M. (2000) Decomposition and carbon cycling of dead trees in tropical forests of the central Amazon. *Oecologia*, **122**, 380–388.

Chandrashekara, U.M. & Ramakrishnan, P.S. (1994) Vegetation and gap dynamics of a tropical wet evergreen forest in the Western Ghats of Kerala, India. *Journal of Tropical Ecology*, **10**, 337–354.

Chanway, C.P., Holl, F.B. & Turkington, R. (1989) Effect of *Rhizobium leguminosarum* biovar *trifolii* on specificity between *Trifolium repens* and *Lolium perenne*. *Journal of Ecology*, **77**, 1150–1160.

Chapin, F.S., Walker, L.R., Fastie, C.L. & Sharman, L.C. (1994) Mechanisms of primary succession following deglaciation at Glacier Bay, Alaska. *Ecological Monographs*, **64**, 149–175.

Chase, J.M. (2003) Community assembly: when should history matter? *Oecologia*, **136**, 489–498.

Chase, J.M. (2010) Stochastic community assembly causes higher biodiversity in more productive environments. *Science*, **328**, 1388–1391.

Chase, J.M. (2014) Spatial scale resolves the niche versus neutral theory debate. *Journal of Vegetation Science*, **25**, 319–322.

Chase, J.M. & Leibold, M.A. (2003) *Ecological Niches: Linking Classical and Contemporary Approaches*. Chicago University Press, Chicago, IL.

Chase, J.M. & Myers, J.A. (2011) Disentangling the importance of ecological niches from stochastic processes across scales. *Philosophical Transactions of the Royal Society of London B*, **366**, 2351–2363.

Chave, J. (2004) Neutral theory and community ecology. *Ecology Letters*, **7**, 241–253.

Chen, J., Franklin, J.F. & Spies, T.A. (1993) Contrasting microclimates among clearcut, edge, and interior of old-growth Douglas-fir forest. *Agricultural and Forest Meteorology*, **63**, 219–237.

Chen, Y., Wright, S.J., Muller-Landau, H.C., Hubbell, S.P., Wang, Y. & Yu, S. (2016) Positive effects of neighborhood complementarity on tree growth in a Neotropical forest. *Ecology*, **97**, 776–785.

Cheng, J.J., Mi, X.C., Nadrowski, K., Ren, H.B., Zhang, J.T. & Ma, K.P. (2012) Separating the effect of mechanisms shaping species-abundance distributions at multiple scales in a subtropical forest. *Oikos*, **121**, 236–244.

Chesson, P. & Huntly, N. (1997) The roles of harsh and fluctuating conditions in the dynamics of ecological communities. *American Naturalist*, **150**, 519–553.

Chesson, P.L. (1985) Coexistence of competitors in spatially and temporally varying environments: a look at the combined effects of different sorts of variability. *Theoretical Population Biology*, **28**, 263–287.

Chesson, P.L. (1990) Geometry, heterogeneity and competition in variable environments. *Philosophical Transactions of the Royal Society of London B*, **330**, 165–173.

Chesson, P.L. (2000a) Mechanisms of maintenance of species diversity. *Annual Review of Ecology and Systematics*, **31**, 343–366.

Chesson, P.L. (2000b) General theory of competitive coexistence in spatially varying environments. *Theoretical Population Biology*, **58**, 211–237.

Chesson, P.L. (2008) Quantifying and testing species coexistence mechanisms. In *Unity in Diversity: Reflections on Ecology after the Legacy of Ramon Margalef* (eds F. Valladares, A. Camacho, A. Elosegi, C. Gracia, M. Estrada, J.C. Senar & J.-P. Gili), pp. 119–164. Fundation Banco Bilbao-Vizcaya Argentaria, Bilbao.

Chesson, P.L., Donahue, M., Melbourne, B. & Sears, A. (2005) Scale transition theory for understanding mechanisms in metacommunities. In *Metacommunities: Spatial Dynamics and Ecological Communities* (eds M. Holyoak, M.A. Leibold & R.D. Holt), pp. 279–306. The University of Chicago Press, Chicago, IL.

Chiarucci, A., Mistral, M., Bonini, I., Anderson, B.J. & Wilson, J.B. (2002) Canopy occupancy: how much of the space in plant communities is filled? *Folia Geobotanica*, **37**, 333–338.

Chisholm, R.A. & Pacala, S.W. (2010) Niche and neutral models predict asymptotically equivalent species abundance distributions in high-diversity ecological communities. *Proceedings of the National Academy of Sciences of the U.S.A.*, **107**, 15821–15825.

Choi, J.H., Fushimi, K., Abe, N., Tanaka, H., Maeda, S., Morita, A., Hara, M., Motohashi, R., Matsunaga, J., Eguchi, Y., Ishigaki, N., Hashizume, D., Koshino, H. & Kawagishi, H. (2010) Disclosure of the "fairy" of fairy-ring-forming fungus *Lepista sordida*. *ChemBioChem*, **11**, 1373–1377.

Christensen, O. (1975) Wood litter fall in relation to abscission, environmental factors, and the decomposition cycle in a Danish oak forest. *Oikos*, **26**, 187–195.

Cipollini, D., Rigsby, C.M. & Barto, E.K. (2012) Microbes as targets and mediators of allelopathy in plants. *Journal of Chemical Ecology*, **38**, 714–727.

Clark, D.B. & Clark, D.A. (1991) The impact of physical damage on canopy tee regeneration in tropical rain forest. *Journal of Ecology*, **79**, 447–457.

Clark, J.S., Dietze, M., Chakraborty, S., Agarwal, P.K., Ibanez, I., LaDeau, S. & Wolosin, M. (2007) Resolving the biodiversity paradox. *Ecology Letters*, **10**, 647–659.

Clarke, P.J. (2002) Experiments on tree and shrub establishment in temperate grassy woodlands: Seedling survival. *Austral Ecology*, **27**, 606–615.

Clarkson, B.R., Schipper, L.A. & Silvester, W.B. (2009) Nutritional niche separation in coexisting bog species demonstrated by ^{15}N-enriched simulated rainfall. *Austral Ecology*, **34**, 377–385.

Clements, C.F., Warren, P.H., Collen, B., Blackburn, T., Worsfold, N. & Petchey, O. (2013) Interactions between assembly order and temperature can alter both short- and long-term community composition. *Ecology and Evolution*, **3**, 5201–5208.

Clements, F.E. (1904) Studies on the vegetation of the state. III. The development and structure of vegetation. *Reports of the Botanical Survey of Nebraska*, **7**, 1–175.

Clements, F.E. (1905) *Research Methods in Ecology*. University Publishing, Lincoln, Nebraska.

Clements, F.E. (1907) *Plant Physiology and Ecology*. Henry Holt, New York.

Clements, F.E. (1916) *Plant Succession: An Analysis of the Development of Vegetation*. Carnegie Institute of Washington, Washington.

Clements, F.E. (1920) *Plant Indicators: The Relation of Plant Communities to Process and Practice*. Carnegie Institution of Washington, Washington.

Clements, F.E. (1929) The phylogeny of climaxes. *Yearbook Carnegie Institute of Washington*, **28**, 202–203.

Clements, F.E. (1934) The relict method in dynamic ecology. *Journal of Ecology*, **22**, 39–68.

Clements, F.E. (1935) Experimental ecology in the public service. *Ecology*, **16**, 342–363.

Clements, F.E. (1936) Nature and structure of the climax. *Journal of Ecology*, **24**, 252–284.

Clements, F.E. & Goldsmith, G.W. (1924) *The Phytometer Method in Ecology*. Carnegie Institution of Washington, Washington.

Clements, F.E. & Shelford, V.E. (1939) *Bio-ecology*. Wiley, New York.

Clements, F.E., Weaver, J.E. & Hanson, H.C. (1929) *Plant Competition: An Analysis of Community Functions*. Carnegie Institute of Washington, Washington.

Clymo, R.S. & Hayward, P.M. (1982) The ecology of Sphagnum. In *Bryophyte Ecology* (ed. A.J.E. Smith), pp. 229–289. Chapman and Hall, London.

Cockayne, L. (1926) *Monograph on The New Zealand Beech Forests. I. The Ecology of the Forests and Taxonomy of the Beeches*. Government Printer, Wellington.

Cody, M.L. (1966) A general theory of clutch size. *Evolution*, **20**, 174–184.

Cody, M.L. (1974) *Competition and the Structure of Bird Communities*. Princeton University Press, Princeton, NJ.

Cody, M.L. (1986) Structural niches in plant communities. In *Community Ecology* (eds J.M. Diamond & T.J. Case), pp. 381–405. Harper and Row, New York.

Cody, M.L. (1989) Discussion: Structure and assembly of communities. In *Perspectives in Ecological Theory* (eds J. Roughgarden, R.M. May & S.A. Levin), pp. 227–241. Princeton University Press, Princeton, NJ.

Cody, M.L. & Prigge, B.A. (2003) Spatial and temporal variations in the timing of leaf replacement in a *Quercus cornelius-mulleri* population. *Journal of Vegetation Science*, **14**, 789–798.

Cohen, J.E. (1968) Alternate derivations of a species-abundance relation. *American Naturalist*, **102**, 165–172.

Cole, D.N. (1995) Experimental trampling of vegetation. II. Predictors of resistance and resilience. *Journal of Applied Ecology*, **32**, 215–224.

Cole, D.N. & Spildie, D.R. (1998) Hiker, horse and llama trampling effects on native vegetation in Montana, USA. *Journal of Environmental Management*, **53**, 61–71.

Cole, D.N. & Trull, S.J. (1992) Quantifying vegetation response to recreational disturbance in the North Cascades, Washington. *Northwest Science*, **66**, 229–236.

Collins, S.L., Bradford, J.A. & Sims, P.L. (1987) Succession and fluctuation in Artemisia dominated grassland. *Vegetatio*, **73**, 89–99.

Collins, S.L., Suding, K.N., Cleland, E.E., Batty, M., Pennings, S.C., Gross, K.L., Grace, J.B., Gough, L., Fargione, J.E. & Clark, C.M. (2008) Rank clocks and plant community dynamics. *Ecology*, **89**, 3534–3541.

Colwell, R.K. & Rangel, T.F. (2009) Hutchinson's duality: the once and future niche. *Proceedings of the National Academy of Sciences of the U.S.A.*, **106**, 19651–19658.

Connell, J.H. (1978) Diversity in tropical rain forests and coral reefs. *Science*, **199**, 1302–1310.

Connell, J.H. & Slatyer, R.O. (1977) Mechanisms of succession in natural communities and their role in community stability and organisation. *American Naturalist*, **111**, 1119–1144.

Connell, J.H. & Sousa, W.P. (1983) On the evidence needed to judge ecological stability or persistence. *American Naturalist*, **121**, 789–824.

Connolly, J. (1997) Substitutive experiments and the evidence for competitive hierarchies in plant communities. *Oikos*, **80**, 179–182.

Connor, E.F. & McCoy, E.D. (1979) The statistics and biology of the species-area relationship. *American Naturalist*, **113**, 791–833.

Connor, E.F., McCoy, E.D. & Cosby, B.J. (1983) Model discrimination and expected slope values in species-area studies. *American Naturalist*, **122**, 789–796.

Coomes, D.A., Allen, R.B., Bentley, W.A., Burrows, L.E., Canham, C.D., Fagan, L., Forsyth, D.M., Gaxiola-Alcantar, A., Parfitt, R.L., Ruscoe, W.A., Wardle, D.A., Wilson, D.J. & Wright, E.F. (2005) The hare, the tortoise and the crocodile: the ecology of angiosperm dominance, conifer persistence and fern filtering. *Journal of Ecology*, **93**, 918–935.

Coomes, D.A. & Grubb, P.J. (2000) Impacts of root competition in forests and woodlands: a theoretical framework and review of experiments. *Ecological Monographs*, **70**, 171–207.

Corbin, J.D. & D'Antonio, C.M. (2004) Competition between native perennial and exotic annual grasses: implications for an historical invasion. *Ecology*, **85**, 1273–1283.

Cornelissen, J.H.C. & Thompson, K. (1997) Functional leaf attributes predict litter decomposition rate in herbaceous plants. *New Phytologist*, **135**, 109–114.

Cornell, H.V. & Lawton, J.H. (1992) Species interactions, local and regional processes, and limits to the richness of ecological communities: a theoretical perspective. *Journal of Animal Ecology*, **61**, 1–12.

Cornwell, W.K. & Ackerly, D.D. (2009) Community assembly and shifts in plant trait distributions across an environmental gradient in coastal California. *Ecological Monographs*, **79**, 109–126.

Correia, M., Montesinos, D., French, K. & Rodríguez-Echeverría, S. (2016) Evidence for enemy release and increased seed production and size for two invasive Australian acacias. *Journal of Ecology*, **104**, 1391–1399.

Cowell, C.M., Hoalst-Pullen, N. & Jackson, M.T. (2010) The limited role of canopy gaps in the successional dynamics of a mature mixed Quercus forest remnant. *Journal of Vegetation Science*, **21**, 201–212.

Cowles, H.C. (1901) The physiographic ecology of Chicago and vicinity; a study of the origin, development, and classification of plant societies. *Botanical Gazette*, **31**, 145–182.

Cowling, R.M., Mustart, P.J., Laurie, H. & Richards, M.B. (1994) Species diversity, functional diversity and functional redundancy in fynbos communities. *South African Journal of Science*, **90**, 333–337.

Cowling, R.M., Ojeda, F., Lamont, B.B., Rundel, P.W. & Lechmere-Oertel, R. (2005) Rainfall reliability, a neglected factor in explaining convergence and divergence of plant traits in fire-prone mediterranean-climate ecosystems. *Global Ecology and Biogeography*, **14**, 509–519.

Cowling, R.M. & Witkowski, E.T.F. (1994) Convergence and non-convergence of plant traits in climatically and edaphically matched sites in Mediterranean Australia and South Africa. *Australian Journal of Ecology*, **19**, 220–232.

Craine, J.M. (2005) Reconciling plant strategy theories of Grime and Tilman. *Journal of Ecology*, **93**, 1041–1052.

Crawley, M.J. (1986) The structure of plant communities. In *Plant Ecology* (ed. M.J. Crawley), pp. 1–50. Blackwell, Oxford.

Crawley, M.J. (1997) Life history and environment. In *Plant Ecology* (ed. M.J. Crawley), pp. 73–131. Blackwell, Oxford.

Crawley, M.J., Brown, S.L., Heard, M.S. & Edwards, G.R. (1999) Invasion-resistance in experimental grassland communities: species richness or species identity? *Ecology Letters*, **2**, 140–148.

Crepy, M.A. & Casal, J.J. (2015) Photoreceptor-mediated kin recognition in plants. *New Phytologist*, **205**, 329–338.

Crews, T.E., Kitayama, K., Fownes, J.H., Riley, R.H., Herbert, D.A., Mueller-Dombois, D. & Vitousek, P.M. (1995) Changes in soil phosphorus fractions and ecosystem dynamics across a long chronosequence in Hawaii. *Ecology*, **76**, 1407–1424.

Csotonyi, J.T. & Addicott, J.F. (2004) Influence of trampling-induced microtopography on growth of the soil crust bryophyte *Ceratodon purpureus* in Jasper National Park. *Canadian Journal of Botany*, **82**, 1382–1392.

Cuesta, B., Villar-Salvador, P., Puertolas, J., Benayas, J.M.R. & Michalet, R. (2010) Facilitation of *Quercus ilex* in Mediterranean shrubland is explained by both direct and indirect interactions mediated by herbs. *Journal of Ecology*, **98**, 687–696.

Culmsee, H., Leuschner, C., Moser, G. & Pitopang, R. (2010) Forest aboveground biomass along an elevational transect in Sulawesi, Indonesia, and the role of Fagaceae in tropical montane rain forests. *Journal of Biogeography*, **37**, 960–974.

Cupper, M.L., Drinnan, A.N. & Thomas, I. (2000) Holocene palaeoenvironments of salt lakes in the Darling Anabranch region, south-western New South Wales, Australia. *Journal of Biogeography*, **27**, 1079–1094.

Curran, L.M. & Leighton, M. (2000) Vertebrate responses to spatiotemporal variation in seed production of mast-fruiting Dipterocarpaceae. *Ecological Monographs*, **70**, 101–128.

Curtis, J.T. & McIntosh, R.P. (1951) An upland forest continuum in the prairie-forest border region of Wisconsin. *Ecology*, **32**, 476–496.

Dahlin, A.S. & Stenberg, M. (2010) Transfer of N from red clover to perennial ryegrass in mixed stands under different cutting strategies. *European Journal of Agronomy*, **33**, 149–156.

Dale, M.B. (2002) Models, measures and messages: an essay on the role for induction. *Community Ecology*, **3**, 191–204.

Dale, M.P. & Causton, D.R. (1992) The ecophysiology of Veronica chamaedrys, V. montana and V. officinalis. I. Light quality and light quantity. *Journal of Ecology*, **80**, 483–492.

Dale, M.R.T. (1984) The contiguity of upslope and downslope boundaries of species in a zoned community. *Oikos*, **42**, 92–96.

Dale, M.R.T. (1985) Graph theoretical methods for comparing phytosociological structures. *Vegetatio*, **63**, 79–88.

Dallimore, W. (1917) Natural grafting of branches and roots. *Bulletin of Miscellaneous Information*, **1917**, 303–306.

Dalling, J.W., Swaine, M.D. & Garwood, N.C. (1998) Dispersal patterns and seed bank dynamics of pioneer trees in moist tropical forest. *Ecology*, **79**, 564–578.

Dansereau, P. (1964) Six problems in New Zealand vegetation. *Bulletin of the Torrey Botanical Club*, **91**, 114–140.

Darnell, R.M. (1970) Evolution and the ecosystem. *American Zoologist*, **10**, 9–15.

Darwin, C. (1859) *On the Origin of Species by Means of Natural Selection, or the Preservation of Favoured Races in the Struggle for Life*, 1st edn. John Murray, London.

Davies, K.F., Cavender-Bares, J. & Deacon, N. (2011) Native communities determine the identity of exotic invaders even at scales at which communities are unsaturated. *Diversity and Distributions*, **17**, 35–42.

Davis, M.A., Grime, J.P. & Thompson, K. (2000) Fluctuating resources in plant communities: a general theory of invasibility. *Journal of Ecology*, **88**, 528–534.

Davis, M.B. (1981) Quaternary history and the stability of forest communities. *Forest Succession: Concepts and Applications* (eds D.C. West, H.H. Shugart & D.B. Botkin), pp. 132–153. Springer, New York.

Dawson, T.E. (1993) Hydraulic lift and water use by plants – implications for water balance, performance and plant-plant interactions. *Oecologia*, **95**, 565–574.

de Bello, F., Price, J.N., Münkemüller, T., Liira, J., Zobel, M., Thuiller, W., Gerhold, P., Götzenberger, L., Lavergne, S., Lepš, J., Zobel, K. & Pärtel, M. (2012) Functional species pool framework to test for biotic effects on community assembly. *Ecology*, **93**, 2263–2273.

de Brouwer, J.F.C. & Stal, L.J. (2002) Daily fluctuations of exopolymers in cultures of the benthic diatoms Cylindrotheca closterium and Nitschia sp. (Bacillariophyceae). *Journal of Phycology*, 38, 464–472.

de Kroon, H. & Kalliola, R. (1995) Shoot dynamics of the giant grass *Gynerium sagittatum* in Peruvian Amazon floodplains, a clonal plant that does show self-thinning. *Oecologia*, **101**, 124–131.

de Kroon, H. & van Groenendael, J. (1997) *The Ecology and Evolution of Clonal Plants*. Backhuys, Leyden.

de Lange, P.J. & Stockley, J.E. (1987) The flora of the Lost World Cavern, Mangapu Caves System, Waitomo, Te Kuiti. *Bulletin of the Wellington Botanical Society*, **43**, 3–6.

de Rooij-van der Goes, P.C.E.M. (1995) The role of plant-parasitic nematodes and soil-borne fungi in the decline of *Ammophila arenaria* (L.) Link. *New Phytologist*, **129**, 661–669.

del Moral, R. (1983) Competition as a control mechanism in subalpine meadows. *American Journal of Botany*, **70**, 232–245.

del Moral, R. (2007) Limits to convergence of vegetation during early primary succession. *Journal of Vegetation Science*, **18**, 479–488.

del Moral, R. (2009) Increasing deterministic control of primary succession on Mount St. Helens, Washington. *Journal of Vegetation Science*, **20**, 1145–1154.

del Moral, R., Saura, J.M. & Emenegger, J.N. (2010) Primary succession trajectories on a barren plain, Mount St. Helens, Washington. *Journal of Vegetation Science*, **21**, 857–867.

DeLonge, M., D'Odorico, P. & Lawrence, D. (2008) Feedbacks between phosphorus deposition and canopy cover: the emergence of multiple stable states in tropical dry forests. *Global Change Biology*, **14**, 154–160.

DeMalach, N., Zaady, E., Weiner, J. & Kadmon, R. (2016) Size asymmetry of resource competition and the structure of plant communities. *Journal of Ecology*, **104**, 899–910.

Den Hartog, C. (1970) *Sea Grasses of the World*. North Holland, Amsterdam.

Deng, J.M., Zuo, W.Y., Wang, Z.Q., Fan, Z.X., Ji, M.F., Wang, G.X., Ran, J.Z., Zhao, C.M., Liu, J.Q., Niklas, K.J., Hammond, S.T. & Brown, J.H. (2012) Insights into plant size-density relationships from models and agricultural crops. *Proceedings of the National Academy of Sciences of the U.S.A.*, **109**, 8600–8605.

DeWoody, Y.D., Swihart, R.K., Craig, B.A. & Goheen, J.R. (2003) Diversity and stability in communities structured by asymmetric resource allocation. *American Naturalist*, **162**, 514–527.

Diamond, J.M. (1975) Assembly of species communities. In *Ecology and Evolution of Communities* (eds M.L. Cody & J.M. Diamond), pp. 342–444. Harvard University Press, Cambridge, MA.

Díaz Barradas, M.C., García Novo, F., Collantes, M. & Zunzunegu, M. (2001) Vertical structure of wet grasslands under grazed and non-grazed conditions in Tierra del Fuego. *Journal of Vegetation Science*, **12**, 385–390.

Díaz, S., Cabido, M. & Casanoves, F. (1998) Plant functional traits and environmental filters at a regional scale. *Journal of Vegetation Science*, **9**, 113–122.

Diaz-Martin, Z., Swamy, V., Terborgh, J., Alvarez-Loayza, P. & Cornejo, F. (2014) Identifying keystone plant resources in an Amazonian forest using a long-term fruit-fall record. *Journal of Tropical Ecology*, **30**, 291–301.

Dice, L.R. (1952) *Natural Communities*. University of Michigan Press, Ann Arbor.

Dickie, I.A., Fukami, T., Wilkie, J.P., Allen, R.B. & Buchanan, P.K. (2012) Do assembly history effects attenuate from species to ecosystem properties? A field test with wood-inhabiting fungi. *Ecology Letters*, **15**, 133–141.

Diekmann, M. & Lawesson, J.E. (1999) Shifts in ecological behaviour of herbaceous forest species along a transect from northern Central to North Europe. *Folia Geobotanica*, **34**, 127–141.

Diez, J.M., Dickie, I., Edwards, G., Hulme, P.E., Sullivan, J.J. & Duncan, R.P. (2010) Negative soil feedbacks accumulate over time for non-native plant species. *Ecology Letters*, **13**, 803–809.

DiMichele, W.A., Phillips, T.L. & Nelson, W.J. (2002) Place vs. time and vegetational persistence: a comparison of four tropical mires from the Illinois Basin during the height of the Pennsylvanian ice age. *International Journal of Coal Geology*, **50**, 43–72.

Dixon, A.F.G. & Kundu, R. (1994) Ecology of host alternation in aphids. *European Journal of Entomology*, **91**, 63–70.

Dodd, A.P. (1940) *The Biological Campaign Against Prickly-Pear*. Commonwealth Prickly Pear Board, Brisbane.

Dodd, J., Heddle, E.M., Pate, J.S. & Dixon, K.W. (1984) Rooting patterns of sandplain plants and their functional significance. In *Kongwan: Plant Life of the Sandplain* (eds J.S. Pate & J.S. Beard), pp. 146–177. University of Western Australia Press, Nedlands.

Dodd, M.E., Silvertown, J., McConway, K., Potts, J. & Crawley, M.J. (1995) Community stability: a 60-year record of trends and outbreaks in the occurrence of species in the Park Grass Experiment. *Journal of Ecology*, **83**, 277–285.

Dornbush, M.E. & Wilsey, B.J. (2010) Experimental manipulation of soil depth alters species richness and co-occurrence in restored tallgrass prairie. *Journal of Ecology*, **98**, 117–125.

Dostal, P. (2011) Plant Competitive Interactions and Invasiveness: Searching for the Effects of Phylogenetic Relatedness and Origin on Competition Intensity. *American Naturalist*, **177**, 655–667.

Drake, J.A. (1990) The mechanics of community assembly and succession. *Journal of Theoretical Biology*, **147**, 213–233.

Drake, J.A. (1991) Community-assembly mechanics and the structure of an experimental species ensemble. *American Naturalist*, **137**, 1–26.

Drake, J.A., Zimmermann, C.R., Purucker, T. & Rojo, C. (1999) of the assembly trajectory. *Assembly Rules and Restoration Ecology. Bridging the Gap Between Theory and Practice*. (eds V.M. Temperton, R.J. Hobbs, T. Nuttle & S. Halle), p. 464. Island Press, Washington DC.

Drew, M.C. (1975) Comparison of the effects of a localized supply of phosphate, nitrate, ammonium and potassium on the growth of the seminal root system, and the shoot, in barley. *New Phytologist*, **75**, 479–490.

Drude, O. (1885) Die Vertheilung und zusammensetzung östlicher pflanzengenossenschaften in der umgebung von Dresden. In *Festschrift der Isis in Dresden*, pp. 75–107. Warnatz and Lehmann, Dresden.

Dublin, H.T. (1995) Vegetation dynamics in the Serengeti-Mara ecosystem: the role of elephants, fire and other factors. In *Serengeti II: Dynamics, Management and Conservation of an Ecosystem* (eds A.R.E. Sinclair & P. Arcese), pp. 71–90. University of Chicago Press, Chicago, IL.

Dublin, H.T., Sinclair, A.R.E. & McGlade, J. (1990) Elephants and fire as causes of multiple stable states in the Serengeti-Mara woodlands. *Journal of Animal Ecology*, **59**, 1147–1164.

Dudley, J.P. (1999) Seed dispersal of Acacia erioloba by African bush elephants in Hwange National Park, Zimbabwe. *African Journal of Ecology*, **37**, 375–385.

Dudley, S.A., Murphy, G.P. & File, A.L. (2013) Kin recognition and competition in plants. *Functional Ecology*, **27**, 898–906.

Dukes, J.S. (2001) Biodiversity and invasibility in grassland microcosms. *Oecologia*, **126**, 563–568.

Dukes, J.S. (2002) Species composition and diversity affect grassland susceptibility and response to invasion. *Ecological Applications*, **12**, 602–617.

Dulloo, M.E., Kell, S.P. & Jones, C.G. (2002) Impact and control of invasive alien species on small islands. *International Forestry Review*, **4**, 277–285.

Dumbrell, A.J., Nelson, M., Helgason, T., Dytham, C. & Fitter, A.H. (2010) Relative roles of niche and neutral processes in structuring a soil microbial community. *ISME Journal*, **4**, 337–345.

Dunbar, M.J. (1960) The evolution of stability in marine environments: natural selection at the level of the ecosystem. *American Naturalist*, **94**, 129–136.

Duncan, R.P. (1993) Flood disturbance and the coexistence of species in a lowland podocarp forest, south Westland, New-Zealand. *Journal of Ecology*, **81**, 403–416.

Dunnett, N.P. & Grime, J.P. (1999) Competition as an amplifier of short-term vegetation responses to climate: an experimental test. *Functional Ecology*, **13**, 388–395.

Dunnett, N.P. & Willis, A.J. (2004) Monitoring of permanent plots in roadside verge vegetation at Bibury, Gloucestershire. *Bulletin of the British Ecological Society*, **35**, 18–21.

Dunnett, N.P., Willis, A.J., Hunt, R. & Grime, J.P. (1998) A 38 year study of relations between the weather and vegetation dynamics in a road verge near Bibury, Gloucestershire. *Journal of Ecology*, **86**, 610–623.

During, H.J. & Willems, J.H. (1986) The impoverishment of the bryophyte and lichen flora of the Dutch chalk grasslands in the thirty years 1953–1983. *Biological Conservation*, **36**, 143–158.

Dybzinski, R. & Tilman, D. (2007) Resource use patterns predict long-term outcomes of plant competition for nutrients and light. *American Naturalist*, **170**, 305–318.

Dyer, A.R., Fenech, A. & Rice, K.J. (2000) Accelerated seedling emergence in interspecific competitive neighbourhoods. *Ecology Letters*, **3**, 523–529.

Eckstein, R.L. & Donath, T.W. (2005) Interactions between litter and water availability affect seedling emergence in four familial pairs of floodplain species. *Journal of Ecology*, **93**, 807–816.

Edwards, G.R. & Crawley, M.J. (1999) Rodent seed predation and seedling recruitment in mesic grassland. *Oecologia*, **118**, 288–296.

Edwards, P.J. (1988) Effects of the fairy ring fungus *Agaricus arvensis* on nutrient availability in grassland. *New Phytologist*, **110**, 377–381.

Egler, F.E. (1954) Vegetation science concepts. I. Initial floristic composition, a factor in old-field vegetation development. *Vegetatio*, **4**, 412–417.

Egler, F.E. (1977) *The Nature of Vegetation: Its Management and Mismanagement*. Aton Forest Connecticut; in cooperation with the Connecticut Conservation Association, Bridgewater.

Ehlers, B.K., David, P., Damgaard, C.F. & Lenormand, T. (2016) Competitor relatedness, indirect soil effects and plant coexistence. *Journal of Ecology*, **104**, 1126–1135.

Ehrenfeld, J.G., Kourtev, P. & Huang, W. (2001) Changes in soil functions following invasions of exotic understory plants in deciduous forests. *Ecological Applications*, **11**, 1287–1300.

Ehrenfeld, J.G., Ravit, B. & Elgersma, K. (2005) Feedback in the plant-soil system. *Annual Review of Environment and Resources*, **30**, 75–115.

Ehrlén, J. (1996) Spatiotemporal variation in predispersal seed predation intensity. *Oecologia*, **108**, 708–713.

Ehrlén, J. & van Groenendael, J.M. (1998) The trade-off between dispersability and longevity – an important aspect of plant species diversity. *Applied Vegetation Science*, **1**, 29–36.

El Mehdawi, A.F., Cappa, J.J., Fakra, S.C., Self, J. & Pilon-Smits, E.A.H. (2012) Interactions of selenium hyperaccumulators and nonaccumulators during cocultivation on seleniferous or nonseleniferous soil – the importance of having good neighbors. *New Phytologist*, **194**, 264–277.

Eldridge, D.J. & Rath, D. (2002) Hip holes: Kangaroo (Macropus spp.) resting sites modify the physical and chemical environment of woodland soils. *Austral Ecology*, **27**, 527–536.

Elmendorf, S.C. & Harrison, S.P. (2011) Is plant community richness regulated over time? Contrasting results from experiments and long-term observations. *Ecology*, **92**, 602–609.

Elton, C.S. (1927) *Animal Ecology*. Sidgwick and Jackson, London.

Emborg, J., Christensen, M. & Heilmann-Clausen, J. (2000) The structural dynamics of Suserup Skov, a near-natural temperate deciduous forest in Denmark. *Forest Ecology and Management*, **126**, 173–189.

Emerman, S.H. & Dawson, T.E. (1996) Hydraulic lift and its influence on the water content of the rhizosphere: an example from sugar maple, *Acer saccharum*. *Oecologia*, **108**, 273–278.

Engelhardt, K.A.M. & Kadlec, J.A. (2001) Species traits, species richness and the resilience of wetlands after disturbance. *Journal of Aquatic Plant Management*, **39**, 36–39.

Enquist, B.J. & Niklas, K.J. (2001) Invariant scaling relations across tree-dominated communities. *Nature*, **410**, 655–660.

Enright, N.J. (1999) Litterfall dynamics in a mixed conifer-angiosperm forest in northern New Zealand. *Journal of Biogeography*, **26**, 149–157.

Eriksson, O. (2005) Game theory provides no explanation for seed size variation in grasslands. *Oecologia*, **144**, 98–105.

Eschenbach, C., Glauner, R., Kleine, M. & Kappen, L. (1998) Photosynthesis rates of selected tree species in lowland dipterocarp rainforest of Sabah, Malaysia. *Trees Structure and Function*, **12**, 356–365.

Esseen, P.-A. (1985) Litter fall of epiphytic macrolichens in two old *Picea abies* forests in Sweden. *Canadian Journal of Botany*, **63**, 980–987.

Evangelista, P.H., Kumar, S., Stohlgren, T.J., Jarnevich, C.S., Crall, A.W., Norman, J.B. & Barnett, D.T. (2008) Modelling invasion for a habitat generalist and a specialist plant species. *Diversity and Distributions*, **14**, 808–817.

Evans, M.N. & Barkham, J.P. (1992) Coppicing and natural disturbance in temperate woodlands, a review. In *Ecology and Management of Coppiced Woodlands* (ed. G.P. Buckley), pp. 79–98. Chapman and Hall, London.

Evans, R.C. & Turkington, R. (1988) Maintenance of morphological variation in a biotically patchy environment. *New Phytologist*, **109**, 369–376.

Eviner, V.T. & Chapin, F.S. (2003) Functional matrix: a conceptual framework for predicting multiple plant effects on ecosystem processes. *Annual Review of Ecology Evolution and Systematics*, **34**, 455–485.

Facelli, J.M. (1994) Multiple indirect effects of plant litter affect the establishment of woody seedlings in old fields. *Ecology*, **75**, 1727–1735.

Facelli, J.M. & Temby, A.M. (2002) Multiple effects of shrubs on annual plant communities in arid lands of South Australia. *Austral Ecology*, **27**, 422–432.

Facelli, J.M., Williams, R., Fricker, S. & Ladd, B. (1999) Establishment and growth of seedlings of *Eucalyptus obliqua*: interactive effects of litter, water, and pathogens. *Australian Journal of Ecology*, **24**, 484–494.

Fahnestock, J.T., Povirk, K.L. & Welker, J.M. (2000) Ecological significance of litter redistribution by wind and snow in arctic landscapes. *Ecography*, **23**, 623–631.

Falcone, P., Keller, R., Le Tacon, F. & Oswald, H. (1986) Facteurs influencant la forme des feuillus en plantations. *Revue Forestiere Francaise*, **38**, 315–323.

Falik, O., Reides, P., Gersani, M. & Novoplansky, A. (2003) Self/non-self discrimination in roots. *Journal of Ecology*, **91**, 525–531.

Fargione, J. & Tilman, D. (2006) Plant species traits and capacity for resource reduction predict yield and abundance under competition in nitrogen- limited grassland. *Functional Ecology*, **20**, 533–540.

Fargione, J.E., Brown, C.S. & Tilman, D. (2003) Community assembly and invasion: an experimental test of neutral versus niche processes. *Proceedings of the National Academy of Sciences of the U.S.A.*, **100**, 8916–8920.

Fargione, J.E. & Tilman, D. (2005a) Diversity decreases invasion via both sampling and complementarity effects. *Ecology Letters*, **8**, 604–611.

Fargione, J.E. & Tilman, D. (2005b) Niche differentiation in phenology and rooting depth promote coexistence with a dominant C4 bunchgrass. *Oecologia*, **143**, 598–606.

Farley, R.A. & Fitter, A.H. (1999) Temporal and spatial variation in soil resources in a deciduous woodland. *Journal of Ecology*, **87**, 688–696.

Fastie, C.L. (1995) Causes and ecosystem consequences of multiple pathways of primary succession at Glacier Bay, Alaska. *Ecology*, **76**, 1899–1916.

Federle, W., Maschwitz, U. & Hölldobler, B. (2002) Pruning of host plant neighbours as defence against enemy ant invasions: Crematogaster ant partners of Macaranga protected by "wax barriers" prune less than their congeners. *Oecologia*, **132**, 264–270.

Felker-Quinn, E., Schweitzer, J.A. & Bailey, J.K. (2013) Meta-analysis reveals evolution in invasive plant species but little support for Evolution of Increased Competitive Ability (EICA). *Ecology and Evolution*, **3**, 739–751.

Fellbaum, C.R., Mensah, J.A., Cloos, A.J., Strahan, G.E., Pfeffer, P.E., Kiers, E.T. & Bucking, H. (2014) Fungal nutrient allocation in common mycorrhizal networks is regulated by the carbon source strength of individual host plants. *New Phytologist*, **203**, 646–656.

Fiala, B., Maschwitz, U., Pong, T.Y. & Helbig, A.J. (1989) Studies of a South East Asian ant-plant association: protection of Macaranga trees by *Crematogaster boreensis*. *Oecologia*, **79**, 463–470.

Field, J.P., Breshears, D.D., Whicker, J.J. & Zou, C.B. (2012) Sediment capture by vegetation patches: implications for desertification and increased resource redistribution. *Journal of Geophysical Research Biogeosciences*, **117**, 1033.

Fine, P.V.A. (2002) The invasibility of tropical forests by exotic plants. *Journal of Tropical Ecology*, **18**, 687–705.

Firestone, J.L. & Jasieniuk, M. (2013) Small population size limits reproduction in an invasive grass through both demography and genetics. *Oecologia*, **172**, 109–117.

Firn, R. (2004) Plant intelligence: an alternative point of view. *Annals of Botany*, **93**, 345–351.

Flory, S.L., Long, F.R. & Clay, K. (2011) Invasive Microstegium populations consistently outperform native range populations across diverse environments. *Ecology*, **92**, 2248–2257.

Fonteyn, P.J. & Mahall, B.E. (1978) Competition among desert perennials. *Nature*, **275**, 544–545.

Fortin, M.-J. & Dale, M.R.T. (2005) *Spatial Data Analysis: A Guide for Ecologists*. Cambridge University Press, Cambridge, UK.

Fortunel, C., Paine, C.E.T., Fine, P.V.A., Kraft, N.J.B. & Baraloto, C. (2014) Environmental factors predict community functional composition in Amazonian forests. *Journal of Ecology*, **102**, 145–155.

Fowler, N.L. (1981) Competition and coexistence in a North Carolina grassland. II. The effects of the experimental removal of species. *Journal of Ecology*, **69**, 843–854.

Fowler, N.L. (1986a) The role of competition in plant communities in arid and semiarid regions. *Annual Review of Ecology and Systematics*, **17**, 89–110.

Fowler, N.L. (1986b) Microsite requirements for germination and establishment of tree grass species. *American Midland Naturalist*, **115**, 131–145.

Fowler, N.L. (1990) Disorderliness in plant communities: comparisons, causes, and consequences. *Perspectives on Plant Competition* (eds J.B. Grace & D. Tilman), pp. 291–306. Academic Press, San Diego.

Franco, M. (1986) The influence of neighbours on the growth of modular organisms with an example from trees. *Philosophical Transactions of the Royal Society of London B*, **313**, 209–225.

Fraser, E.C., Lieffers, V.J. & Landhäusser, S.M. (2006) Carbohydrate transfer through root grafts to support shaded trees. *Tree Physiology*, **26**, 1019–1023.

Freckleton, R.P. & Watkinson, A.R. (2001) Predicting competition coefficients for plant mixtures: reciprocity, transitivity and correlations with life-history traits. *Ecology Letters*, 348–357

Freckleton, R.P., Watkinson, A.R. & Rees, M. (2009) Measuring the importance of competition in plant communities. *Journal of Ecology*, **97**, 379–384.

Fridley, J.D. & Sax, D.F. (2014) The imbalance of nature: revisiting a Darwinian framework for invasion biology. *Global Ecology and Biogeography*, **23**, 1157–1166.

Fuentes, E.R. (1976) Ecological convergence of lizard communities in Chile and California. *Ecology*, **57**, 3–17.

Fuentes, M. (2004) Slight differences among individuals and the unified neutral theory of biodiversity. *Theoretical Population Biology*, **66**, 199–203.

Fujinuma, R., Bockheim, J. & Balster, N. (2005) Base-cation cycling by individual tree species in old-growth forests of Upper Michigan, USA. *Biogeochemistry*, **74**, 357–376.

Fukami, T. (2004) Community assembly along a species pool gradient: implications for multiple-scale patterns of species diversity. *Population Ecology*, **46**, 137–147.

Fukami, T., Bezemer, T.M., Mortimer, S.R. & van der Putten, W.H. (2005) Species divergence and trait convergence in experimental plant community assembly. *Ecology Letters*, **8**, 1283–1290.

Fukami, T., Dickie, I.A., Wilkie, J.P., Paulus, B.C., Park, D., Roberts, A., Buchanan, P.K. & Allen, R.B. (2010) Assembly history dictates ecosystem functioning: evidence from wood decomposer communities. *Ecology Letters*, **13**, 675–684.

Fuller, J.L. (1998) Ecological impact of the mid-holocene hemlock decline in southern Ontario, Canada. *Ecology*, **79**, 2337–2351.

Fuller, M.M. & Enquist, B.J. (2012) Accounting for spatial autocorrelation in null models of tree species association. *Ecography*, **35**, 510–518.

Furness, N.H. & Upadhyaya, M.K. (2002) Differential susceptibility of agricultural weeds to ultraviolet-B radiation. *Canadian Journal of Plant Science*, **82**, 789–796.

Futuyma, D.J. & Wasserman, S.S. (1980) Resource concentration and herbivory in oak forests. *Science*, **210**, 920–922.

Fynn, R.W.S., Morris, C.D. & Krikman, K.P. (2005) Plant strategies and trait trade-offs influence trends in competitive ability along gradients of soil fertility and disturbance. *Journal of Ecology*, **93**, 384–394.

García-Palacios, P., Shaw, E.A., Wall, D.H. & Hättenschwiler, S. (2016) Temporal dynamics of biotic and abiotic drivers of litter decomposition. *Ecology Letters*, **19**, 554–563.

Gartner, T.B. & Cardon, Z.G. (2004) Decomposition dynamics in mixed species leaf litter. *Oikos*, **104**, 230–246.

Gause, G.F. (1934) *The Struggle for Existence*. Williams and Wilkins, Baltimore.

Gentry, A.H. (1988) Changes in plant community diversity and floristic composition on environmental and geographical gradients. *Annals of the Missouri Botanic Garden*, **75**, 1–34.

Genung, M.A., Bailey, J.K. & Schweitzer, J.A. (2012) Welcome to the neighbourhood: interspecific genotype by genotype interactions in Solidago influence above- and belowground biomass and associated communities. *Ecology Letters*, **15**, 65–73.

Germino, M.J. & Smith, W.K. (1999) Sky exposure, crown architecture, and low-temperature photoinhibition in conifer seedlings at alpine treeline. *Plant Cell and Environment*, **22**, 407–415.

Gigon, A. (1997) Fluctuations of dominance and the coexistence of plant species in limestone grasslands. *Phytocoenologia*, **27**, 275–287.

Gilbert, B. & Lechowicz, M.J. (2004) Neutrality, niches, and dispersal in a temperate forest understory. *Proceedings of the National Academy of Sciences of the U.S.A.*, **101**, 7651–7656.

Gilbert, G.S. (2002) Evolutionary ecology of plant diseases in natural ecosystems. *Annual Review of Phytopathology*, **40**, 13–43.

Gilbert, G.S. & Parker, I.M. (2010) Rapid evolution in a plant-pathogen interaction and the consequences for introduced host species. *Evolutionary Applications*, **3**, 144–156.

Gill, D.E., Chao, L., Perkins, S.L. & Wolf, J.B. (1995) Genetic mosaicism in plants and clonal animals. *Annual Review of Ecology and Systematics*, **26**, 423–444.

Gillett, J.B. (1962) Pest pressure, an underestimated factor in evolution. *Systematics Association Publications No. 4*, pp. 37–46. Oxford.

Gillman, L.N. (2016) Seedling mortality from litterfall increases with decreasing latitude. *Ecology*, **97**, 530–535.

Gilmore, R.G. & Snedaker, S.C. (1993) Mangrove forests. *Biodiversity of the Southeastern United States* (ed. W.H. Martin), pp. 165–199. Wiley, New York.

Gilpin, M.E. & Case, T.J. (1976) Multiple domains of attraction in competition communities. *Nature*, **261**, 39–42.

Gitay, H. & Wilson, J.B. (1995) Post-fire changes in community structure of tall tussock grasslands: a test of alternative models of succession. *Journal of Ecology*, **83**, 775–782.

Gleason, H.A. (1917) The structure and development of the plant association. *Bulletin of the Torrey Botanical Club*, **44**, 463–481.

Gleason, H.A. (1926) The individualistic concept of the plant association. *Bulletin of the Torrey Botanical Club*, **53**, 7–26.

Gleason, H.A. (1927) Further views on the succession-concept. *Ecology*, **8**, 299–326.

Gleason, H.A. (1936) Is the synusia an association? *Ecology*, **17**, 444–451.

Gleason, H.A. (1939) The individualistic concept of the plant association. *American Midland Naturalist*, **21**, 92–110.

Goldberg, D.E. & Landa, K. (1991) Competitive effect and response – hierarchies and correlated traits in the early stages of competition. *Journal of Ecology*, **79**, 1013–1030.

Goldberg, D.E., Rajaniemi, T.K., Gurevitch, J. & Stewart-Oaten, A. (1999) Empirical approaches to quantifying interaction intensity: competition and facilitation along productivity gradients. *Ecology*, **80**, 1118–1131.

Goldberg, D.E. & Werner, P.A. (1983) Equivalence of competitors in plant communities: a null hypothesis and a field experimental approach. *American Journal of Botany*, **70**, 1098–1104.

Goldblum, D. (1997) The effects of treefall gaps on understory vegetation in New York State. *Journal of Vegetation Science*, **8**, 125–132.

Golubski, A.J., Gross, K.L. & Mittelbach, G.G. (2010) Recycling-mediated facilitation and co-existence based on plant size. *American Naturalist*, **176**, 588–600.

Gómez-Aparicio, L., Canham, C.D. & Martin, P.H. (2008) Neighbourhood models of the effects of the invasive Acer platanoides on tree seedling dynamics: linking impacts on communities and ecosystems. *Journal of Ecology*, **96**, 78–90.

Gómez-Aparicio, L., Gómez, J.M., Zamora, R. & Boettinger, J.L. (2005) Canopy vs. soil effects of shrubs facilitating tree seedlings in Mediterranean montane ecosystems. *Journal of Vegetation Science*, **16**, 191–198.

Goodall, D.W. (1954) Vegetational classification and vegetational continua. *Angewandte Pflanzensoziologie*, **1**, 168–182.

Goodall, D.W. (1963) The continuum and the individualistic association. *Vegetatio*, **11**, 297–316.

Goodall, D.W. (1966) The nature of the mixed community. *Proceedings of the Ecological Society of Australia*, **1**, 84–96.

Gotelli, N.J. & Graves, G.R. (1996) *Null Models in Ecology*. Smithsonian Institution Press, Washington.

Grace, J. (1983) *Plant-Atmosphere Relationships*. Chapman and Hall, London.

Grace, J.B. (1991) A clarification of the debate between Grime and Tilman. *Functional Ecology*, **5**, 583–587.

Grace, J.B. & Wetzel, R.G. (1981) Habitat partitioning and competitive displacement in cattails (Typha): experimental field studies. *American Naturalist*, **118**, 463–474.

Grady, K.C., Wood, T.E., Kolb, T.E., Hersch-Green, E., Shuster, S.M., Gehring, C.A., Hart, S.C., Allan, G.J. & Whitham, T.G. (2016) Local biotic adaptation of trees and shrubs to plant neighbors. *Oikos*, 583–593.

Graham, R.W. & Grimm, E.C. (1990) Effects of global climate change on the patterns of terrestrial biological communities. *Trends in Ecology and Evolution*, **5**, 289–292.

Grant, K., Kreyling, J., Heilmeier, H., Beierkuhnlein, C. & Jentsch, A. (2014) Extreme weather events and plant plant interactions: shifts between competition and facilitation among grassland species in the face of drought and heavy rainfall. *Ecological Research*, **29**, 991–1001.

Gravel, D., Canham, C.D., Beaudet, M. & Messier, C. (2010) Shade tolerance, canopy gaps and mechanisms of coexistence of forest trees. *Oikos*, **119**, 475–484.

Graves, J.D., Press, M.C., Smith, S. & Stewart, G.R. (1992) The carbon canopy economy of the association between cowpea and the parasitic angiosperm *Striga gesnerioides*. *Plant Cell and Environment*, **15**, 283–288.

Grayston, S.J., Wang, S., Campbell, C.D. & Edwards, A.C. (1998) Selective influence of plant species on microbial diversity in the rhizosphere. *Soil Biology and Biochemistry*, **30**, 369–378.

Green, P.T., O'Dowd, D.J. & Lake, P.S. (2008) Recruitment dynamics in a rainforest seedling community: context-independent impact of a keystone consumer. *Oecologia*, **156**, 373–385.

Greig-Smith, P. (1952) Ecological observations on degraded and secondary forest in Trinidad, British West Indies. *Journal of Ecology*, **40**, 283–330.

Greig-Smith, P. (1983) *Quantitative Plant Ecology*, 3rd edn. Blackwell, Oxford.

Grigg, A.H. & Mulligan, D.R. (1999) Litterfall from two eucalypt woodlands in central Queensland. *Australian Journal of Ecology*, **24**, 662–664.

Griggs, R.F. (1914) Observations on the behavior of some species at the edges of their ranges. *Bulletin of the Torrey Botanical Club*, **41**, 25–49.

Grime, J.P. (1973) Control of species density in herbaceous vegetation. *Journal of Environmental Management*, **1**, 151–167.

Grime, J.P. (1974) Vegetation classification by reference to strategies. *Nature*, **250**, 26–31.

Grime, J.P. (1977) Evidence for the existence of three primary strategies in plants and its relevance to ecological and evolutionary theory. *American Naturalist*, **111**, 1169–1194.

Grime, J.P. (1979) *Plant Strategies and Vegetation Processes*. Wiley, Chichester.

Grime, J.P. (1988) The C-S-R model of primary plant strategies – origins, implications and tests. In *Plant Evolutionary Biology* (eds L.D. Gottlieb & S.K. Jain), pp. 371–393. Chapman and Hall, London.

Grime, J.P. (2001) *Plant Strategies, Vegetation Processes, and Ecosystem Properties*, 2nd edn. Wiley, Chichester.

Grime, J.P. (2006) Trait convergence and trait divergence in herbaceous plant communities: mechanisms and consequences. *Journal of Vegetation Science*, **17**, 255–260.

Grime, J.P., Hodgson, J.G. & Hunt, R. (2007) *Comparative Plant Ecology: A Functional Approach to Common British Species*, 2nd edn. Castlepoint Press, Dalbeattie.

Grime, J.P. & Hunt, R. (1975) Relative growth rate: its range and adaptive significance in a local flora. *Journal of Ecology*, **63**, 393–422.

Grime, J.P., Hunt, R. & Krzanowski, W.J. (1987) Evolutionary physiological ecology of plants. *Evolutionary Physiological Ecology* (ed. P. Calow), pp. 105–125. Cambridge University Press, Cambridge, UK.

Grime, J.P., Thompson, K., Hunt, R., Hodgson, J.G., Cornelissen, J.H.C., Rorison, I.H., Hendry, G.A.F., Ashenden, T.W., Askew, A.P., Band, S.R., Booth, R.E., Bossard, C.C., Campbell, B.D., Cooper, J.E.L., Davison, A.W., Gupta, P.L., Hall, W., Hand, D.W., Hannah, M.A., Hillier, S.H., Hodkinson, D.J., Jalili, A., Liu, Z., Mackey, J.M.L., Matthews, N., Mowforth, M.A., Neal, A.M., Reader, R.J., Reiling, K., Ross-Fraser, W., Spencer, R.E., Sutton, F., Tasker, D.E., Thorpe, P.C. & Whitehouse, J. (1997) Integrated screening validates primary axes of specialisation in plants. *Oikos*, **79**, 259–281.

Grime, J.P., Willis, A.J., Hunt, R. & Dunnett, N.P. (1994) Climate-vegetation relationships in the Bibury road verge experiments. In *Long-Term Experiments in Agricultural and Ecological Sciences* (eds R.A. Leigh & A.E. Johnston), pp. 271–285. CAB International, Wallingford.

Grinnell, J. (1904) The origin and distribution of the chestnut-backed chickadee. *Auk*, **21**, 375–377.

Groom, M.J. (1998) Allee effects limit population viability of an annual plant. *American Naturalist*, **151**, 487–496.

Gross, C.L., Nelson, P.A., Haddadchi, A. & Fatemi, M. (2012) Somatic mutations contribute to genotypic diversity in sterile and fertile populations of the threatened shrub, *Grevillea rhizomatosa* (Proteaceae). *Annals of Botany*, **109**, 331–342.

Gross, K., Cardinale, B.J., Fox, J.W., Gonzalez, A., Loreau, M., Polley, H.W., Reich, P.B. & van Ruijven, J. (2014) Species richness and the temporal stability of biomass production: a new analysis of recent biodiversity experiments. *American Naturalist*, **183**, 1–12.

Grostal, P. & O'Dowd, D.J. (1994) Plants, mites and mutualism – leaf domatia and the abundance and reproduction of mites on *Viburnum tinus* (Caprifoliaceae). *Oecologia*, **97**, 308–315.

Grover, J.P. (1994) Assembly rules for communities of nutrient-limited plants and specialist herbivores. *American Naturalist*, **143**, 258–282.

Grubb, P.J. (1982) Control of relative abundance in roadside Arrhenatheretum: results of a long-term garden experiment. *Journal of Ecology*, **70**, 845–861.

Grubb, P.J. (1985) Plant populations and vegetation in relation to habitat, disturbance and competition: problems of generalization. In *The Population Structure of Vegetation* (ed. J. White), pp. 595–621. Junk, Dordrecht.

Grubb, P.J. (1987) Some generalizing ideas about colonisation and succession in green plants and fungi. In *Colonization, Succession and Stability* (eds A.J. Gray, M.J. Crawley & P.J. Edwards), pp. 81–102. Blackwell, Oxford.

Gruntman, M. & Novoplansky, A. (2004) Physiologically mediated self/non-self discrimination in roots. *Proceedings of the National Academy of Sciences of the U.S.A.*, **101**, 3863–3867.

Gulmon, S., Rundel, P.W., Ehleringer, J.R. & Mooney, H.A. (1979) Spatial relationships and competition in a Chilean desert cactus. *Oecologia*, **44**, 40–43.

Gulmon, S.L. (1977) A comparative study of the grassland of California and Chile. *Flora*, **166**, 261–278.

Gundale, M.J., Kardol, P., Nilsson, M.C., Nilsson, U., Lucas, R.W. & Wardle, D.A. (2014) Interactions with soil biota shift from negative to positive when a tree species is moved outside its native range. *New Phytologist*, **202**, 415–421.

Gurevitch, J. & Unnasch, R.S. (1989) Experimental removal of a dominant species at two levels of soil fertility. *Canadian Journal of Botany*, **67**, 3470–3477.

Haase, P. (1990) Environmental and floristic gradients in Westland, New Zealand, and the discontinuous distribution of Nothofagus. *New Zealand Journal of Botany*, **28**, 25–40.

Hacker, S.D. & Bertness, M.D. (1999) Experimental evidence for factors maintaining plant species diversity in a New England salt marsh. *Ecology*, **80**, 2064–2073.

Hairston, N.G. (1981) An experimental test of a guilds: salamander competition. *Ecology*, **62**, 65–72.

Halkett, J.C. (1991) *The Native Forests of New Zealand*. GP Publications, Wellington, New Zealand.

Hall, S.J. & Raffaelli, D.G. (1993) Food webs: theory and reality. *Advances in Ecological Research*, **24**, 187–239.

Hallett, J.G. (1982) Habitat selection and the community matrix of a desert small-mammal fauna. *Ecology*, **63**, 1400–1410.

Handa, I.T., Harmsen, R. & Jefferies, R.L. (2002) Patterns of vegetation change and the recovery potential of degraded areas in a coastal marsh system of the Hudson Bay lowlands. *Journal of Ecology*, **90**, 86–99.

Hanley, M.E. & Sykes, R.J. (2009) Impacts of seedling herbivory on plant competition and implications for species coexistence. *Annals of Botany*, **103**, 1347–1353.

Hansen, G.J.A., Ives, A.R., Vander Zanden, M.J. & Carpenter, S.R. (2013) Are rapid transitions between invasive and native species caused by alternative stable states, and does it matter? *Ecology*, **94**, 2207–2219.

Hanya, G. (2005) Comparisons of dispersal success between the species fruiting prior to and those at the peak of migrant frugivore abundance. *Plant Ecology*, **181**, 167–177.

Hara, T. & Srutek, M. (1995) Shoot growth and mortality patterns of *Urtica dioica*, a clonal forb. *Annals of Botany*, **76**, 235–243.

Harberd, D.J. (1962) Some observations on natural clones in *Festuca ovina*. *New Phytologist*, **61**, 85–100.

Hargeby, A., Blindow, I. & Hansson, L.-A. (2004) Shifts between clear and turbid states in a shallow lake: multi-causal stress from climate, nutrients and biotic interactions. *Archiv für Hydrobiologie*, **161**, 433–454.

Harper, J.L. (1967) A darwinian approach to plant ecology. *Journal of Ecology*, **55**, 247–270.

Harper, J.L. (1977) *Population Biology of Plants*. Academic Press, London.

Harper, J.L. & Ogden, J. (1970) The reproductive strategy of higher plants. I. The concept of strategy with special reference to Senecio vulgaris L. *Journal of Ecology*, **58**, 681–698.

Harpole, W.S. & Tilman, D. (2006) Non-neutral patterns of species abundance in grassland communities. *Ecology Letters*, **9**, 15–23.

Harries, J.H. & Norrington-Davies, J. (1977) Competition studies in diploid and tetraploid varieties of *Lolium perenne*. II. The inhibition of germination. *Journal of Agricultural Science*, **88**, 411–415.

Harrison, K.A. & Bardgett, R.D. (2010) Influence of plant species and soil conditions on plant-soil feedback in mixed grassland communities. *Journal of Ecology*, **98**, 384–395.

Hart, S.P., Schreiber, S.J. & Levine, J.M. (2016) How variation between individuals affects species coexistence. *Ecology Letters*, **19**, 825–838.

Hautier, Y., Hector, A., Vojtech, E., Purves, D. & Turnbull, L.A. (2010) Modelling the growth of parasitic plants. *Journal of Ecology*, **98**, 857–866.

Hautier, Y., Niklaus, P.A. & Hector, A. (2009) Competition for light causes plant biodiversity after eutrophication. *Science*, **324**, 636–638.

Haydon, D. (1994) Pivotal assumptions determining the relationship between stability and complexity – an analytical synthesis of the stability-complexity debate. *American Naturalist*, **144**, 14–29.

He, Q. & Bertness, M.D. (2014) Extreme stresses, niches, and positive species interactions along stress gradients. *Ecology*, **95**, 1437–1443.

He, Q., Bertness, M.D. & Altieri, A.H. (2013) Global shifts towards positive species interactions with increasing environmental stress. *Ecology Letters*, **16**, 695–706.

He, W.M., Feng, Y.L., Ridenour, W.M., Thelen, G.C., Pollock, J.L., Diaconu, A. & Callaway, R.M. (2009) Novel weapons and invasion: biogeographic differences in the competitive effects of *Centaurea maculosa* and its root exudates ($\pm$)-catechin. *Oecologia*, **159**, 803–815.

He, X.H., Critchley, C., Ng, H. & Bledsoe, C. (2004) Reciprocal N ($^{15}NH_4+$ or $^{15}NO_3$) transfer between non-N_2-fixing Eucalyptus maculata and N_2-fixing Casuarina cunninghamiana linked by the ectomycorrhizal fungus Pisolithus sp. *New Phytologist*, **163**, 629–640.

Heads, M.J. (1989) Integrating earth and life sciences in New Zealand natural history: the parallel arcs model. *New Zealand Journal of Zoology*, **16**, 549–585.

Hector, A. (2006) Overyielding and stable species coexistence. *New Phytologist*, **172**, 1–3.

Hector, A., Hautier, Y., Saner, P., Wacker, L., Bagchi, R., Joshi, J., Schererlorenzen, M., Spehn, E.M., Bazeleywhite, E., Weilenmann, M., Caldeira, M.C., Dimitrakopoulos, P.G., Finn, J.A., Hussdanell, K., Jumpponen, A., Mulder, C.P.H., Palmborg, C., Pereira, J.S., Siamantziouras, A.S.D., Terry, A.C., Troumbis, A.Y., Schmid, B. & Loreau, M. (2010) General stabilizing effects of plant diversity on grassland productivity through population asynchrony and over-yielding. *Ecology*, **91**, 2213–2220.

Hegarty, E.E. (1991) Vine-host interactions. In *The Biology of Vines* (eds F.E. Putz & H.A. Mooney), pp. 357–375. Cambridge University Press, Cambridge, UK.

Hellström, G.B. & Lubke, R.A. (1993) Recent changes to a climbing-falling dune system on the Robberg Peninsula, Southern Cape Coast, South Africa. *Journal of Coastal Research*, **9**, 647–653.

Herben, T. & Goldberg, D.E. (2014) Community assembly by limiting similarity vs. competitive hierarchies: testing the consequences of dispersion of individual traits. *Journal of Ecology*, **102**, 156–166.

Herben, T., Krahulec, F., Hadincová, V. & Kovářová, M. (1990) Fine scale dynamics in a mountain grassland. In *Spatial Processes in Plant Communities* (eds F. Krahulec, A.D.Q. Agnew, S. Agnew & J.H. Willems), pp. 173–184. SPB Academic Publishing, The Hague.

Herben, T., Krahulec, F., Hadincová, V., Pecháčková, S. & Wildová, R. (2003) Year-to-year variation in plant competition in a mountain grassland. *Journal of Ecology*, **91**, 103–113.

Herbert, D.A., Fownes, J.H. & Vitousek, P.M. (1999) Hurricane damage to a Hawaiian forest: nutrient supply rate affects resistance and resilience. *Ecology*, **80**, 908–920.

Herrera, C.M., Jordano, P., Lopez-Soria, L. & Amat, J.A. (1994) Recruitment of a mast-fruiting, bird-dispersed tree-bridging frugivore activity and seedling establishment. *Ecological Monographs*, **64**, 315–344.

Hill, S.B. & Kotanen, P.M. (2009) Evidence that phylogenetically novel non-indigenous plants experience less herbivory. *Oecologia*, **161**, 581–590.

Himanen, S.J., Blande, J.D., Klemola, T., Pulkkinen, J., Heijari, J. & Holopainen, J.K. (2010) Birch (Betula spp.) leaves adsorb and re-release volatiles specific to neighbouring plants – a mechanism for associational herbivore resistance? *New Phytologist*, **186**, 722–732.

Hirose, T. & Werger, M.J.A. (1995) Canopy structure and photon flux partitioning among species in a herbaceous plant community. *Ecology*, **76**, 466–474.

Hirota, M., Holmgren, M., VanNes, E.H. & Scheffer, M. (2011) Global resilience of tropical forest and savanna to critical transitions. *Science*, **334**, 232–235.

Hoagland, B.W. & Collins, S.L. (1997) Gradient models, gradient analysis, and hierarchical structure in plant communities. *Oikos*, **78**, 23–30.

Hochberg, M.E., Menaut, J.C. & Gignoux, J. (1994) Influences of tree biology and fire in the spatial structure of the West African savannah. *Journal of Ecology*, **82**, 217–226.

Hodgkinson, K.C. & Baas Becking, H.G. (1977) Effect of defoliation on root growth of some arid zone perennial plants. *Australian Journal of Agricultural Research*, **29**, 31–42.

Hodgson, J.G., Wilson, P.J., Hunt, R., Grime, J.P. & Thompson, K. (1999) Allocating C-S-R plant functional types: a soft approach to a hard problem. *Oikos*, **85**, 282–294.

Hoffmann, W.A., Adasme, R., Haridasan, M., de Carvalho, M.T., Geiger, E.L., Pereira, M.A.B., Gotsch, S.G. & Franco, A.C. (2009) Tree topkill, not mortality, governs the dynamics of savanna-forest boundaries under frequent fire in central Brazil. *Ecology*, **90**, 1326–1337.

Hofgaard, A. (1993) Structure and regeneration patterns in a virgin *Picea abies* forest in northern Sweden. *Journal of Vegetation Science*, **4**, 601–608.

Holdredge, C. & Bertness, M.D. (2011) Litter legacy increases the competitive advantage of invasive *Phragmites australis* in New England wetlands. *Biological Invasions*, **13**, 423–433.

Holt, G. & Chesson, P. (2014) Variation in moisture duration as a driver of coexistence by the storage effect in desert annual plants. *Theoretical Population Biology*, **92**, 36–50.

Holzapfel, C. & Mahall, B.E. (1999) Bidirectional facilitation and interference between shrubs and annuals in the Mojave desert. *Ecology*, **80**, 1747–1761.

Hooper, D.U. & Dukes, J.S. (2004) Overyielding among plant functional groups in a long-term experiment. *Ecology Letters*, **7**, 95–105.

Hooper, D.U. & Dukes, J.S. (2010) Functional composition controls invasion success in a California serpentine grassland. *Journal of Ecology*, **98**, 764–777.

Houlahan, J.E., Currie, D.J., Cottenie, K., Cumming, G.S., Ernest, S.K., Findlay, C.S., Fuhlendorf, S.D., Gaedke, U., Legendre, P., Magnuson, J.J., McArdle, B.H., Muldavin, E.H., Noble, D., Russell, R., Stevens, R.D., Willis, T.J., Woiwod, I.P. & Wondzell, S.M. (2007) Compensatory dynamics are rare in natural ecological communities. *Proceedings of the National Academy of Sciences of the U.S.A.*, **104**, 3273–3277.

Howard, K.S.C., Eldridge, D.J. & Soliveres, S. (2012) Positive effects of shrubs on plant species diversity do not change along a gradient in grazing pressure in an arid shrubland. *Basic and Applied Ecology*, **13**, 159–168.

Hubbell, S.P. (2001) *The Unified Neutral Theory of Biodiversity and Biogeography*. Princeton University Press, Princeton, NJ.

Hubbell, S.P. (2005) Neutral theory in community ecology and the hypothesis of functional equivalence. *Functional Ecology*, **19**, 166–172.

Hubbell, S.P. & Foster, R.B. (1986) Biology, chance, and history and the structure of tropical rain forest tree communities. In *Community Ecology* (eds J.M. Diamond & T.J. Case), pp. 314–329. Harper and Row, New York.

Hughes, J.B. & Roughgarden, J. (1998) Aggregate community properties and the strength of species' interactions. *Proceedings of the National Academy of Sciences of the U.S.A.*, **95**, 6837–6842.

Hughes, J.B. & Roughgarden, J. (2000) Species diversity and biomass stability. *American Naturalist*, **155**, 618–627.

Hunt, R. & Hope-Simpson, J.F. (1990) Growth of *Pyrola rotundfolia* ssp. *maritima* in relation to shade. *New Phytologist*, **114**, 129–137.

Hunter, A.F. & Aarssen, L.W. (1988) Plants helping plants. *BioScience*, **38**, 34–40.

Hutchinson, G.E. (1941) Ecological aspects of succession in natural populations. *American Naturalist*, **75**, 406–418.

Hutchinson, G.E. (1951) Copepodology for the ornithologist. *Ecology*, **32**, 571–577.

Hutchinson, G.E. (1957) Concluding remarks. *Cold Spring Harbour Symposium on Quantitative Biology*, **22**, 415–427.

Hutchinson, G.E. (1959) Homage to Santa Rosalia, or why are there so many kinds of animals. *American Naturalist*, **93**, 145–159.

Hutchinson, G.E. (1961) The paradox of the plankton. *American Naturalist*, **95**, 137–145.

Hyatt, L.A., Rosenberg, M.S., Howard, T.G., Bole, G., Fang, W., Anastasia, J., Brown, K., Grella, R., Hinman, K., Kurdziel, J. & Gurevitch, J. (2003) The distance dependence prediction of the Janzen-Connell hypothesis: a meta-analysis. *Oikos*, **103**, 590–602.

Inderjit, Dakshini, K.M.M. & Einhellig, F.A. (1994) *Allelopathy: Organisms, Processes, and Applications*. American Chemical Society, Washington.

Inderjit & Mallik, A.U. (1997) Effects of Ledum groenlandicum amendments on black spruce seedling growth. *Plant Ecology*, **133**, 29–36.

Innes, L., Hobbs, P.J. & Bardgett, R.D. (2004) The impacts of individual plant species on rhizosphere microbial communities in soils of different fertility. *Biology and Fertility of Soils*, **40**, 7–13.

Inouye, R.S., Allison, T.D. & Johnson, N.C. (1987) Old-Field succession on a Minnesota sand plain. *Ecology*, **68**, 12–26.

Isbell, F.I., Polley, H.W. & Wilsey, B.J. (2009) Biodiversity, productivity and the temporal stability of productivity: patterns and processes. *Ecology Letters*, **12**, 443–451.

Ives, A.R. & Hughes, J.B. (2002) General relationships between species diversity and stability in competitive systems. *American Naturalist*, **159**, 388–395.

Jabot, F. & Chave, J. (2011) Analyzing tropical forest tree species abundance distributions using a nonneutral model and through approximate Bayesian inference. *American Naturalist*, **178**, 37–47.

Jackson, S.T. & Williams, J.W. (2004) Modern analogs in quaternary paleoecology: here today, gone yesterday, gone tomorrow? *Annual Review of Earth and Planetary Science*, **32**, 495–537.

Jacobsen, A.L., Esler, K.J., Pratt, R.B. & Ewers, F.W. (2009) Water stress tolerance of shrubs in mediterranean-type climate regions: convergence of fynbos and succulent karoo communities with California shrub communities. *American Journal of Botany*, **96**, 1445–1453.

Janse, J.H. (1997) A model of nutrient dynamics in shallow lakes in relation to multiple stable states. *Hydrobiologia*, **342**, 1–8.

Jogesh, T., Carpenter, D. & Cappuccino, N. (2008) Herbivory on invasive exotic plants and their non-invasive relatives. *Biological Invasions*, **10**, 797–804.

Johnson, D.J., Beaulieu, W.T., Bever, J.D. & Clay, K. (2012) Conspecific negative density dependence and forest diversity. *Science*, **336**, 904–907.

Johnson, E.A. (1992) *Fire and Vegetation Dynamics: Studies from the North American Boreal Forest*. Cambridge University Press, Cambridge, UK.

Jokinen, K. (1991) Yield and competition in barley variety mixtures. *Journal of Agricultural Science in Finland*, **63**, 287–305.

Jones, C.G., Lawton, J.H. & Shachak, M. (1994) Organisms as ecosystem engineers. *Oikos*, **69**, 373–386.

Jones, J.I. & Sayer, C.D. (2003) Does the fish-invertebrate-periphyton cascade precipitate plant loss in shallow lakes? *Ecology*, **84**, 2155–2167.

Jones, M.G. (1933) Grassland management and its effect on the sward. *Journal of the Royal Agricultural Society of England*, **94**, 21–41.

Jonzén, N., Nolet, B.A., Santamaría, J. & Svensson, M. (2002) Seasonal herbivory and mortality compensation in a swan-pondweed system. *Ecological Modelling*, **147**, 209–219.

Jordán, F., Takács-Sánta, A. & Molnár, I. (1999) A reliability theoretical quest for keystones. *Oikos*, **86**, 453–462.

Joshi, J. & Vrieling, K. (2005) The enemy release and EICA hypothesis revisited: incorporating the fundamental difference between specialist and generalist herbivores. *Ecology Letters*, **8**, 704–714.

Jump, A.S. & Woodward, E.I. (2003) Seed production and population density decline approaching the range-edge of Cirsium species. *New Phytologist*, **160**, 349–358.

Kadmon, R. & Shmida, A. (1990) Competition in a variable environment: an experimental study in a desert annual plant population. *Israel Journal of Botany*, **39**, 403–412.

Kaneko, N. & Salamanca, E.F. (1999) Mixed leaf litter effects on decomposition rates and soil microarthropod communities in an oak-pine stand in Japan. *Ecological Research*, **14**, 131–138.

Karban, R., Shiojiri, K., Ishizaki, S., Wetzel, W.C. & Evans, R.Y. (2013) Kin recognition affects plant communication and defence. *Proceedings of the Royal Society of London B*, **280**, 20123062.

Kazmierczak, E., van der Maarel, E. & Noest, V. (1995) Plant communities in kettle-holes in central Poland: chance occurrence of species? *Journal of Vegetation Science*, **6**, 863–874.

Keddy, P.A. (1983) Shoreline vegetation in Axe Lake, Ontario: effects of exposure on vegetation patterns. *Ecology*, **64**, 331–344.

Keddy, P.A. (1992) Assembly and response rules: two goals for predictive community ecology. *Journal of Vegetation Science*, **3**, 157–164.

Keddy, P.A. (2007) *Plants and Vegetation. Origins. Processes. Consequences.* Cambridge University Press, Cambridge, UK.

Keddy, P.A., Fraser, L.H. & Wisheu, I.C. (1998) A comparative approach to examine competitive response of 48 wetland plant species. *Journal of Vegetation Science*, **9**, 777–786.

Keeley, J.E. & Fotheringham, C.J. (1997) Trace gas emissions and smoke-induced seed germination. *Science*, **276**, 1248–1250.

Keith, S.A., Newton, A.C., Morecroft, M.D., Golicher, D.J. & Bullock, J.M. (2011) Plant metacommunity structure remains unchanged during biodiversity loss in English woodlands. *Oikos*, **120**, 302–310.

Keller, E. & Steffen, K.L. (1995) Increased chilling tolerance and altered carbon metabolism in tomato leaves following application of mechanical stress. *Physiologia Plantarum*, **93**, 519–525.

Kellman, M. & Tackaberry, R. (1993) Disturbance and tree species coexistence in tropical riparian forest fragments. *Global Ecology and Biogeography Letters*, **3**, 1–9.

Kelly, B.J., Wilson, J.B. & Mark, A.F. (1989) Causes of the species/area relation: a study of islands in Lake Manapouri, New Zealand. *Journal of Ecology*, **71**, 1021–1028.

Kelly, C.K., Blundell, S.J., Bowler, M.G., Fox, G.A., Harvey, P.H., Lomas, M.R. & Woodward, F.I. (2011) The statistical mechanics of community assembly and species distribution. *New Phytologist*, **191**, 819–827.

Kelly, C.K., Venable, D.L. & Zimmerer, K. (1988) Host specialization in *Cuscuta costaricensis*: an assessment of host use relative to host availability. *Oikos*, **53**, 315–320.

Kenkel, N.C., McIlraith, A.L., Burchill, C.A. & Jones, G. (1991) Competition and the response of three plant species to a salinity gradient. *Canadian Journal of Botany*, **69**, 2497–2502.

Kielland, K. (1994) Amino acid absorption by arctic plants: implications for plant nutrition and nitrogen cycling. *Ecology*, **75**, 2373–2383.

Kikvidze, Z., Khetsuriani, L. & Kikodze, D. (2005) Small-scale guild proportions and niche complementarity in a Caucasian subalpine hay meadow. *Journal of Vegetation Science*, **16**, 565–570.

Kikvidze, Z., Khetsuriani, L., Kikodze, D. & Callaway, R.M. (2006) Seasonal shifts in competition and facilitation in subalpine plant communities of the central Caucasus. *Journal of Vegetation Science*, **17**, 77–82.

King, T.J. (1977) The plant ecology of ant-hills in calcareous grasslands. I. Patterns of species in relation to ant-hills in southern England. *Journal of Ecology*, **65**, 235–256.

King, W.M. & Wilson, J. (2006) Differentiation between native and exotic plant species from a dry grassland: fundamental responses to resource availability, and growth rates. *Austral Ecology*, **31**, 996–1004.

Kissel, R., Wilson, J.B., Bannister, P. & Mark, A.F. (1987) Water relations of some native and exotic shrubs of New Zealand. *New Phytologist*, **107**, 29–37.

Kitajima, K. (1994) Relative importance of photosynthetic traits and allocation patterns as correlates of seedling shade tolerance of 13 tropical trees. *Oecologia*, **98**, 419–428.

Klekowski, E.J. & Godfrey, P.J. (1989) Ageing and mutation in plants. *Nature*, **340**, 389–391.

Kleyer, M. (2002) Validation of plant functional types across two contrasting landscapes. *Journal of Vegetation Science*, **13**, 167–178.

Klimeš, L., Jongepier, J.W. & Jongepierová, I. (1995) Variability in species richness and guild structure in two species-rich grasslands. *Folia Geobotanica et Phytotaxonomica*, **30**, 243–253.

Klinger, L.F. (1996) The myth of the classic hydrosere model of bog succession. *Arctic and Alpine Research*, **28**, 1–9.

Kloeppel, B.D. & Abrams, M.D. (1995) Ecophysiological attributes of the native *Acer saccharum* and the exotic *Acer platanoides* in urban oak forests in Pennsylvania, USA. *Tree Physiology*, **15**, 739–746.

Knight, B., Zhao, F.J., McGrath, S.P. & Shen, Z.G. (1997) Zinc and cadmium uptake by the hyperaccumulator Thlaspi caerulescens in contaminated soils and its effects on the concentration and chemical speciation of metals in soil solution. *Plant and Soil*, **197**, 71–78.

Knoop, W.T. & Walker, B.H. (1985) Interactions of woody and herbaceous vegetation in a southern African savanna. *Journal of Ecology*, **73**, 235–253.

Kohls, S.J., Baker, D.D., van Kessel, C. & Dawson, J.O. (2003) An assessment of soil enrichment by actinorhizal N_2 fixation using delta-^{15}N values in a chronosequence of deglaciation at Glacier Bay, Alaska. *Plant and Soil*, **254**, 11–17.

Kohyama, T. & Takada, T. (2009) The stratification theory for plant coexistence promoted by one-sided competition. *Journal of Ecology*, **97**, 463–471.

Konno, M., Iwamoto, S. & Seiwa, K. (2011) Specialization of a fungal pathogen on host tree species in a cross-inoculation experiment. *Journal of Ecology*, **99**, 1394–1401.

Kooijman, A.M. & Bakker, C. (1994) The acidification capacity of wetland bryophytes as influenced by simulated clean and polluted rain. *Aquatic Botany*, **48**, 133–144.

Korhola, A. (1996) Initiation of a sloping mire complex in southwestern Finland: autogenic versus allogenic controls. *Ecoscience*, **3**, 216–222.

Körner, C. (2003a) *Alpine Plant Life: Functional Plant Ecology of High Mountain Ecosystems*, 2nd edn. Springer, Berlin.

Körner, C. (2003b) Limitation and stress: always or never? *Journal of Vegetation Science*, **14**, 141–143.

Körner, C., Stöcklin, J., Reuther-Thiébaud, L. & Pelaez-Riedl, S. (2008) Small differences in arrival time influence composition and productivity of plant communities. *New Phytologist*, **177**, 698–705.

Körner, S. (2001) Development of submerged macrophytes in shallow Lake Muggelsee (Berlin, Germany) before and after its switch to the phytoplankton-dominated state. *Archiv für Hydrobiologie*, **152**, 395–409.

Korstian, C.F. & Stickel, P.W. (1927) The natural replacement of blight-killed chestnut in the hardwood forests of the Northeast. *Journal of Agricultural Research*, **34**, 631–648.

Kraft, N.J.B. & Ackerly, D.D. (2010) Functional trait and phylogenetic tests of community assembly across spatial scales in an Amazonian forest. *Ecological Monographs*, **80**, 401–422.

Kraft, N.J.B., Valencia, R. & Ackerly, D.D. (2008) Functional traits and niche-based tree community assembly in an Amazonian forest. *Science*, **322**, 580–582.

Král, K., McMahon, S.M., Janik, D., Adam, D. & Vrska, T. (2014) Patch mosaic of developmental stages in central European natural forests along vegetation gradient. *Forest Ecology and Management*, **330**, 17–28.

Kraus, M., Fusseder, A. & Beck, E. (1987) Development and replenishment of the P-depletion zone around the primary root of maize during the vegetation period. *Plant and Soil*, **101**, 247–255.

Krebs, C.J. (1978) *Ecology, the Experimental Analysis of Distribution and Abundance*. Harper and Row, New York.

Kuang, J.J. & Chesson, P. (2010) Interacting coexistence mechanisms in annual plant communities: frequency-dependent predation and the storage effect. *Theoretical Population Biology*, **77**, 56–70.

Kubota, Y. (2000) Spatial dynamics of regeneration in a conifer/broad-leaved forest in northern Japan. *Journal of Vegetation Science*, **11**, 633–640.

Kubota, Y., Konno, Y. & Hiura, T. (1994) Stand structure and growth patterns of understorey trees in a coniferous forest, Taisetsuzan National Park, northern Japan. *Ecological Research*, **9**, 333–341.

Kucbel, S., Jaloviar, P., Saniga, M., Vencurik, J. & Klimaš, V. (2010) Canopy gaps in an old-growth fir-beech forest remnant of Western Carpathians. *European Journal of Forest Research*, **129**, 249–259.

Kueffer, C., Kronauer, L. & Edwards, P.J. (2009) Wider spectrum of fruit traits in invasive than native floras may increase the vulnerability of oceanic islands to plant invasions. *Oikos*, **118**, 1327–1234.

Kuhry, P., Nicholson, B.J., Gignac, L.D. & Bayley, S.E. (1993) Development of Sphagnum dominated peatlands in boreal continental Canada. *Canadian Journal of Botany*, **71**, 10–22.

Kuiters, A.T. & Slim, P.A. (2003) Tree colonisation of abandoned arable land after 27 years of horse-grazing: the role of bramble as a facilitator of oak wood regeneration. *Forest Ecology and Management*, **181**, 239–251.

Kunin, W.E. (1998) Biodiversity at the edge: a test of the importance of spatial "mass effects" in the Rothamsted Park Grass experiments. *Proceedings of the National Academy of Sciences of the U.S.A.*, **95**, 207–212.

La Peyre, M.K.G., Grace, J.B., Hahn, E. & Mendelssohn, I.A. (2001) The importance of competition in regulating plant species abundance along a salinity gradient. *Ecology*, **82**, 62–69.

Laird, R.A. & Schamp, B.S. (2006) Competitive intransitivity promotes species coexistence. *American Naturalist*, **168**, 182–193.

Laland, K.N., Odling-Smee, F.J. & Feldman, M.W. (1996) The evolutionary consequences of niche construction: a theoretical investigation using two-locus theory. *Journal of Evolutionary Biology*, **9**, 293–316.

Laland, K.N., Odling-Smee, F.J. & Feldman, M.W. (1999) Evolutionary consequences of niche construction and their implications for ecology. *Proceedings of the National Academy of Sciences of the U.S.A.*, **96**, 10242–10247.

Laliberte, E., Turner, B.L., Costes, T., Pearse, S.J., Wyrwoll, K.H., Zemunik, G. & Lambers, H. (2012) Experimental assessment of nutrient limitation along a 2-million-year dune chronosequence in the south-western Australia biodiversity hotspot. *Journal of Ecology*, **100**, 631–642.

Lambers, J.H.R., Harpole, W.S., Tilman, D. & Knops, J. (2004) Mechanisms responsible for the positive diversity-productivity relationship in Minnesota grasslands. *Ecology Letters*, **7**, 661–668.

Lambers, J.H.R., Yelenik, S.G., Colman, B.P. & Levine, J.M. (2010) California annual grass invaders: the drivers or passengers of change? *Journal of Ecology*, **98**, 1147–1156.

Lamont, B.B., Klinkhamer, P.G.L. & Witkowski, E.T.F. (1993) Population fragmentation may reduce fertility to zero in *Banksia goodii* – a demonstration of the Allee effect. *Oecologia*, **94**, 446–450.

Lamprey, H. (1964) Estimation of the large mammal densities, biomass and energy exchange in the Tarangire Game Reserve and the Maasai Steppe in Tanzania. *East African Wildlife Journal*, **2**, 1–46.

Landres, P.B. & MacMahon, J.A. (1980) Guilds and community organization: analysis of an oak woodland avifauna in Sonora, Mexico. *Auk*, **97**, 351–365.

Lankau, R.A. (2012) Coevolution between invasive and native plants driven by chemical competition and soil biota. *Proceedings of the National Academy of Sciences of the U.S.A.*, **109**, 11240–11245.

Lankau, R.A. & Strauss, S.Y. (2007) Mutual feedbacks maintain both genetic and species diversity in a plant community. *Science*, **317**, 1561–1563.

Larson, J.E. & Funk, J.L. (2016) Regeneration: an overlooked aspect of trait-based plant community assembly models. *Journal of Ecology*, **104**, 1284–1298.

Law, R. & Morton, R.D. (1996) Permanence and the assembly of ecological communities. *Ecology*, **77**, 762–775.

Lawton, J.H. (1991) Ecology as she is done, and could be done. *Oikos*, **61**, 289–290.

Lawton, J.H. (1999) Are there general laws in ecology? *Oikos*, **84**, 177–192.

Leake, J.R. (1994) The biology of myco-heterotrophic ('saprophytic') plants. *New Phytologist*, **127**, 171–216.

Lee, M.R., Flory, S.L. & Phillips, R.P. (2012) Positive feedbacks to growth of an invasive grass through alteration of nitrogen cycling. *Oecologia*, **170**, 457–465.

Lee, W.G., Wilson, J.B., Meurk, C.D. & Kennedy, P.C. (1991) Invasion of the subantarctic Auckland Islands, New Zealand, by the asterad tree Olearia lyallii and its interaction with a resident myrtaceous tree Metrosideros umbellata. *Journal of Biogeography*, **18**, 493–508.

Leeflang, L. (1999) Are stoloniferous plants able to avoid neighbours in response to low R:FR ratios in reflected light? *Plant Ecology*, **141**, 59–65.

Leger, E.A. & Rice, K.J. (2003) Invasive Californian poppies (*Eschscholzia californica* Cham.) grow larger than native individuals under reduced competition. *Ecology Letters*, **6**, 257–264.

Lehman, C.L. & Tilman, D. (2000) Biodiversity, stability, and productivity in competitive communities. *American Naturalist*, **156**, 534–552.

Leibold, M.A. & Mikkelson, G.M. (2002) Coherence, species turnover, and boundary clumping: elements of meta-community structure. *Oikos*, **97**, 237–250.

Lenton, T.M. & van Oijen, M. (2002) Gaia as a complex adaptive system. *Philosophical Transactions of the Royal Society of London B*, **357**, 683–695.

Lepik, A., Abakumova, M., Zobel, K. & Semchenko, M. (2012) Kin recognition is density-dependent and uncommon among temperate grassland plants. *Functional Ecology*, **26**, 1214–1220.

Leschen, R.A.B., Buckley, T.R., Harman, H.M. & Shulmeister, J. (2008) Determining the origin and age of the Westland beech (Nothofagus) gap, New Zealand, using fungus beetle genetics. *Molecular Ecology*, **17**, 1256–1276.

Levin, S.A. (1974) Dispersion and population interactions. *American Naturalist*, **108**, 207–228.

Levine, J.M. (2001) Local interactions, dispersal, and native and exotic plant diversity along a California stream. *Oikos*, **95**, 397–408.

Levins, R. & Culver, D. (1971) Regional coexistence of species and competition between rare species. *Proceedings of the National Academy of Sciences of the U.S.A.*, **68**, 1246–1248.

Levins, R. & Lewontin, R. (1985) *The Dialectical Biologist*. Harvard University Press, Cambridge, MA.

Lewis, M.W., Leslie, M.E. & Liljegren, S.J. (2006) Plant separation: 50 ways to leave your mother. *Current Opinion in Plant Biology*, **9**, 59–65.

Lewontin, R.C. (1969) The meaning of stability. *Diversity and Stability in Ecological Systems, Brookhaven Symposia in Biology No. 22*, pp. 13–23. Brookhaven National Laboratory, New York.

Li, L., Li, S.-M., Sun, J.-H., Zhou, L.-L., Bao, X.-G., Zhang, H.-G. & Zhang, F.-S. (2007) Diversity enhances agricultural productivity via rhizosphere phosphorus facilitation on phosphorus-deficient soils. *Proceedings of the National Academy of Sciences of the U.S.A.*, **104**, 11192–11196.

Liancourt, P., Callaway, R.M. & Michalet, R. (2005) Stress tolerance and competitive-response ability determine the outcome of biotic interactions. *Ecology*, **86**, 1611–1618.

Lichtenthaler, H.K. (1996) Vegetation stress: an introduction to the stress concept in plants. *Journal of Plant Physiology*, **148**, 4–14.

Lieberman, M., Peralta, R. & Hartshorn, G.S. (1995) Canopy closure and the distribution of tropical forest tree species at La-Selva, Costa-Rica. *Journal of Tropical Ecology*, **11**, 161–178.

Lind, E.M. & Morrison, M.E.S. (1974) *East African Vegetation*. Longman, London.

Lippmaa, T. (1939) The unistratal concept of plant communities (the unions). *American Midland Naturalist*, **21**, 111–145.

Liu, X.B., Liang, M.X., Etienne, R.S., Wang, Y.F., Staehelin, C. & Yu, S.X. (2012) Experimental evidence for a phylogenetic Janzen-Connell effect in a subtropical forest. *Ecology Letters*, **15**, 111–118.

Lloyd, K.M., Lee, W.G., Fenner, M. & Loughnan, A.E. (2003) Vegetation change after artificial disturbance in an alpine Chionochloa pallens grassland in New Zealand. *New Zealand Journal of Ecology*, **27**, 31–36.

Lloyd, K.M., McQueen, A.A.M., Lee, B.J., Wilson, R.C.B., Walker, S. & Wilson, J.B. (2000) Evidence on ecotone concepts from switch, environmental and anthropogenic ecotones. *Journal of Vegetation Science*, **11**, 903–910.

Lockwood, J.A. & Lockwood, D.R. (1993) Catastrophe theory: a unified paradigm for rangeland ecosystem dynamics. *Journal of Range Management*, **46**, 282–288.

Long, J.N. & Smith, F.W. (1992) Volume increment in *Pinus contorta* var. *latifolia*: the influence of stand development and crown dynamics. *Forest Ecology and Management*, **53**, 53–64.

Long, Z.T., Carson, W.P. & Peterson, C.J. (1998) Can disturbance create refugia from herbivores? An example with hemlock regeneration on treefall mounds. *Journal of the Torrey Botanical Society*, **125**, 165–168.

Long, Z.T., Mohler, C.L. & Carson, W.P. (2003) Extending the resource concentration hypothesis to plant communities: effects of litter and herbivores. *Ecology*, **84**, 652–665.

Longworth, J.B., Mesquita, R.C., Bentos, T.V., Moreira, M.P., Massoca, P.E. & Williamson, G.B. (2014) Shifts in dominance and species assemblages over two decades in alternative successions in central amazonia. *Biotropica*, **46**, 529–537.

Lord, J.M., Wilson, J.B., Steel, J.B. & Anderson, B.J. (2000) Community reassembly: a test using limestone grassland in New Zealand. *Ecology Letters*, **3**, 213–218.

Loreau, M. & Behera, N. (1999) Phenotypic diversity and stability of ecosystem processes. *Theoretical Population Biology*, **56**, 29–47.

Loreau, M. & Hector, A. (2001) Partitioning selection and complementarity in biodiversity experiments. *Nature*, **412**, 72–76.

Loreau, M., Mouquet, N. & Gonzalez, A. (2003) Biodiversity as spatial insurance in heterogeneous landscapes. *Proceedings of the National Academy of Sciences of the U.S.A.*, **100**, 12765–12770.

Loreau, M., Sapijanskas, J., Isbell, F. & Hector, A. (2012) Niche and fitness differences relate the maintenance of diversity to ecosystem function: comment. *Ecology*, **93**, 1482–1487.

Lortie, C.J., Brooker, R.W., Choler, P., Kikvidze, Z., Michalet, R., Pugnaire, F.I. & Callaway, R.M. (2004) Rethinking plant community ecology. *Oikos*, **107**, 433–438.

Lovelock, J.E. (1979) *Gaia: A New Look at Life on Earth*. Oxford University Press, Oxford.

Luh, H.-K. & Pimm, S.L. (1993) The assembly of ecological communities: a minimalist approach. *Journal of Animal Ecology*, **62**, 749–765.

Luo, W.B., Xie, Y.H., Chen, X.S., Li, F. & Qin, X.Y. (2010) Competition and facilitation in three marsh plants in response to a water-level gradient. *Wetlands*, **30**, 525–530.

Lüscher, A., Connolly, J. & Jacquard, P. (1992) Neighbour specificity between *Lolium perenne* and *Trifolium repens* from a natural pasture. *Oecologia*, **91**, 404–409.

Lusk, C. & Ogden, J. (1992) Age structure and dynamics of a podocarp broadleaf forest in Tongariro-National-Park, New-Zealand. *Journal of Ecology*, **80**, 379–393.

Lusk, C.H., Chazdon, R.L., Hofmann, G. & Memmott, J. (2006) A bounded null model explains juvenile tree community structure along light availability gradients in a temperate rain forest. *Oikos*, **112**, 131–137.

Lutz, H.J. (1943) Injuries to trees caused by Celastrus and Vitis. *Bulletin of the Torrey Botanical Club*, **70**, 436–439.

Lyons, K.G. & Schwartz, M.W. (2001) Rare species loss alters ecosystem function – invasion resistance. *Ecology Letters*, **4**, 358–365.

Maalouf, J.P., Le Bagousse-Pinguet, Y., Marchand, L., Touzard, B. & Michalet, R. (2012) The interplay of stress and mowing disturbance for the intensity and importance of plant interactions in dry calcareous grasslands. *Annals of Botany*, **110**, 821–828.

MacArthur, R. & Levins, R. (1967) The limiting similarity, convergence, and divergence of coexisting species. *American Naturalist*, **101**, 377–385.

MacArthur, R.H. (1957) On the relative abundance of bird species. *Proceedings of the National Academy of Sciences of the U.S.A.*, **43**, 293–295.

MacArthur, R.H. (1972) *Geographical Ecology: Patterns in the Distribution of Species*. Harper and Row, New York.

MacArthur, R.H. & Wilson, E.O. (1963) An equilibrium theory of insular zoogeography. *Evolution*, **17**, 373–387.

MacArthur, R.H. & Wilson, E.O. (1967) *The Theory of Island Biogeography*. Princeton University Press, Princeton, NJ.

MacDonald, I.A.W. & Cooper, J. (1995) Insular Lessons for Global Biodiversity Conservation with Particular Reference to Alien Invasions. In *Islands: Biological Diversity and Ecosystem Function* (eds P. Vitousek, L.L. Loope & H. Andersen), pp. 189–202. Springer-Verlag, Berlin.

MacNally, R. (2000) Coexistence of a locally undifferentiated foraging guild: avian snatchers in a southeastern Australian forest. *Austral Ecology*, **25**, 69–82.

Madon, O. & Médail, F. (1997) The ecological significance of annuals on a Mediterranean grassland (Mt Ventoux, France). *Plant Ecology*, **129**, 189–199.

Madritch, M., Donaldson, J.R. & Lindroth, R.L. (2006) Genetic identity of Populus tremuloides litter influences decomposition and nutrient release in a mixed forest stand. *Ecosystems*, **9**, 528–537.

Maestre, F.T. & Cortina, J. (2004) Do positive interactions increase with abiotic stress? A test from a semi-arid steppe. *Biology Letters*, **271**, 331–333.

Maestre, F.T., Valladares, F. & Reynolds, J.F. (2005) Is the change of plant–plant interactions with abiotic stress predictable? A meta-analysis of field results in arid environments. *Journal of Ecology*, **93**, 748–757.

Mahall, B.E. & Callaway, R.M. (1991) Root communication among desert shrubs. *Proceedings of the National Academy of Sciences of the U.S.A.*, **88**, 874–876.

Mahall, B.E. & Callaway, R.M. (1992) Root communication mechanisms and intracommunity distributions of two Mojave Desert shrubs. *Ecology*, **73**, 2145–2151.

Makita, A. (1996) Density regulation during the regeneration of two monocarpic bamboos: self-thinning or intra clonal regulation? *Journal of Vegetation Science*, **7**, 281–288.

Mallik, A.U. (2003) Conifer regeneration problems in boreal and temperate forests with ericaceous understory: role of disturbance, seedbed limitation, and keystone species change. *Critical Reviews in Plant Sciences*, **22**, 341–366.

Malloch, A.J.C. (1997) Influence of salt spray on dry coastal vegetation. In *Dry Coastal Ecosystems: General Aspects* (ed. E. van der Maarel), pp. 411–420. Elsevier, Amsterdam.

Malmer, N., Albinsson, C., Svensson, B.M. & Wallén, B. (2003) Interferences between Sphagnum and vascular plants: effects on plant community structure and peat formation. *Oikos*, **100**, 469–482.

Malmer, N., Svensson, B.M. & Wallén, B. (1994) Interactions between Sphagnum mosses and field layer vascular plants in the development of peat-forming systems. *Folia Geobotanica et Phytotaxonomica*, **29**, 483–496.

Malmstrom, C.M., McCullough, A.J., Johnson, H.A., Newton, L.A. & Borer, E.T. (2005) Invasive annual grasses indirectly increase virus incidence in California native perennial bunchgrasses. *Oecologia*, **145**, 153–164.

Manders, P.T. & Richardson, D.M. (1992) Colonisation of Cape Fynbos communities by forest species. *Forest Ecology and Management*, **48**, 277–293.

Manders, P.T. & Smith, R.E. (1992) Effects of watering regime on growth and competitive ability of nursery grown Cape Fynbos and forest plants. *South African Journal of Botany*, **58**, 188–194.

Mangan, S.A., Herre, E.A. & Bever, J.D. (2010) Specificity between Neotropical tree seedlings and their fungal mutualists leads to plant-soil feedback. *Ecology*, **91**, 2594–2603.

Mangla, S., Inderjit & Callaway, R.M. (2008) Exotic invasive plant accumulates native soil pathogens which inhibit native plants. *Journal of Ecology*, **96**, 58–67.

Mark, A.F. & Wilson, J.B. (2005) Tempo and mode of vegetation dynamics over 50 years in a New Zealand alpine cushion/tussock community. *Journal of Vegetation Science*, **16**, 227–236.

Marks, C.O. & Muller-Landau, H.C. (2007) Comment on "From plant traits to plant communities: a statistical mechanistic approach to biodiversity". *Science*, **316**, 1425.

Maron, J.L., Marler, M., Klironomos, J.N. & Cleveland, C.C. (2011) Soil fungal pathogens and the relationship between plant diversity and productivity. *Ecology Letters*, **14**, 36–41.

Marquard, E., Weigelt, A., Temperton, V.M., Roscher, C., Schumacher, J., Buchmann, N., Fischer, M., Weisser, W.W. & Schmid, B. (2009) Plant species richness and functional composition drive overyielding in a six-year grassland experiment. *Ecology*, **90**, 3290–3302.

Marr, J.W. (1977) The development and movement of tree islands near the upper limit of tree growth in the southern Rocky Mountains. *Ecology*, **58**, 1159–1164.

Marrs, R.H. & Hicks, M.J. (1986) Study of vegetation change at Lakenheath Warren: a re-examination of A.S. Watt's theories of bracken dynamics in relation to succession and management. *Journal of Applied Ecology*, **23**, 1029–1046.

Marshall, C. (1996) Sectoriality and physiological organisation in herbaceous plants: an overview. *Vegetatio*, **127**, 9–16.

Martin, M.M. & Harding, J. (1981) Evidence for the evolution of competition between two species of annual plants. *Evolution*, **35**, 975–987.

Martin, P.H., Canham, C.D. & Kobe, R.K. (2010) Divergence from the growth-survival trade-off and extreme high growth rates drive patterns of exotic tree invasions in closed-canopy forests. *Journal of Ecology*, **98**, 778–789.

Martin, P.H. & Marks, P.L. (2006) Intact forests provide only weak resistance to a shade-tolerant invasive Norway maple (*Acer platanoides* L.). *Journal of Ecology*, **94**, 1070–1079.

Mason, N.W.H. & Wilson, J.B. (2006) Mechanisms of species coexistence in a lawn community: mutual corroboration between two independent assembly rules. *Community Ecology*, **7**, 109–116.

Mason, N.W.H., Wilson, J.B. & Steel, J.B. (2007) Are alternative stable states more likely in high stress environments? Logic and available evidence do not support Didham et al. 2005. *Oikos*, **116**, 353–357.

Matista, A.A. & Silk, W.K. (1997) An electronic device for continuous, in vivo measurement of forces exerted by twining vines. *American Journal of Botany*, **84**, 1164–1168.

Matsui, T., Dougherty, N.J., Loughnan, A.E., Swaney, J.K., Laurence, B.L., Lloyd, K.M. & Wilson, J.B. (2002) Local texture convergence within three communities in Fiordland, New Zealand. *New Zealand Journal of Ecology*, **26**, 15–22.

May, F., Giladi, I., Ristow, M., Ziv, Y. & Jeltsch, F. (2013) Metacommunity, mainland-island system or island communities? Assessing the regional dynamics of plant communities in a fragmented landscape. *Ecography*, **36**, 842–853.

May, R.M. (1972) Will a large complex system be stable? *Nature*, **238**, 413–414.

May, R.M. (1973) *Stability and Complexity in Model Ecosystems*. Princeton University Press, Princeton, NJ.

May, R.M. (1975) Patterns of species abundance and diversity. In *Ecology and Evolution of Communities* (eds M.L. Cody & J.M. Diamond), pp. 81–120. Harvard University Press, Cambridge, MA.

Maynard Smith, J. & Price, G.R. (1973) The logic of animal conflict. *Nature*, **246**, 15–18.

Mazza, C.A., Gimenez, P.I., Kantolic, A.G. & Ballare, C.L. (2013) Beneficial effects of solar UV-B radiation on soybean yield mediated by reduced insect herbivory under field conditions. *Physiologia Plantarum*, **147**, 307–315.

McCarthy-Neumann, S. & Kobe, R.K. (2010) Conspecific plant-soil feedbacks reduce survivorship and growth of tropical tree seedlings. *Journal of Ecology*, **98**, 396–407.

McCune, B. & Allen, T.F.H. (1985) Will similar forests develop on similar sites? *Canadian Journal of Botany*, **63**, 367–376.

McDonald, A.K., Kinucan, R.J. & Loomis, L.E. (2009) Ecohydrological interactions within banded vegetation in the northeastern Chihuahuan Desert, USA. *Ecohydrology*, **2**, 66–71.

McGinley, M.A., Dhillion, S.S. & Neumann, J.C. (1994) Environmental heterogeneity and seedling establishment – ant-plant-microbe interactions. *Functional Ecology*, **8**, 607–615.

McInerny, G.J. & Etienne, R.S. (2012) Ditch the niche – is the niche a useful concept in ecology or species distribution modelling? *Journal of Biogeography*, **39**, 2096–2102.

McKee, A., La Roi, G. & Frankilin, J.F. (1982) Structure, composition and reproductive behaviour of terrace forests, South Fork Hoh River, Olympic National Park. In *Ecological Research in National Parks of the Pacific Northwest* (eds E.E. Starkey, J.F. Franklin & J.W. Matthews), pp. 22–28. National Park Service Cooperative Park Studies Unit, Corvallis.

McNeilly, T. & Roose, M.L. (1996) Co-adaptation between neighbours? A case study with Lolium perenne genotypes. *Euphytica*, **92**, 121–128.

McPherson, G.R., Wright, H.A. & Wester, D.B. (1988) Patterns of shrub invasion in semiarid Texas grasslands. *American Midland Naturalist*, **120**, 391–397.

McSorley, R. & Dickson, D.W. (1995) Effect of tropical rotation crops on *Meloidogyne incognita* and other plant-parasitic nematodes. *Journal of Nematology*, **27**, 535–544.

Meier, C.L., Keyserling, K. & Bowman, W.D. (2009) Fine root inputs to soil reduce growth of a neighbouring plant via distinct mechanisms dependent on root carbon chemistry. *Journal of Ecology*, **97**, 941–949.

Meijer, W. (1976) A note on *Podostemon ceratophyllum* Michx. as an indicator of clean streams in and around the Appalachian Mountains. *Castanea*, **41**, 319–324.

Meiners, S.J., Cadenasso, M.L. & Pickett, S.T.A. (2007) Succession on the Piedmont of New Jersey and its implications for ecological restoration. In *Old fields: Dynamics and Restoration of Abandoned Farmland* (eds V.A. Cramer & R.J. Hobbs), pp. 145–161. Island Press, Washington DC.

Mensah, R.K. & Khan, M. (1997) Use of *Medicago sativa* (L.) interplantings/trap crops in the management of the green mirid, *Creontiades dilutus* (Stål) in commercial cotton in Australia. *International Journal of Pest Management*, **43**, 197–202.

Mergeay, J., De Meester, L., Eggermont, H. & Verschuren, D. (2011) Priority effects and species sorting in a long paleoecological record of repeated community assembly through time. *Ecology*, **92**, 2267–2275.

Merritt, D.M., Nilsson, C. & Jansson, R. (2010) Consequences of propagule dispersal and river fragmentation for riparian plant community diversity and turnover. *Ecological Monographs*, **80**, 609–626.

Michalet, R., Le Bagousse-Pinguet, Y., Maalouf, J.P. & Lortie, C.J. (2014) Two alternatives to the stress-gradient hypothesis at the edge of life: the collapse of facilitation and the switch from facilitation to competition. *Journal of Vegetation Science*, **25**, 609–613.

Middendorp, R.S., Vlam, M., Rebel, K.T., Baker, P.J., Bunyavejchewin, S. & Zuidema, P.A. (2013) Disturbance history of a seasonal tropical forest in Western Thailand: a spatial dendroecological analysis. *Biotropica*, **45**, 578–586.

Midgley, J.J., Cameron, M.C. & Bond, W.J. (1995) Gap characteristics and replacement patterns in the Knysna Forest, South Africa. *Journal of Vegetation Science*, **6**, 29–36.

Midgley, J.J., Cowling, R.M., Hendricks, H., Desmet, P.G., Esler, K. & Rundel, P.W. (1997) Population ecology of tree succulents (Aloe and Pachypodium) in the arid western Cape: decline of keystone species. *Biodiversity and Conservation*, **6**, 869–876.

Mihail, J.D., Alexander, H.M. & Taylor, S.J. (1998) Interactions between root-infecting fungi and plant density in an annual legume, *Kummerowia stipulacea. Journal of Ecology*, **86**, 739–748.

Miller, A.D. & Chesson, P. (2009) Coexistence in disturbance-prone communities: how a resistance-resilience trade-off generates coexistence via the storage effect. *American Naturalist*, **173**, 30–43.

Miller, A.E., Bowman, W.D. & Suding, K.N. (2007) Plant uptake of inorganic and organic nitrogen: neighbour identity matters. *Ecology*, **88**, 1832–1840.

Miller, G., Mangan, J., Pollard, D., Thompson, S., Felzer, B. & Magee, J. (2005) Sensitivity of the Australian monsoon to insolation and vegetation: implications for human impact on continental moisture balance. *Geology*, **33**, 65–68.

Miller, M.F. (1995) Acacia seed survival, seed germination and seedling growth following pod consumption by large herbivores and seed chewing rodents. *African Journal of Ecology*, **33**, 194–210.

Miller, T.E. & Werner, P.A. (1987) Competitive effects and responses between plant species in a first-year old-field community. *Ecology*, **68**, 1201–1210.

Millington, W.F. & Chaney, W.R. (1973) Shedding of shoots and branches. In *Shedding of Plant Parts* (ed. T.T. Kozlowski), pp. 149–204. Academic Press, New York.

Minchinton, T.E., Simpson, J.C. & Bertness, M.D. (2006) Mechanisms of exclusion of native coastal marsh plants by an invasive grass. *Journal of Ecology*, **94**, 342–354.

Mirams, R.V. (1957) Aspects of the natural regeneration of the kauri (*Agathis australis* Salisb.). *Transactions of the Royal Society of New Zealand*, **84**, 661–680.

Mirkin, B.M. (1994) Which plant communities do exist? *Journal of Vegetation Science*, **5**, 283–284.

Mitamura, M., Yamamura, Y. & Nakano, T. (2009) Large-scale canopy opening causes decreased photosynthesis in the saplings of shade-tolerant conifer, *Abies veitchii*. *Tree Physiology*, **29**, 137–145.

Mitchell, C., Tilman, D. & Groth, J. (2002) Effects of grassland plant species diversity, abundance, and composition on foliar fungal disease. *Ecology*, **83**, 1713–1726.

Mitchell, C.E. (2003) Trophic control of grassland production and biomass by pathogens. *Ecology Letters*, **6**, 147–155.

Mitchell, C.E. & Power, A.G. (2003) Release of invasive plants from fungal and viral pathogens. *Nature*, **421**, 625–627.

Mitchell, C.E. & Power, A.G. (2006) Disease dynamics in plant communities. *Disease Ecology: Community Structure and Pathogen Dynamics* (eds S.K. Collinge & C. Ray), pp. 58–72. Oxford University Press, Oxford.

Mitchell, C.E., Reich, P.B., Tilman, D. & Groth, J.V. (2003) Effects of elevated CO_2, nitrogen deposition, and decreased species diversity on foliar fungal plant disease. *Global Change Biology*, **9**, 438–451.

Mitchell, S. (1989) Primary production in a shallow eutrophic lake dominated alternately by phytoplankton and by submerged macrophytes. *Aquatic Botany*, **33**, 101–110.

Mitchley, J. & Grubb, P. (1986) Control of relative abundance of perennials in chalk grassland in southern England. I. Constancy of rank order and results of pot- and field-experiments on the role of interference. *Journal of Ecology*, **74**, 1139–1166.

Mittelbach, G., Steiner, C., Scheiner, S., Gross, K., Reynolds, H., Waide, R., Willig, M., Dodson, S. & Gough, L. (2001) What is the observed relationship between species richness and productivity? *Ecology*, **82**, 2381–2396.

Moeller, D. (2004) Facilitative interactions among plants via shared pollinators. *Ecology*, **85**, 3289–3301.

Mohler, C. (1990) Co-occurrence of oak subgenera: implications for niche differentiation. *Bulletin of the Torrey Botanical Club*, **117**, 247–255.

Molloy, B., Partridge, T. & Thomas, W. (1991) Decline of tree lupin (*Lupinus arboreus*) on Kaitorete Spit, Canterbury, New-Zealand, 1984–1990. *New Zealand Journal of Botany*, **29**, 349–352.

Moog, D., Kahmen, S. & Poschlod, P. (2005) Application of C-S-R- and LHS-strategies for the distinction of differently managed grasslands. *Basic and Applied Ecology*, **6**, 133–143.

Moore, J.C., de Ruiter, P.C. & Hunt, H.W. (1993) Influence of productivity on the stability of real and modelled ecosystems. *Science*, **261**, 906–908.

Moore, J.L., Mouquet, N., Lawton, J.H. & Loreau, M. (2001) Coexistence, saturation and invasion resistance in simulated plant assemblages. *Oikos*, **94**, 303–314.

Moore, R.M. & Williams, J.D. (1976) A study of a subalpine woodland-grassland boundary. *Australian Journal of Ecology*, **1**, 145–153.

Morrison, J.A. & Mauck, K. (2007) Experimental field comparison of native and non-native maple seedlings: natural enemies, ecophysiology, growth and survival. *Journal of Ecology*, **95**, 1036–1049.

Motzkin, G., Orwig, D.A. & Foster, D.R. (2002) Vegetation and disturbance history of a rare dwarf pitch pine community in western New England. *USA Journal of Biogeography*, **29**, 1455–1467.

Mouillot, D., Mason, N.W.H. & Wilson, J. (2007) Is the abundance of species determined by their functional traits? A new method with a test using plant communities. *Oecologia*, **152**, 729–737.

Mouillot, D. & Wilson, J.B. (2002) Can we tell how a community was constructed? A comparison of five indices for their ability to identify theoretical models of community construction. *Theoretical Population Biology*, **61**, 141–151.

Mouquet, N., Leadley, P., Mériguet, J. & Loreau, M. (2004) Immigration and local competition in herbaceous plant communities: a three-seed-sowing experiment. *Oikos*, **104**, 77–90.

Mueller-Dombois, D. & Ellenberg, H. (1974) *Aims and Methods of Vegetation Ecology*. John Wiley and Sons, New York.

Mukhortova, L.V., Kirdyanov, A.V., Myglan, V.S. & Guggenberger, G. (2009) Wood transformation in dead-standing trees in the forest-tundra of Central Siberia. *Biology Bulletin [Russian Academy of Sciences]*, **36**, 58–65.

Muko, S. & Iwasa, Y. (2003) Incomplete mixing promotes species coexistence in a lottery model with permanent spatial heterogeneity. *Theoretical Population Biology*, **64**, 359–368.

Muler, A.L., Oliveira, R.S., Lambers, H. & Veneklaas, E.J. (2014) Does cluster-root activity benefit nutrient uptake and growth of co-existing species? *Oecologia*, **174**, 23–31.

Munday, P.L. (2004) Competitive coexistence of coral-dwelling fishes: the lottery hypothesis revisited. *Ecology*, **85**, 623–628.

Murillo, N., Laterra, P. & Monterubbianesi, G. (2007) Post-dispersal granivory in a tall-tussock grassland: a positive feedback mechanism of dominance? *Journal of Vegetation Science*, **18**, 799–806.

Muro-Pérez, G., Jurado, E., Flores, J., Sánchez-Salas, J., García-Pérez, J. & Estrada, E. (2012) Positive effects of native shrubs on three specially protected cacti species in Durango, Mexico. *Plant Species Biology*, **27**, 53–58.

Murphy, G.P. & Dudley, S.A. (2009) Kin recognition: competition and cooperation in Impatiens (Balsaminaceae). *American Journal of Botany*, **96**, 1990–1996.

Murray, B.R. (1998) Density-dependent germination and the role of seed leachate. *Australian Journal of Ecology*, **23**, 411–418.

Murrell, D.J. & Law, R. (2003) Heteromyopia and the spatial coexistence of similar competitors. *Ecology Letters*, **6**, 48–59.

Musselman, L.J. & Press, M.C. (1995) Introduction to parasitic plants. In *Parasitic Plants* (eds M.C. Press & J.D. Graves), pp. 1–13. Chapman & Hall, London.

Muth, C.C. & Bazzaz, F.A. (2002) Tree seedling canopy responses to conflicting photosensory cues. *Oecologia*, **132**, 197–204.

Myers, B.J. & Talsma, T. (1992) Site water-balance and tree water status in irrigated and fertilised stands of Pinus radiata. *Forest Ecology and Management*, **52**, 17–42.

Myers, J.A., Chase, J.M., Jimenez, I., Jorgensen, P.M., Araujo-Murakami, A., Paniagua-Zambrana, N. & Seidel, R. (2013) Beta-diversity in temperate and tropical forests reflects dissimilar mechanisms of community assembly. *Ecology Letters*, **16**, 151–157.

Myers, J.A. & Harms, K.E. (2009) Local immigration, competition from dominant guilds, and the ecological assembly of high-diversity pine savannas. *Ecology*, **90**, 2745–2754.

Nadkarni, N.M. & Matelson, T.J. (1992) Biomass and nutrient dynamics of epiphytic litterfall in a neotropical montane forest, Costa Rica. *Biotropica*, **24**, 24–30.

Naeem, S. & Li, S. (1997) Biodiversity enhances ecosystem reliability. *Nature*, **390**, 507–509.

Nagashima, H. & Hikosaka, K. (2011) Plants in a crowded stand regulate their height growth so as to maintain similar heights to neighbours even when they have potential advantages in height growth. *Annals of Botany*, **108**, 207–214.

Nantel, P. & Gagnon, D. (1999) Variability in the dynamics of northern peripheral versus southern populations of two clonal plant species, *Helianthus divaricatus* and *Rhus aromatica*. *Journal of Ecology*, **87**, 748–760.

Nason, J.D., Herre, E.A. & Hamrick, J.L. (1998) The breeding structure of a tropical keystone plant resource. *Nature*, **391**, 685–687.

Nee, S. & May, R.M. (1992) Dynamics of metapopulations: habitat destruction and competitive coexistence. *Journal of Animal Ecology*, **61**, 37–40.

Newman, E.I. (1978) Allelopathy: adaptation or accident? In *Biochemical Aspects of Plant and Animal Coevolution* (ed. J.B. Harborne), pp. 327–342. Academic Press, London.

Newman, E.I. & Eason, W.R. (1993) Rates of phosphorus transfer within and between ryegrass *(Lolium perenne)* plants. *Functional Ecology*, **7**, 242–248.

Newman, E.I. & Ritz, K. (1986) Evidence on the pathways of phosphorus transfer between vesicular-arbuscular mycorrhizal plants. *New Phytologist*, **104**, 77–87.

Newman, E.I. & Rovira, A.D. (1975) Allelopathy among some British grassland species. *Journal of Ecology*, **63**, 727–737.

Niinemets, U. (1996) Changes in foliage distribution with relative irradiance and tree size: differences between the saplings of *Acer platanoides* acid *Quercus robur. Ecological Research*, **11**, 269–281.

Niinemets, U. (1997) Role of foliar nitrogen in light harvesting and shade tolerance of four temperate deciduous woody species. *Functional Ecology*, **11**, 518–531.

Niklas, K.J. (1998) Effects of vibration on mechanical properties and biomass allocation pattern of *Capsella bursa-pastoris* (Cruciferae). *Annals of Botany*, **82**, 147–156.

Niklas, K.J. & Hammond, S.T. (2013) Biophysical effects on plant competition and coexistence. *Functional Ecology*, **27**, 854–864.

Nippert, J.B. & Knapp, A.K. (2007) Linking water uptake with rooting patterns in grassland species. *Oecologia*, **153**, 261–272.

Nobel, P.S. (1997) Root distribution and seasonal production in the northwestern Sonoran Desert for a C3 subshrub, a C4 bunchgrass, and a CAM leaf succulent. *American Journal of Botany*, **84**, 949–955.

Nord, E.A., Zhang, C.C. & Lynch, J.P. (2011) Root responses to neighbouring plants in common bean are mediated by nutrient concentration rather than self/non-self recognition. *Functional Plant Biology*, **38**, 941–952.

Northup, R.R., Dahlgren, R.A. & McColl, J.G. (1998) Polyphenols as regulators of plant-litter-soil interactions in northern California's pygmy forest: a positive feedback? *Biogeochemistry*, **42**, 189–220.

Norton, D.A. (1997) Host specificity and spatial distribution patterns of mistletoes. In *New Zealand's Loranthaceous Mistletoes* (eds P.J. de Lange & D.A. Norton), pp. 105–109. Department of Conservation, Wellington, NZ.

Novoplansky, A., Cohen, D. & Sachs, T. (1990) How Portulaca seedlings avoid their neighbours. *Oecologia*, **82**, 490–493.

Novotny, V. & Basset, Y. (2005) Host specificity of insect herbivores in tropical forests. *Proceedings of the Royal Society of London B*, **272**, 1083–1090.

O'Brien, K.L. (1996) Tropical deforestation and climate change. *Progress in Physical Geography*, **20**, 311–335.

O'Brien, T.A., Moorby, J. & Whittington, W.J. (1967) The effect of management and competition on the uptake of ^{32}phosphorus by ryegrass, meadow fescue and their natural hybrid. *Journal of Applied Ecology*, **4**, 513–520.

O'Connor, I. & Aarssen, L.W. (1988) Species association patterns in abandoned sand quarries. *Vegetatio*, **73**, 101–109.

Odling-Smee, F.J. (1988) Niche-constructing phenotypes. In *The Role of Behaviour in Evolution* (ed. H.C. Plotkin), pp. 73–132. MIT Press, Cambridge, MA.

Odling-Smee, F.J., Laland, K.N. & Feldman, M.W. (1996) Niche construction. *American Naturalist*, **147**, 641–648.

Odling-Smee, F.J., Laland, K.N. & Feldman, M.W. (2003) *Niche Construction: The Neglected Process in Evolution*. Princeton University Press, Princeton, NJ.

Odum, H.T. (1971) *Environment, Power and Society*. Wiley-Interscience, New York.

Ogden, J. (1985) Past, present and future: studies on the population dynamics of some long-lived trees. In *Studies on Plant Demography* (ed. J. White), pp. 3–16. Academic Press, London.

Okuyama, T. & Holland, J.N. (2008) Network structural properties mediate the stability of mutualistic communities. *Ecology Letters*, **11**, 208–216.

Olff, H., Hoorens, B., de Goede, R.G.M., van der Putten, W.H. & Gleichman, J.M. (2000) Small-scale shifting mosaics of two dominant grassland species: the possible role of soil-borne pathogens. *Oecologia*, **125**, 45–54.

Oliver, L.R. & Schreiber, M.V. (1974) Competition for CO_2 in a heteroculture. *Weed Science*, **22**, 125–130.

Onipchenko, V.G., Blinnikov, M.S., Gerasimova, M.A., Volkova, E.V. & Cornelissen, J.H.C. (2009) Experimental comparison of competition and facilitation in alpine communities varying in productivity. *Journal of Vegetation Science*, **20**, 718–727.

Orians, G.H. (1975) Diversity, stability and maturity in natural ecosystems. In *Unifying Concepts in Ecology* (eds W.H. van Dobben & R.H. Lowe-McConnell), pp. 139–150. Junk, The Hague.

Orians, G.H., Dirzo, R. & Cushman, J.H. (1996) *Biodiversity and Ecosystem Processes in Tropical Forests*. Springer, Berlin.

Orians, G.H. & Paine, R.T. (1983) Convergent evolution at the community level. In *Coevolution* (eds D.J. Futuyma & M. Slatkin), pp. 431–458. Sinauer, Sunderland.

Orians, G.H. & Solbrig, O.T. (1977) A cost-income model of leaves and roots with special reference to arid and semiarid areas. *American Naturalist*, **111**, 677–690.

Orrock, J.L. & Christopher, C.C. (2010) Density of intraspecific competitors determines the occurrence and benefits of accelerated germination. *American Journal of Botany*, **97**, 694–699.

Orrock, J.L., Witter, M.S. & Reichman, O.J. (2008) Apparent competition with an exotic plant reduces native plant establishment. *Ecology*, **89**, 1168–1174.

Otto-Bliesner, B.L. & Upchurch, G.R. (1997) Vegetation-induced warming of high-latitude regions during the late Cretaceous period. *Nature*, **385**, 804–807.

Ozinga, W.A., Schaminée, J.H.J., Bekker, R.M., Bonn, S., Poschlod, P., Tackenberg, O., Bakker, J. & van Groenendael, J. (2005) Predictability of plant species composition from environmental conditions is constrained by dispersal limitation. *Oikos*, **108**, 555–561.

Padovan, A., Keszei, A., Foley, W.J. & Külheim, C. (2013) Differences in gene expression within a striking phenotypic mosaic Eucalyptus tree that varies in susceptibility to herbivory. *BMC Plant Biology*, **13**, 1–12.

Page, R.R., da Vinha, S.G. & Agnew, A.D.Q. (1985) The reaction of some sand dune plant species to experimentally imposed environmental change. *Vegetatio*, **61**, 105–114.

Paine, R.T. (1969) A note on trophic complexity and community stability. *American Naturalist*, **103**, 91–93.

Pake, C.E. & Venable, D.L. (1996) Seed banks in desert annuals: implications for persistence and coexistence in variable environments. *Ecology*, **77**, 1427–1435.

Palmer, M.W. (1994) Variation in species richness: towards a unification of hypotheses. *Folia Geobotanica et Phytotaxonomica*, **29**, 511–530.

Paquette, A., Fontaine, B., Berninger, F., Dubois, K., Lechowicz, M.J., Messier, C., Posada, J.M., Valladares, F. & Brisson, J. (2012) Norway maple displays greater seasonal growth and phenotypic plasticity to light than native sugar maple. *Tree Physiology*, **32**, 1339–1347.

Parfitt, R.L., Ross, D.J., Coomes, D.A., Richardson, S.J., Smale, M.C. & Dahlgren, R.A. (2005) N and P in New Zealand soil chronosequences and relationships with foliar N and P. *Biogeochemistry*, **75**, 305–328.

Parker, J.D. & Hay, M.E. (2005) Biotic resistance to plant invasions? Native herbivores prefer non-native plants. *Ecology Letters*, **8**, 959–967.

Parmesan, C. (2000) Unexpected density-dependent effects of herbivory in a wild population of the annual *Collinsia torreyi*. *Journal of Ecology*, **88**, 392–400.

Parrish, J.A.D. & Bazzaz, F.A. (1976) Underground niche separation in successional plants. *Ecology*, **57**, 1281–1288.

Parrish, J.A.D. & Bazzaz, F.A. (1979) Difference in pollination niche relationships in early and late successional plant communities. *Ecology*, **60**, 597–610.

Paterson, D.M. (1989) Short-term changes in the erodibility of intertidal cohesive sediments related to the migratory behavior of epipelic diatoms. *Limnology and Oceanography*, 34, 223–234.

Pauchard, A., García, R.A., Peña, E., González, C., Cavieres, L.A. & Bustamante, R.O. (2008) Positive feedbacks between plant invasions and fire regimes: *Teline monspessulana* (L.) K. Koch (Fabaceae) in central Chile. *Biological Invasions*, **10**, 547–553.

Pavlick, R., Drewry, D.T., Bohn, K., Reu, B. & Kleidon, A. (2013) The Jena Diversity-Dynamic Global Vegetation Model (JeDi-DGVM): a diverse approach to representing terrestrial bio-geography and biogeochemistry based on plant functional trade-offs. *Biogeosciences*, **10**, 4137–4177.

Pearse, I.S., Hughes, K., Shiojiri, K., Ishizaki, S. & Karban, R. (2013) Interplant volatile signaling in willows: revisiting the original talking trees. *Oecologia*, **172**, 869–875.

Peco, B. (1989) Modelling Mediterranean pasture dynamics. *Vegetatio*, **83**, 269–276.

Peh, K.S.H., Lewis, S.L. & Lloyd, J. (2011) Mechanisms of monodominance in diverse tropical tree-dominated systems. *Journal of Ecology*, **99**, 891–898.

Pelletier, B., Fyles, J.W. & Dutilleul, P. (1999) Tree species control and spatial structure of forest floor properties in a mixed-species stand. *Ecoscience*, **6**, 79–91.

Pellino, M., Hojsgaard, D., Schmutzer, T., Scholz, U., Horandl, E., Vogel, H. & Sharbel, T.F. (2013) Asexual genome evolution in the apomictic Ranunculus auricomus complex: examining the effects of hybridization and mutation accumulation. *Molecular Ecology*, **22**, 5908–5921.

Peltzer, D.A., Wardle, D.A., Allison, V.J., Baisden, W.T., Bardgett, R.D., Chadwick, O.A., Condron, L.M., Parfitt, R.L., Porder, S., Richardson, S.J., Turner, B.L., Vitousek, P.M., Walker, J. & Walker, L.R. (2010) Understanding ecosystem retrogression. *Ecological Monographs*, **80**, 509–529.

Peltzer, D.A. & Wilson, S.D. (2001) Competition and environmental stress in temperate grasslands. In *Competition and Succession in Pastures* (eds P.G. Tow & A. Lazenby), pp. 193–212. CAB International, Wallingford.

Peltzer, D.A., Wilson, S.D. & Gerry, A.K. (1998) Competition intensity along a productivity gradient in a low-diversity grassland. *American Naturalist*, **151**, 465–485.

Penfould, W.T. (1964) The relation of grazing to plant succession in the tallgrass prairie. *Journal of Range Management*, **17**, 256–260.

Pennings, S.C., Grant, M.-B. & Bertness, M.D. (2005) Plant zonation in low-latitude salt marshes: disentangling the roles of flooding, salinity and competition. *Journal of Ecology*, **93**, 159–167.

Peppler-Lisbach, C. & Kleyer, M. (2009) Patterns of species richness and turnover along the pH gradient in deciduous forests: testing the continuum hypothesis. *Journal of Vegetation Science*, **20**, 984–995.

Perez-Salicrup, D.R. & Barker, M.G. (2000) Effect of liana cutting on water potential and growth of adult *Senna multijuga* (Caesalpinoideae) trees in a Bolivian tropical forest. *Oecologia*, **124**, 469–475.

Petermann, J.S., Fergus, A.J.F., Roscher, C., Turnbull, L.A., Weigelt, A. & Schmid, B. (2010) Biology, chance, or history? The predictable reassembly of temperate grassland communities. *Ecology*, **91**, 408–421.

Petermann, J.S., Fergus, A.J.F., Turnbull, L.A. & Schmid, B. (2008) Janzen-Connell effects are widespread and strong enough to maintain diversity in grasslands. *Ecology*, **89**, 2399–2406.

Peterson, C.J. & Pickett, S.T.A. (1995) Forest reorganization: a case study in an old-growth forest catastrophic blowdown. *Ecology*, **76**, 763–774.

Petraitis, P.S. & Hoffman, C. (2010) Multiple stable states and relationship between thresholds in processes and states. *Marine Ecology Progress Series*, **413**, 189–200.

Petritan, A.M., Nuske, R.S., Petritan, I.C. & Tudose, N.C. (2013) Gap disturbance patterns in an old-growth sessile oak (*Quercus petraea* L.)-European beech (*Fagus sylvatica* L.) forest remnant in the Carpathian Mountains, Romania. *Forest Ecology and Management*, **308**, 67–75.

Pfisterer, A.B. & Schmid, B. (2002) Diversity-dependent production can decrease the stability of ecosystem functioning. *Nature*, **416**, 84–86.

Phillips, J. (1935) Succession, development, the climax, and the complex organism: an analysis of concepts. III. The complex organism: conclusions. *Journal of Ecology*, **23**, 488–508.

Phillips, O.L., Martínez, R.V., Mendoza, A.M., Baker, T.R. & Vargas, N.M. (2005) Large lianas as hyperdynamic elements of the tropical forest canopy. *Ecology*, **86**, 1250–1258.

Pianka, E.R. (1980) Guild structure in desert lizards. *Oikos*, **35**, 194–201.

Pianka, E.R. & Horn, H.S. (2005) Ecology's legacy from Robert MacArthur. In *Ecological Paradigms Lost: Routes of Theory Change* (eds K. Cuddington & B.E. Beisner), pp. 213–232. Elsevier, Amsterdam.

Pickett, S.T.A. (1976) Succession: an evolutionary interpretation. *American Naturalist*, **110**, 107–119.

Pickett, S.T.A. & Bazzaz, F.A. (1978) Organization of an assemblage of early successional species on a soil moisture gradient. *Ecology*, **59**, 1248–1255.

Pielou, E.C. (1975) *Ecological Diversity*. Wiley, New York.

Pielou, E.C. (1978) Latitudinal overlap of seaweed species: evidence for quasi-sympatric speciation. *Journal of Biogeography*, **5**, 227–238.

Pielou, E.C. & Routledge, R.D. (1976) Salt marsh vegetation: latitudinal gradients in the zonation patterns. *Oecologia*, **24**, 311–321.

Pierik, R. & de Wit, M. (2014) Shade avoidance: phytochrome signalling and other aboveground neighbour detection cues. *Journal of Experimental Botany*, **65**, 2815–2824.

Pierik, R., Mommer, L. & Voesenek, L. (2013) Molecular mechanisms of plant competition: neighbour detection and response strategies. *Functional Ecology*, **27**, 841–853.

Pieterse, A.H. & Murphy, K. (1993) *Aquatic Weeds: the Ecology and Management of Nuisance Aquatic Vegetation*. Oxford University Press, Oxford.

Pigliucci, M. & Murren, C.J. (2003) Perspective: genetic assimilation and a possible evolutionary paradox: Can macroevolution sometimes be so fast as to pass us by? *Evolution*, **57**, 1455–1464.

Pigott, C.D. (1970) The response of plants to climate and climatic change. In *The Flora of a Changing Britain* (ed. F.H. Perring), pp. 32–44. Classey, Faringdon.

Pigott, C.D. (1980) [Review of Grime (1979), Plant strategies and vegetation processes]. *Journal of Ecology*, **68**, 704–706.

Pigott, C.D. & Huntley, J.P. (1981) Factors controlling the distribution of *Tilia cordata* at the northern limits of its geographical range. III. Nature and causes of seed sterility. *New Phytologist*, **87**, 817–839.

Platt, W.J. & Connell, J.H. (2003) Natural disturbances and directional replacement of species. *Ecological Monographs*, **73**, 507–522.

Pleasants, J.M. (1980) Competition for bumblebee pollinators in Rocky Mountain plant communities. *Ecology*, **61**, 1446–1459.

Pleasants, J.M. (1990) Null-model tests for competitive displacement: the fallacy of not focusing on the whole community. *Ecology*, **71**, 1078–1084.

Ploeg, A.T. & Maris, P.C. (1999) Effect of temperature on suppression of Meloidogyne incognita by Tagetes cultivars. *Journal of Nematology*, **31**, 709–714.

Plumptre, A.J. (1994) The effects of trampling damage by herbivores on the vegetation of the Parc National des Volcans, Rwanda. *African Journal of Ecology*, **32**, 115–129.

Pocock, M.J.O., Evans, D.M. & Memmott, J. (2012) The robustness and restoration of a network of ecological networks. *Science*, **335**, 973–977.

Pokon, R., Novotny, V. & Samuelson, G.A. (2005) Host specialization and species richness of root-feeding chrysomelid larvae (Chrysomelidae, Coleoptera) in a New Guinea rain forest. *Journal of Tropical Ecology*, **21**, 595–604.

Poli, C.H.E.C., Hodgson, J., Cosgrove, G.P. & Arnold, G.C. (2006) Selective behaviour in cattle grazing pastures of strips of birdsfoot trefoil and red clover. 1. The effects of relative sward area. *Journal of Agricultural Science*, **144**, 165–171.

Poorter, L., Bongers, F., Sterck, F.J. & Woll, H. (2005) Beyond the regeneration phase: differentiation of height-light trajectories among tropical tree species. *Journal of Ecology*, **93**, 256–267.

Post, D.M. & Palkovacs, E.P. (2009) Eco-evolutionary feedbacks in community and ecosystem ecology: interactions between the ecological theatre and the evolutionary play. *Philosophical Transactions of the Royal Society B*, **364**, 1629–1640.

Potts, A.J., Midgley, J.J., Child, M.F., Larsen, C. & Hempson, T. (2011) Coexistence theory in the Cape Floristic Region: revisiting an example of leaf niches in the Proteaceae. *Austral Ecology*, **36**, 212–219.

Poulin, B., Wright, S.J., Lefebvre, G. & Calderon, O. (1999) Interspecific synchrony and asynchrony in the fruiting phenologies of congeneric bird-dispersed plants in Panama. *Journal of Tropical Ecology*, **15**, 213–227.

Poulson, T.L. & Platt, W.J. (1996) Replacement patterns of beech and sugar maple in Warren Woods, Michigan. *Ecology*, **77**, 1234–1253.

Pound, R. & Clements, F.E. (1900) *The Phytogeography of Nebraska*, 2nd edn. Botanical Seminar, Lincoln, NE.

Power, A.G. & Mitchell, C.E. (2004) Pathogen spillover in disease epidemics. *American Naturalist*, **164**, 79–89.

Presley, S.J., Higgins, C.L. & Willig, M.R. (2010) A comprehensive framework for the evaluation of metacommunity structure. *Oikos*, **119**, 908–917.

Press, M.C. (1998) Dracula or Robin Hood: a functional role for root hemiparasites in nutrient poor ecosystems. *Oikos*, **82**, 609–611.

Preston, C.D., Pearman, D.A. & Dines, T.D. (2002) *New Atlas of the British & Irish Flora: An Atlas of the Vascular Plants of Britain, Ireland, the Isle of Man and the Channel Islands.* Oxford University Press, Oxford.

Preston, F.W. (1948) The commonness and rarity of species. *Ecology*, **29**, 254–283.

Preston, F.W. (1962) The canonical distribution of commonness and rarity. *Ecology*, **43**, 185–215.

Prider, J., Watling, J. & Facelli, J.M. (2009) Impacts of a native parasitic plant on an introduced and a native host species: implications for the control of an invasive weed. *Annals of Botany*, **103**, 107–115.

Prins, H.H.T. & van der Jeugd, H.P. (1993) Herbivore population crashes and woodland structure in East Africa. *Journal of Ecology*, **81**, 305–314.

Prokopy, R.J., Diehl, S.R. & Cooley, S.S. (1988) Behavioral evidence for host races in Rhagoletis pomonella flies. *Oecologia*, **76**, 138–147.

Putz, F.E. (1984) The natural history of lianas on Barro Colorado Island, Panama. *Ecology*, **65**, 1713–1724.

Putz, F.E. (1991) Silvicultural effects of lianas. In *The Biology of Vines* (eds F.E. Putz & H.A. Mooney), pp. 493–501. Cambridge University Press, Cambridge, UK.

Putz, F.E. (1995) Vines in treetops: consequences of mechanical dependence. In *Forest Canopies* (eds M.D. Lowman & N.M. Nadkari), pp. 311–323. Academic Press, San Diego.

Putz, F.E., Parker, G.G. & Archibald, R.M. (1984) Mechanical abrasion and intercrown spacing. *American Midland Naturalist*, **112**, 24–28.

Pyne, S.J. (1991) *Burning Bush: a Fire History of Australia.* Allen and Unwin, Sydney.

Rabinowitz, D., Rapp, J.K. & Dixon, P.M. (1984) Competitive abilities of sparse grass species: means of persistence or cause of abundance? *Ecology*, **65**, 1144–1154.

Rajaniemi, T.K. (2003) Evidence for size asymmetry of belowground competition. *Basic and Applied Ecology*, **4**, 239–247.

Rastetter, E.B., Gough, L., Hartley, A.E., Herbert, D.A., Nadelhoffer, K.J. & Williams, M. (1999) A revised assessment of species redundancy and ecosystem reliability. *Conservation Biology*, **13**, 440–443.

Rathcke, B. & Lacey, E.P. (1985) Phenological patterns of terrestrial plants. *Annual Review of Ecology and Systematics*, **16**, 179–214.

Ratnadass, A., Fernandes, P., Avelino, J. & Habib, R. (2012) Plant species diversity for sustainable management of crop pests and diseases in agroecosystems: a review. *Agronomy for Sustainable Development*, **32**, 273–303.

Raunkiaer, C. (1909) Formationsundersøgelse og Formationsstatistik. *Botanisk Tidsskrift*, **30**, 20–132.

Raventos, J., Wiegand, T. & Deluis, M. (2010) Evidence for the spatial segregation hypothesis: a test with nine-year survivorship data in a Mediterranean shrubland. *Ecology*, **91**, 2110–2120.

Rebele, F. (2000) Competition and coexistence of rhizomatous perennial plants along a nutrient gradient. *Plant Ecology*, **147**, 77–94.

Redmond, M.D., Wilbur, R.B. & Wilbur, H.M. (2012) Recruitment and dominance of *Quercus rubra* and *Quercus alba* in a previous Oak-Chestnut forest from the 1980s to 2008. *American Midland Naturalist*, **168**, 427–442.

Reich, P.B. (2014) The world-wide 'fast-slow' plant economics spectrum: a traits manifesto. *Journal of Ecology*, **102**, 275–301.

Reich, P.B., Tilman, D., Isbell, F., Mueller, K., Hobbie, S.E., Flynn, D.F.B. & Eisenhauer, N. (2012) Impacts of biodiversity loss escalate through time as redundancy fades. *Science*, **336**, 589–592.

Reich, P.B., Wright, I.J., Cavender-Bares, J., Craine, J.M., Oleksyn, J. & Westoby, M. (2003) The evolution of plant functional variation: traits, spectra, and strategies. *Journal of Plant Sciences* **164**, 143–164.

Reichle, D.E., O'Neill, R.V. & Harris, W.F. (1975) Principles of material and energy exchange in ecosystems. In *Unifying Concepts in Ecology* (eds W.H. van Dobben & R.H. Lowe-McConnell), pp. 27–43. Junk, The Hague.

Reicosky, D.C. (1989) Diurnal and seasonal trends in carbon dioxide concentrations in corn and soybean canopies as affected by tillage and irrigation. *Agricultural and Forest Meteorology*, **48**, 285–303.

Reinhart, K.O. & Callaway, R.M. (2004) Soil biota facilitate exotic Acer invasions in Europe and North America. *Ecological Applications*, **14**, 1737–1745.

Reinhart, K.O., Gurnee, J., Tirado, R. & Callaway, R.M. (2006) Invasion through quantitative effects: intense shade drives native decline and invasive success. *Ecological Applications*, **16**, 1821–1831.

Reintam, E., Trukmann, K. & Kuht, J. (2008) Effect of *Cirsium arvense* L. on soil physical properties and crop growth. *Agricultural and Food Science*, **17**, 153–164.

Rejmánek, M. (1996) Species richness and resistance to invasions. In *Biodiversity and Ecosystem Processes in Tropical Forests* (eds G.H. Orians, R. Dirzo & J.H. Cushman), pp. 153–172. Springer, Berlin.

Rejmánek, M. (1999) Holocene invasions: finally the resolution ecologists were waiting for. *Trends in Ecology and Evolution*, **14**, 8–10.

Reuschel, D., Mattheck, C. & Althaus, C. (1998) The mechanical effect of climbing plants upon the host tree. *Allgemeine Forst und Jagdzeitung*, **169**, 87–91.

Revilla, T. & Weissing, F.J. (2008) Nonequilibrium coexistence in a competition model with nutrient storage. *Ecology*, **89**, 865–877.

Richards, P.W. (1996) *The Tropical Rain Forest: An Ecological Study*, 2nd edn. Cambridge University Press, Cambridge, UK.

Richardson, A.E., Barea, J.M., McNeill, A.M. & Prigent-Combaret, C. (2009) Acquisition of phosphorus and nitrogen in the rhizosphere and plant growth promotion by microorganisms. *Plant and Soil*, **321**, 305–339.

Richardson, S.J., Peltzer, D.A., Allen, R.B., McGlone, M.S. & Parfitt, R.L. (2004) Rapid development of phosphorus limitation in temperate rainforest along the Franz Josef soil chronosequence. *Oecologia*, **139**, 267–276.

Rietkerk, M., van den Bosch, F. & van de Koppel, J. (1997) Site-specific properties and irreversible vegetation changes in semi-arid grazing systems. *Oikos*, **80**, 241–252.

Robberecht, R., Mahall, B.E. & Nobel, P.S. (1983) Experimental removal of intraspecific competitors – effect on water relations and productivity of a desert bunchgrass, *Hilaria rigida*. *Oecologia*, **60**, 21–24.

Robertson, G.P., Hutson, M.A., Evans, F.C. & Tiedje, J.M. (1988) Spatial variability in a successional plant community: patterns of nitrogen availability. *Ecology*, **69**, 1517–1524.

Robinson, J.V. & Dickerson, J.E. (1987) Does invasion sequence affect community structure? *Ecology*, **68**, 587–595.

Rodríguez, M.A. & Gómezal-Sal, A. (1994) Stability may decrease with diversity in grassland communities – empirical-evidence from the 1986 Cantabrian mountains (Spain) drought. *Oikos*, **71**, 177–180.

Rodwell, J.S. (1991–2000) *British Plant Com0munities*. Cambridge University Press, Cambridge, UK.

Rogers, R.S. (1983) Small-area coexistence of vernal forest herbs: does functional similarity of plants matter. *American Naturalist*, **121**, 834–850.

Roggy, J.C., Moiroud, A., Lensi, R. & Domenach, A.M. (2004) Estimating N transfers between N_2-fixing actinorhizal species and the non-N_2-fixing *Prunus avium* under partially controlled conditions. *Biology and Fertility of Soils*, **39**, 312–319.

Root, R.B. (1967) The niche exploitation pattern of the blue-gray gnatcatcher. *Ecological Monographs*, **37**, 317–350.

Roques, K.G., O'Connor, T.G. & Watkinson, A.R. (2001) Dynamics of shrub encroachment in an African savannah: relative influences of fire, herbivory, rainfall and density dependence. *Journal of Applied Ecology*, **38**, 268–280.

Roscher, C., Weigelt, A., Proulx, R., Marquard, E., Schumacher, J., Weisser, W.W. & Schmid, B. (2011) Identifying population- and community-level mechanisms of diversity-stability relationships in experimental grasslands. *Journal of Ecology*, **99**, 1460–1469.

Rosindell, J., Hubbell, S.P., He, F.L., Harmon, L.J. & Etienne, R.S. (2012) The case for ecological neutral theory. *Trends in Ecology and Evolution*, **27**, 203–208.

Rousseaux, M.C., Hall, A.J. & Sánchez, R.A. (1996) Far-red enrichment and photosynthetically active radiation level influence leaf senescence in field-grown sunflower. *Physiologia Plantarum*, **96**, 217–224.

Roxburgh, S.H. & Mokany, K. (2007) Comment on "From plant traits to plant communities: a statistical mechanistic approach to biodiversity". *Science*, **316**, 1425.

Roxburgh, S.H. & Mokany, K. (2010) On testing predictions of species relative abundance from maximum entropy optimisation. *Oikos*, **119**, 583–590.

Roxburgh, S.H., Shea, K. & Wilson, J.B. (2004) The Intermediate Disturbance Hypothesis: patch dynamics and mechanisms of species coexistence. *Ecology*, **85**, 359–371.

Roxburgh, S.H., Watkins, A.J. & Wilson, J.B. (1993) Lawns have vertical stratification. *Journal of Vegetation Science*, **4**, 699–704.

Roxburgh, S.H. & Wilson, J.B. (2000a) Stability and coexistence in a lawn community: mathematical prediction of stability using a community matrix with parameters derived from competition experiments. *Oikos*, **88**, 395–408.

Roxburgh, S.H. & Wilson, J.B. (2000b) Stability and coexistence in a lawn community: experimental assessment of the stability of the actual community. *Oikos*, **88**, 409–423.

Russ, G.R. (1982) Overgrowth in a marine epifaunal community: competitive hierarchies and competitive networks. *Oecologia*, **53**, 12–19.

Ryel, R.J., Caldwell, M.M., Leffler, A.J. & Yoder, C.K. (2003) Rapid soil moisture recharge to depth by roots in a stand of Artemisia tridentata. *Ecology*, **84**, 757–764.

Ryser, P. (1993) Influences of neighbouring plants on seedling establishment in limestone grassland. *Journal of Vegetation Science*, **4**, 195–202.

Sage, R.B. (1999) Weed competition in willow coppice crops: the cause and extent of yield losses. *Weed Research*, **39**, 399–411.

Sale, P.F. (1977) Maintenance of high diversity in coral reef fish communities. *American Naturalist*, **111**, 337–359.

Šamonil, P., Dolezelova, P., Vasickova, I., Adam, D., Valtera, M., Kral, K., Janik, D. & Sebkova, B. (2013a) Individual-based approach to the detection of disturbance history through spatial scales in a natural beech-dominated forest. *Journal of Vegetation Science*, **24**, 1167–1184.

Šamonil, P., Schaetzl, R.J., Valtera, M., Goliáš, V., Baldrian, P., Vaší?ková, I., Adam, D., Janík, D. & Hort, L. (2013b) Crossdating of disturbances by tree uprooting: can treethrow micro-topography persist for 6000 years? *Forest Ecology and Management*, **307**, 123–135.

Samuels, C.L. & Drake, J.A. (1997) Divergent perspectives on community convergence. *Trends in Ecology and Evolution*, **12**, 427–432.

Sanders, N.J., Gotelli, N.J., Heller, N.E. & Gordon, D.M. (2003) Community disassembly by an invasive species. *Proceedings of the National Academy of Sciences of the U.S.A.*, **100**, 2474–2477.

Sarmiento, L., Llambí, L.D., Escalona, A. & Marquez, N. (2003) Vegetation patterns, regeneration rates and divergence in an old-field succession of the high tropical Andes. *Plant Ecology*, **166**, 63–74.

Sartori, F., Lal, R., Ebinger, M.H. & Miller, R.O. (2007) Tree species and wood ash affect soil in Michigan's Upper Peninsula. *Plant and Soil*, **298**, 125–144.

Sato, T. (1994) Stand structure and dynamics of wave-type *Abies sachalinensis* coastal forest. *Ecological Research*, **9**, 77–84.

Sayer, E.J. (2006) Using experimental manipulation to assess the roles of leaf litter in the functioning of forest ecosystems. *Biological Reviews*, **81**, 1–31.

Scheffer, M., Hosper, S.H., Meijer, M.-L., Moss, B. & Jeppesen, E. (1993) Alternative equilibria in shallow lakes. *Trends in Ecology and Evolution*, **8**, 275–279.

Schimper, A.F.W. (1898) *Pflanzen-Geographie auf Physiologischer Grundlage*. Fischer, Jena.

Schimper, A.F.W. (1903) *Plant-Geography upon a Physiological Basis*. Oxford University Press, Oxford.

Schlesinger, W.H., Raikes, J.A., Hartley, A.E. & Cross, A.E. (1996) On the spatial pattern of soil nutrients in desert ecosystems. *Ecology*, **77**, 364–374.

Schliephake, E., Graichen, K. & Rabenstein, F. (2000) Investigations on the vector transmission of the Beet mild yellowing virus (BMYV) and the Turnip yellows virus (TuYV). *Zeitschrift für Pflanzenkrankheiten und Pflanzenschutz*, **107**, 81–87.

Schluter, D. (1990) Species-for-species matching. *American Naturalist*, **136**, 560–568.

Schmitt, J. & Wulff, R.D. (1993) Light spectral quality, phytochrome and plant competition. *Trends in Ecology and Evolution*, **8**, 47–48.

Schmitz, O.J. (1997) Press perturbations and the predictability of ecological interactions in a food web. *Ecology*, **78**, 55–69.

Schmitz, O.J. (2004) Perturbation and abrupt shift in trophic control of biodiversity and productivity. *Ecology Letters*, **7**, 403–409.

Schnitzer, S.A. & Carson, W.P. (2000) Have we forgotten the forest because of the trees? *Trends in Ecology and Evolution*, **15**, 375–376.

Schnitzer, S.A., Kuzee, M.E. & Bongers, F. (2005) Disentangling above- and below-ground competition between lianas and trees in a tropical forest. *Journal of Ecology*, **93**, 1115–1125.

Schoener, T.W. (1986) Resource partitioning. In *Community Ecology: Pattern and Process* (eds J. Kikkawa & D.J. Anderson), pp. 91–126. Blackwell, Melbourne.

Scholte, K. & Vos, J. (2000) Effects of potential trap crops and planting date on soil infestation with potato cyst nematodes and root-knot nematodes. *Annals of Applied Biology*, **137**, 153–164.

Schoonhoven, L.M., Jermy, T. & van Loon, J.J.A. (1998) *Insect-Plant Biology: From Physiology to Evolution*. Chapman and Hall, London.

Schrage, L.J. & Downing, J.A. (2004) Pathways of increased water clarity after fish removal from Ventura Marsh; a shallow, eutrophic wetland. *Hydrobiologia*, **511**, 215–231.

Schröder, A., Persson, L. & De Roos, A.M. (2005) Direct experimental evidence for alternative stable states: a review. *Oikos*, **110**, 3–19.

Schulz, J.P. (1960) *Ecological Studies on Rain Forest in Northern Suriname*. Noord-Hollandsche, Amsterdam.

Scofield, D.G. (2006) Medial pith cells per meter in twigs as a proxy for mitotic growth rate in the apical meristem. *American Journal of Botany*, **93**, 1740–1747.

Scott, P.A. & Hansell, R.I.C. (2002) Development of white spruce tree islands in the shrub zone of the forest-tundra. *Arctic*, **55**, 238–246.

Scott-Phillips, T.C., Laland, K.N., Shuker, D.M., Dickins, T.E. & West, S.A. (2014) The niche construction perspective: a critical appraisal. *Evolution*, **68**, 1231–1243.

Scunthorpe, C.D. (1967) *The Biology of Aquatic Vascular Plants*. Arnold, London.

Sears, A.L.W. & Chesson, P. (2007) New methods for quantifying the spatial storage effect: an illustration with desert annuals. *Ecology*, **88**, 2240–2247.

Šebková, B., Šamonil, P., Valtera, M., Adam, D. & Janík, D. (2012) Interaction between tree species populations and windthrow dynamics in natural beech-dominated forest, Czech Republic. *Forest Ecology and Management*, **280**, 9–19.

Seifan, M., Seifan, T., Ariza, C. & Tielbörger, K. (2010) Facilitating an importance index. *Journal of Ecology*, **98**, 356–361.

Seifert, R.P. & Seifert, F.H. (1976) A community matrix analysis of Heliconia insect communities. *American Naturalist*, **110**, 461–483.

Semchenko, M., Lepik, M., Gotzenberger, L. & Zobel, K. (2012) Positive effect of shade on plant growth: amelioration of stress or active regulation of growth rate? *Journal of Ecology*, **100**, 459–466.

Shea, K. & Chesson, P.L. (2002) Community ecology theory as a framework for biological invasions. *Trends in Ecology and Evolution*, **17**, 170–176.

Shea, K., Roxburgh, S.H. & Rauschert, E.S.J. (2004) Moving from pattern to process: coexistence mechanisms under intermediate disturbance regimes. *Ecology Letters*, **7**, 491–508.

Shepherd, V.A. (1999) Bioelectricity and the rhythms of sensitive plants – the biophysical research of Jagadis Chandra Bose. *Current Science*, **77**, 189–195.

Shimizu, Y. & Tabata, H. (1985) Invasion of Pinus lutchuensis and its influence on the native forest of a Pacific island. *Journal of Biogeography*, **12**, 195–207.

Shipley, B. & Keddy, P.A. (1987) The individualistic and community-unit concepts as falsifiable hypotheses. *Vegetatio*, **69**, 47–56.

Shipley, B., Vile, D. & Garnier, E. (2006) From plant traits to plant communities: a statistical mechanistic approach to biodiversity. *Science*, **314**, 812–814.

Shmida, A. (1981) Mediterranean vegetation in California and Israel: similarities and differences. *Israel Journal of Botany*, **30**, 105–123.

Shnerb, N.M., Sarah, P., Lavee, H. & Solomon, S. (2003) Reactive glass and vegetation patterns. *Physical Review Letters*, **90**, 38101.

Shreve, F. (1942) The desert vegetation of North America. *Botanical Review*, **8**, 195–246.

Shuker, D.M. (2014) Genetic variation in niche construction: a comment on Satz and Nuzhdin. *Trends in Ecology and Evolution*, **29**, 303–304.

Shulski, M.D., Walter-Shea, E., Hubbard, K. & Horst, L.G. (2004) Penetration of photosynthetically active and ultraviolet radiation into alfalfa and tall fescue canopies. *Agronomy Journal*, **96**, 1562–1571.

Shure, D.J. & Phillips, D.L. (1987) Litterfall patterns within different sized disturbance patches in a Southern Appalachian Mountain forest. *American Midland Naturalist*, **118**, 348–357.

Siitonen, J. (2001) Forest management, coarse woody debris and saproxylic organisms: Fennoscandian boreal forests as an example. *Ecological Bulletins*, **49**, 11–42.

Siitonen, J., Martikainen, P., Punttila, P. & Rauh, J. (2000) Coarse woody debris and stand characteristics in mature managed and old-growth boreal mesic forests in southern Finland. *Forest Ecology and Management*, **128**, 211–225.

Silander, J.A. & Antonovics, J. (1982) Analysis of interspecific interactions in a coastal plant community – a perturbation approach. *Nature*, **298**, 557–560.

Silver, M. & Di Paolo, E. (2006) Spatial effects favour the evolution of niche construction. *Theoretical Population Biology*, **70**, 387–400.

Silvertown, J. (1987) Ecological stability: a test case. *American Naturalist*, **130**, 807–810.

Silvertown, J., Dodd, M.E., Gowing, D.J.G. & Mountford, J.O. (1999) Hydrologically-defined niches reveal a basis for species richness in plant communities. *Nature*, **400**, 61–63.

Silvertown, J., Holtier, S., Johnson, J. & Dale, P. (1992) Cellular automaton models of interspecific competition for space: the effect of pattern on process. *Journal of Ecology*, **80**, 527–534.

Silvertown, J., Lines, C.E.M. & Dale, M.P. (1994) Spatial competition between grasses – rates of mutual invasion between four species and the interaction with grazing. *Journal of Ecology*, **82**, 31–38.

Simard, S.W. & Durall, D.M. (2004) Mycorrhizal networks: a review of their extent, function, and importance. *Canadian Journal of Botany*, **82**, 1140–1165.

Simberloff, D. (1980) A succession of paradigms in ecology: essentialism to materialism and probabilism. *Synthese*, **43**, 3–39.

Singh, H.P., Batisha, D.R. & Kohlia, R.K. (1999) Autotoxicity: concept, organisms, and ecological significance. *Critical Reviews in Plant Sciences*, **18**, 757–772.

Šizling, A.L., Storch, D., Sizlingova, E., Reif, J. & Gaston, K.J. (2009) Species abundance distribution results from a spatial analogy of central limit theorem. *Proceedings of the National Academy of Sciences of the U.S.A.*, **106**, 6691–6695.

Skellam, J.G. (1951) Random dispersal in theoretical populations. *Biometrika*, **38**, 196–218.

Skinner, R.H. & Simmons, S.R. (1993) Modulation of leaf elongation, tiller appearance and tiller senescence in spring barley by far-red light. *Plant Cell and Environment*, **16**, 555–562.

Smith, B., Moore, S., Grove, P., Harris, N., Mann, S. & Wilson, J. (1994) Vegetation texture as an approach to community structure: community-level convergence in a New Zealand temperate rainforest. *New Zealand Journal of Ecology*, **18**, 41–50.

Smith, B. & Wilson, J.B. (2002) Community convergence: ecological and evolutionary. *Folia Geobotanica*, **37**, 171–183.

Smith, T.W. & Lundholm, J.T. (2010) Variation partitioning as a tool to distinguish between niche and neutral processes. *Ecography*, **33**, 648–655.

Snaydon, R.W. & Davies, T.M. (1982) Rapid divergence of plant populations in response to recent changes in soil conditions. *Evolution*, **36**, 289–297.

Soliveres, S., Smit, C. & Maestre, F. (2014) Moving forward on facilitation research: response to changing environments and effects on the diversity, functioning and evolution of plant communities. *Biological Reviews*, **90**, 297–313.

Spatharis, S., Orfanidis, S., Panayotidis, P. & Tsirtsis, G. (2011) Assembly processes in upper subtidal macroalgae: the effect of wave exposure. *Estuarine Coastal and Shelf Science*, **91**, 298–305.

Srivastava, D.S. & Jefferies, R.L. (1995) Mosaics of vegetation and soil salinity: a consequence of goose foraging in an arctic salt marsh. *Canadian Journal of Botany*, **73**, 75–83.

Stansfield, J.H., Perrow, M.R., Tench, D.L., Jowitt, A.J.D. & Taylor, A.A.L. (1997) Submerged macrophytes as refuges for grazing Cladocera against fish predation: observations on seasonal changes in relation to macrophyte cover and predation pressure. *Hydrobiologia*, **342**, 229–240.

Stastny, M., Schaffner, U.R.S. & Elle, E. (2005) Do vigour of introduced populations and escape from specialist herbivores contribute to invasiveness? *Journal of Ecology*, **93**, 27–37.

Steel, J.B., Wilson, J.B., Anderson, B.J., Lodge, R.H.E. & Tangney, R.S. (2004) Are bryophyte communities different from higher-plant communities?: Abundance relations. *Oikos*, **104**, 479–486.

Steinger, T., Körner, C. & Schmid, B. (1996) Long-term persistence in a changing climate: DNA analysis suggests very old ages of clones of alpine *Carex curvula*. *Oecologia*, **105**, 94–99.

Stenberg, J.A., Heijari, J., Holopainen, J.K. & Ericson, L. (2007) Presence of *Lythrum salicaria* enhances the bodyguard effects of the parasitoid *Asecodes mento* for *Filipendula ulmaria*. *Oikos*, **116**, 482–490.

Stern, W.R. & Donald, C.M. (1962) The influence of leaf area and radiation on growth of clover in swards. *Australian Journal of Agricultural Research*, **13**, 615–623.

Stewart, W.S. & Bannister, P. (1973) Seasonal changes in carbohydrate content of three Vaccinium spp. with particular reference to *V. uliginosum* L. and its distribution in the British Isles. *Flora*, **162**, 134–155.

Stiles, F.G. (1977) Coadapted competitors: the flowering seasons of hummingbird-pollinated plants in a tropical forest. *Science*, **198**, 1177–1178.

Stiles, F.G. (1979) Reply to Poole & Rathcke. *Science*, **203**, 471.

Stokes, C.J. & Archer, S.R. (2010) Niche differentiation and neutral theory: an integrated perspective on shrub assemblages in a parkland savanna. *Ecology*, **91**, 1152–1162.

Stokke, S. (1999) Sex differences in feeding-patch choice in a megaherbivore: elephants in Chobe National Park, Botswana. *Canadian Journal of Zoology*, **77**, 1723–1732.

Stoll, P. & Newbery, D.M. (2005) Evidence of species-specific neighborhood effects in the Dipterocarpaceae of a Bornean rain forest. *Ecology*, **86**, 3048–3062.

Stoll, P. & Prati, D. (2001) Intraspecific aggregation alters competitive interactions in experimental plant communities. *Ecology*, **82**, 319–327.

Stone, L., Gabric, A. & Berman, T. (1996) Ecosystem resilience, stability and productivity: seeking a relationship. *American Naturalist*, **148**, 892–903.

Stone, L. & Roberts, A. (1991) Conditions for a species to gain advantage from the presence of competitors. *Ecology*, **72**, 1964–1972.

Stoutjesdijk, P.H. & Barkman, J.J. (1992) *Microclimate, Vegetation and Fauna*. Opulus, Knivsta.

Strauss, S.Y., Webb, C.O. & Salamin, N. (2006) Exotic taxa less related to native species are more invasive. *Proceedings of the National Academy of Sciences of the U.S.A.*, **103**, 5841–5845.

Strong, D.R., Jr. (1977) Epiphytic tree loads, tree falls and perennial forest disruption: a mechanism for maintaining higher tree species richness in the tropics without animals. *Journal of Biogeography*, **4**, 215–218.

Strong, D.R., Jr. (1983) Natural variability and the manifold mechanisms of ecological communities. *American Naturalist*, **122**, 636–660.

Strong, D.R., Jr., Lzyska, L.A. & Simberloff, D.S. (1979) Tests of community-wide character displacement against null hypotheses. *Evolution*, **33**, 897–913.

Stubbs, W.J. & Wilson, J.B. (2004) Evidence for limiting similarity in a sand dune community. *Journal of Ecology*, **92**, 557–567.

Stylinski, C.D. & Allen, E.B. (1999) Lack of native species recovery following severe exotic disturbance in southern Californian shrublands. *Journal of Applied Ecology*, **36**, 544–554.

Sugihara, G. (1980) Minimal community structure: an explanation of species abundance patterns. *American Naturalist*, **116**, 770–787.

Sutherland, J.P. (1974) Multiple stable points in natural communities. *American Naturalist*, **108**, 859–873.

Suzuki, S.N. & Suzuki, R.O. (2011) Distance-dependent shifts in net effects by an unpalatable nettle on a palatable plant species. *Acta Oecologica*, **37**, 386–392.

Sykes, M.T., van der Maarel, E., Peet, R.K. & Willems, J.H. (1994) High species mobility in species-rich plant communities – an intercontinental comparison. *Folia Geobotanica et Phytotaxonomica*, **29**, 439–448.

Symstad, A.J. (2000) A test of the effects of functional group richness and composition on grassland invasibility. *Ecology*, **81**, 99–109.

Tahvanainen, J.O. & Root, R.B. (1972) The influence of vegetational diversity on the population ecology of a specialized herbivore, *Phyllotreta cruciferae* (Coleoptera: Chrysomelidae). *Oecologia*, **10**, 321–346.

Takeuchi, Y. & Nakashizuka, T. (2007) Effect of distance and density on seed/seedling fate of two dipterocarp species. *Forest Ecology and Management*, **247**, 167–174.

Tamm, C.O. (1964) Growth of *Hylocomium splendens* in relation to tree canopy. *Bryologist*, **67**, 423–426.

Tamme, R., Gazol, A., Price, J., Hiiesalu, I. & Pärtel, M. (2016) Co-occurring grassland species vary in their responses to fine-scale soil heterogeneity. *Journal of Vegetation Science*, **27**, 1012–1022.

Tan, J.Q., Pu, Z.C., Ryberg, W.A. & Jiang, L. (2012) Species phylogenetic relatedness, priority effects, and ecosystem functioning. *Ecology*, **93**, 1164–1172.

Tangney, R.S., Wilson, J.B. & Mark, A.F. (1990) Bryophyte island biogeography: a study in Lake Manapouri, New Zealand. *Oikos*, **59**, 21–26.

Tansley, A.G. (1917) On competition between *Galium saxatile* L. (*G. hercynicum* Weig.) and *Galium sylvestre* Poll. (*G. asperum* Schreb.) on different types of soil. *Journal of Ecology*, **5**, 173–179.

Tansley, A.G. (1920) The classification of vegetation and the concept of development. *Journal of Ecology*, **8**, 118–149.

Tansley, A.G. (1939) *The British Islands and Their Vegetation*. Cambridge University Press, Cambridge, UK.

Tansley, A.G. & Adamson, R.S. (1925) Studies of the vegetation of the English chalk: III. The chalk grasslands of Hampshire-Sussex border. *Journal of Ecology*, **13**, 177–223.

Tarroux, E. & DesRochers, A. (2011) Effect of natural root grafting on growth response of Jack Pine (*Pinus banksiana*; pinaceae). *American Journal of Botany*, **98**, 967–974.

Taylor, P.J. (1988) The construction and turnover of complex community models having generalized Lotka-Volterra dynamics. *Journal of Theoretical Biology*, **135**, 569–588.

Temperton, V.M., Hobba, R.J., Nuttle, T., Fattorini, M. & Halle, S. (1999) The search for ecological assembly rules and its relevance to restoration ecology. In *Assembly Rules and Restoration Ecology. Bridging the Gap Between Theory and Practice*. (eds V.M. Temperton, R.J. Hobbs, T. Nuttle & S. Halle), p. 464. Island Press, Washington DC.

terHorst, C.P., Miller, T.E. & Powell, E. (2010) When can competition for resources lead to ecological equivalence? *Evolutionary Ecology Research*, **12**, 843–854.

Teste, F.P., Simard, S.W., Durall, D.M., Guy, R.D. & Berch, S.M. (2010) Net carbon transfer between *Pseudotsuga menziesii* var. *glauca* seedlings in the field is influenced by soil disturbance. *Journal of Ecology*, **98**, 429–439.

Thébaud, C. & Simberloff, D. (2001) Are plants really larger in their introduced ranges? *American Naturalist*, **157**, 231–236.

Theimer, T.C. & Gehring, C.A. (1999) Effects of a litter-disturbing bird species on tree seedling germination and survival in an Australian tropical rain forest. *Journal of Tropical Ecology*, **15**, 737–749.

Thies, W. & Kalko, E.K.V. (2004) Phenology of neotropical pepper plants (Piperaceae) and their association with their main dispersers, two short-tailed fruit bats, *Carollia perspicillata* and *C. castanea* (Phyllostomidae). *Oikos*, **104**, 362–376.

Thijs, K.W., Brys, R., Verboven, H.A.F. & Hermy, M. (2012) The influence of an invasive plant species on the pollination success and reproductive output of three riparian plant species. *Biological Invasions*, **14**, 355–365.

Thomas, K.R., Munro, C.L., Graeme, C.B., Steel, J.B. & Wilson, J.B. (1999) Application and evaluation of Dale's non-parametric method for detecting community structure through zonation. *Oikos*, **84**, 261–265.

Thomas, S.C. & Bazzaz, F.A. (1999) Asymptotic height as a predictor of photosynthetic characteristics in Malaysian rain forest trees. *Ecology*, **80**, 1607–1622.

Thomas, W.R. & Pomerantz, M.J. (1981) Feasibility and stability in community dynamics. *American Naturalist*, **117**, 381–385.

Thompson, J.N. & Willson, M.F. (1979) Evolution of temperate fruit/bird interactions: phenological strategies. *Evolution*, **33**, 973–982.

Thompson, K., Hodgson, J.G., Grime, J.P. & Burke, M.J.W. (2001) Plant traits and temporal scale: evidence from a 5-year invasion experiment using native species. *Journal of Ecology*, **89**, 1054–1060.

Thompson, L. & Harper, J.L. (1988) The effect of grasses on the quality of transmitted radiation and its influence on the growth of white clover *Trifolium repens*. *Oecologia*, **75**, 343–347.

Thórhallsdóttir, T.E. (1990) The dynamics of five grasses and white clover in a simulated mosaic sward. *Journal of Ecology*, **78**, 909–923.

Thorpe, A.S., Thelen, G.C., Diaconu, A. & Callaway, R.M. (2009) Root exudate is allelopathic in invaded community but not in native community: field evidence for the novel weapons hypothesis. *Journal of Ecology*, **97**, 641–645.

Thrall, P.H. & Jarosz, A.M. (1994) Host-pathogen dynamics in experimental populations of *Silene alba* and Ustilago violacea. II. Experimental tests of theoretical models. *Journal of Ecology*, **82**, 561–570.

Tilman, D. (1977) Resource competition between planktonic algae: an experimental and theoretical approach. *Ecology*, **58**, 338–348.

Tilman, D. (1981) Tests of resource competition theory using four species of Lake Michigan algae. *Ecology*, **62**, 802–815.

Tilman, D. (1982) *Resource Competition and Community Structure*. Princeton University Press, Princeton, NJ.

Tilman, D. (1986) Nitrogen-limited growth in plants from different successional stages. *Ecology*, **67**, 555–563.

Tilman, D. (1987) Secondary succession and the pattern of plant dominance along experimental nitrogen gradients. *Ecological Monographs*, **57**, 189–214.

Tilman, D. (1988) *Plant Strategies and the Dynamics and Structure of Plant Communities*. Princeton University Press, Princeton, NJ.

Tilman, D. (1993) Species richness of experimental productivity gradients: how important is colonization limitation? *Ecology*, **74**, 2179–2191.

Tilman, D. (1994) Competition and biodiversity in spatially structured habitats. *Ecology*, **75**, 2–16.

Tilman, D. (1997) Community invasibility, recruitment limitation, and grassland biodiversity. *Ecology*, **78**, 81–92.

Tilman, D. & Cowan, M.L. (1989) Growth of old field herbs on a nitrogen gradient. *Functional Ecology*, **3**, 425–438.

Tilman, D. & Lehman, C. (2001) Human-caused environmental change: impacts on plant diversity and evolution. *Proceedings of the National Academy of Sciences of the U.S.A.*, **98**, 5433–5440.

Tilman, D., Reich, P.B., Knops, J., Wedin, D., Mielke, T. & Lehman, C. (2001) Diversity and productivity in a long-term grassland experiment. *Science*, **294**, 843–845.

Tilman, D., Reich, P.B. & Knops, J.M.H. (2006) Biodiversity and ecosystem stability in a decade-long grassland experiment. *Nature*, **441**, 629–632.

Tilman, D. & Wedin, D. (1991a) Plant traits and resource reduction for five grasses growing on a nitrogen gradient. *Ecology*, **72**, 685–700.

Tilman, D. & Wedin, D. (1991b) Dynamics of nitrogen competition between successional grasses. *Ecology*, **72**, 1038–1049.

Timberlake, J.R. & Calvert, G.M. (1993) Preliminary root atlas of Zimbabwe. *Zimbabwe Bulletin of Forestry Research*, **10**, 1–90.

Titman [Tilman], D. (1976) Ecological competition between algae: experimental confirmation of resource-based competition theory. *Science*, **192**, 463–465.

Tokeshi, M. (1996) Species coexistence and abundance: patterns and processes. In *Biodiversity: An Ecological Perspective* (eds T. Abe, S.A. Levin & M. Higashi), pp. 35–55. Springer, New York.

Tokita, K. & Yasutomi, A. (2003) Emergence of a complex and stable network in a model ecosystem with extinction and mutation. *Theoretical Population Biology*, **63**, 131–146.

Tongway, C.H., Valentin, C. & Seghieri, J. (2001) *Banded Vegetation Patterning in Arid and Semi-Arid Environments*. Springer, New York.

Torti, S.D., Coley, P.D. & Kursar, T.A. (2001) Causes and consequences of monodominance in tropical lowland forests. *American Naturalist*, **157**, 141–153.

Touchan, R., Swetnam, T.W. & Grissino-Mayer, H.D. (1995) Effects of livestock grazing on pre-settlement fire regimes in New Mexico. In *Proceedings: Symposium on Fire in Wilderness and Park Management* (eds J.K. Brown, R.W. Mutsch, C.W. Spoon & R.H. Wakimoto), pp. 268–272. Intermountain Research Station, Ogden.

Tregonning, K. & Roberts, A. (1979) Complex systems which evolve towards homeostasis. *Nature*, **281**, 563–564.

Tucker, C.M. & Fukami, T. (2014) Environmental variability counteracts priority effects to facilitate species coexistence: evidence from nectar microbes. *Proceedings of the Royal Society of London B*, **281**, 32637.

Tuomi, J., Augner, M. & Nilsson, P. (1994) A dilemma of plant defences: is it really worth killing the herbivore? *Journal of Theoretical Biology*, **170**, 427–430.

Turkington, R. & Harper, J.L. (1979a) The growth, distribution and neighbour relationships of *Trifolium repens* in a permanent pasture. I. Ordination, pattern and contact. *Journal of Ecology*, **67**, 201–218.

Turkington, R. & Harper, J.L. (1979b) The growth, distribution and neighbour relationships of *Trifolium repens* in a permanent pasture. IV. Fine-scale biotic differentiation. *Journal of Ecology*, **67**, 245–254.

Turkington, R. & Mehrhoff, L.A. (1990) The role of competition in structuring pasture communities. In *Perspectives on Plant Competition* (eds J.B. Grace & D. Tilman), pp. 307–340. Academic Press, San Diego.

Turley, M.C. & Ford, E.D. (2011) Detecting bimodality in plant size distributions and its significance for stand development and competition. *Oecologia*, **167**, 991–1003.

Turnbull, L.A., Rees, M. & Crawley, M.J. (1999) Seed mass and the competition/colonization trade-off: a sowing experiment. *Journal of Ecology*, **87**, 899–912.

Turnbull, M.H. & Yates, D.J. (1993) Seasonal variation in the red far-red ratio and photon flux density in an Australian sub-tropical rainforest. *Agricultural and Forest Meteorology*, **64**, 111–127.

Uhl, C. (1982) Tree dynamics in a species rich tierra firme forest in Amazonia, Venezuela. *Acta Cientifica Venezolana*, **33**, 71–77.

Ulrich, W. (2004) Species co-occurrences and neutral models: reassessing J. M. Diamond's assembly rules. *Oikos*, **107**, 603–609.

Underwood, G.J.C. & Paterson, D.M. (1993) Recovery of intertidal benthic diatoms after biocide treatment and associated sediment dynamics. *Journal of the Marine Biological Association of the United Kingdom*, 73, 25–45.

Urban, M.C. (2011) The evolution of species interactions across natural landscapes. *Ecology Letters*, **14**, 723–732.

Uriarte, M., Condit, R., Canham, C.D. & Hubbell, S.P. (2004) A spatially explicit model of sapling growth in a tropical forest: does the identity of neighbours matter? *Journal of Ecology*, **92**, 348–360.

Usher, M.B. (1987) Modelling successional processes in ecosystems. In *Colonization, Succession and Stability* (eds A.J. Gray, M.J. Crawley & P.J. Edwards), pp. 31–56. Blackwell, Oxford.

Usinowicz, J., Wright, S.J. & Ives, A.R. (2012) Coexistence in tropical forests through asynchronous variation in annual seed production. *Ecology*, **93**, 2073–2084.

Vaidyanathan, L.V., Drew, M.C. & Nye, P.H. (1968) The measurement and mechanism of ion diffusion in soils. IV. The concentration dependence of diffusion concentrations of potassium in soils at a range of moisture levels and a method for the estimation of the differential diffusion coefficient at any concentration. *Journal of Soil Science*, **19**, 94–107.

Valiente-Banuet, A. & Verdú, M. (2008) Temporal shifts from facilitation to competition occur between closely related taxa. *Journal of Ecology*, **96**, 489–494.

Valone, T.J. & Barber, N.A. (2008) An empirical evaluation of the insurance hypothesis in diversity-stability models. *Ecology*, **89**, 522–531.

Valone, T.J., Meyer, M., Brown, J.H. & Chew, R.M. (2002) Timescale of perennial grass recovery in desertified arid grasslands following livestock removal. *Conservation Biology*, **16**, 995–1002.

van de Koppel, J., Herman, P.M.J., Thoolen, P. & Heip, C.H.R. (2001) Do alternate stable states occur in natural ecosystems? Evidence from a tidal flat. *Ecology*, **82**, 3449–3461.

van de Koppel, J. & Prins, H.H.T. (1998) The importance of herbivore interactions for the dynamics of African savanna woodlands: an hypothesis. *Journal of Tropical Ecology*, **14**, 565–576.

van de Koppel, J., Rietkerk, M., van Langevelde, F., Kumar, L., Klausmeier, C.A., Fryxell, J.M., Hearne, J.W., van Andel, J., de Ridder, N., Skidmore, A., Stroosnijder, L. & Prins, H.H. (2002) Spatial heterogeneity and irreversible vegetation change in semiarid grazing systems. *American Naturalist*, **159**, 209–218.

van den Bergh, J.P. & Elberse, W.T. (1962) Competition between *Lolium perenne* L. and *Anthoxanthum odoratum* L. at two levels of phosphate and potash. *Journal of Ecology*, **50**, 87–95.

van der Heide, T., Smolders, A., Rijkens, B., van Nes, E.H., van Katwijk, M.M. & Roelofs, J. (2008) Toxicity of reduced nitrogen in eelgrass (*Zostera marina*) is highly dependent on shoot density and pH. *Oecologia*, **158**, 411–419.

van der Heide, T., van Nes, E.H., van Katwijk, M.M., Scheffer, M., Hendriks, A.J. & Smolders, A.J.P. (2010) Alternative stable states driven by density-dependent toxicity. *Ecosystems*, **13**, 841–850.

van der Maarel, E. & Sykes, M.T. (1997) Rates of small-scale species mobility in alvar limestone grassland. *Journal of Vegetation Science*, **8**, 199–208.

Van der Putten, W.H. & Peters, B.A.M. (1997) How soil pathogens may affect plant competition. *Ecology*, **78**, 1785–1795.

Van der Stoel, C.D., Van Der Putten, W.H. & Duyts, H. (2002) Development of a negative soil plant feedback in the expansion zone of the clonal grass *Ammophila arenaria* following root formation and nematode colonisation. *Journal of Ecology*, **90**, 978–988.

van Donk, E. & van de Bund, W.J. (2002) Impact of submerged macrophytes including charophytes on phyto- and zooplankton communities: allelopathy versus other mechanisms. *Aquatic Botany*, **72**, 261–274.

van Gardingen, P. & Grace, J. (1991) Plants and wind. *Advances in Botanical Research*, **18**, 192–248.

van Kleunen, M. & Fischer, M. (2009) Release from foliar and floral fungal pathogen species does not explain the geographic spread of naturalized North American plants in Europe. *Journal of Ecology*, **97**, 385–392.

van Nes, E.H. & Scheffer, M. (2004) Large species shifts triggered by small forces. *American Naturalist*, **164**, 255–266.

van Nes, E.H., Scheffer, M., van den Berg, M.S. & Coops, H. (2002) Dominance of charophytes in eutrophic shallow lakes - when should we expect it to be an alternative stable state? *Aquatic Botany*, **72**, 275–296.

van Ruijven, J. & Berendse, F. (2007) Contrasting effects of diversity on the temporal stability of plant populations. *Oikos*, **116**, 1323–1330.

Van Ruijven, J. & Berendse, F. (2009) Long-term persistence of a positive plant diversity-productivity relationship in the absence of legumes. *Oikos*, **118**, 101–106.

van Ruijven, J., De Deyn, G.B. & Berendse, F. (2003) Diversity reduces invasibility in experimental plant communities: the role of plant species. *Ecology Letters*, **6**, 910–918.

van Steenis, C.G.G.T. (1981) *Rheophytes of the World*. Sijhoff and Noordhoff, Alphen.

Vance, R.R. (1984) Interference competition and the coexistence of two competitors on a single limiting resource. *Ecology*, **65**, 1349–1357.

Vandermeer, J.H. (1969) The competitive structure of communities: an experimental approach with protozoa. *Ecology*, **50**, 362–371.

Vandermeer, J.H. (1977) Notes on density dependence in *Welfia georgii* Wendl. ex Burnett (Palmae): a lowland rainforest species in Costa Rica. *Brenesia*, **10**, 9–15.

Vandermeer, J.H., de la Cerda, I.G., Perfecto, I., Boucher, D., Ruiz, J. & Kaufmann, A. (2004) Multiple basins of attraction in a tropical forest: evidence for non-equilibrium community structure. *Ecology*, **85**, 575–579.

Vanderwall, S.B. (2001) The evolutionary ecology of nut dispersal. *Botanical Review*, **67**, 74–117.

Vanelslander, B., Dewever, A., Vanoostende, N., Kaewnuratchadasorn, P., Vanormelingen, P., Hendrickx, F., Sabbe, K. & Vyverman, W. (2009) Complementarity effects drive positive diversity effects on biomass production in experimental benthic diatom biofilms. *Journal of Ecology*, **97**, 1075–1082.

Vanhinsberg, A. & Vantienderen, P. (1997) Variation in growth form in relation to spectral light quality (red/far-red ratio) in *Plantago lanceolata* L. in sun and shade populations. *Oecologia*, **111**, 452–459.

Vannette, R.L. & Fukami, T. (2014) Historical contingency in species interactions: towards niche-based predictions. *Ecology Letters*, **17**, 115–124.

Vasquez, E.C. & Meyer, G.A. (2011) Relationships among leaf damage, natural enemy release, and abundance in exotic and native prairie plants. *Biological Invasions*, **13**, 621–633.

Vasseur, D.A., Amarasekare, P., Rudolf, V.H.W. & Levine, J.M. (2011) Eco-evolutionary dynamics enable coexistence via neighbor-dependent selection. *American Naturalist*, **178**, 96–109.

Vázquez-Yanes, C., Orozco-Segovia, A., Rincón, E., Sánchez-Coronado, M.E., Huante, P., Toledo, J.R. & Barradas, V.L. (1990) Light beneath the litter in a tropical forest: effect on seed germination. *Ecology*, **71**, 1952–1958.

Veblen, T.T. & Stewart, G.H. (1980) Comparison of forest structure and regeneration on Bench and Stewart Islands, New Zealand. *New Zealand Journal of Ecology*, **3**, 50–68.

Veblen, T.T. & Stewart, G.H. (1982) On the conifer regeneration gap in New Zealand: dynamics of *Libocedrus bidwillii* stands on South Island. *Journal of Ecology*, **70**, 413–436.

Vellend, M. (2010) Conceptual synthesis in community ecology. *Quarterly Review of Biology*, **85**, 183–206.

Vellend, M. (2016) *The Theory of Ecological Communities*. Princeton University Press, Princeton, NJ.

Veloz, S.D., Williams, J.W., Blois, J.L., He, F., OttoBliesner, B. & Liu, Z.Y. (2012) No-analog climates and shifting realized niches during the late quaternary: implications for 21st-century predictions by species distribution models. *Global Change Biology*, **18**, 1698–1713.

Veresoglou, D.S. & Fitter, A.H. (1984) Spatial and temporal patterns of growth and nutrient uptake of five co-existing grasses. *Journal of Ecology*, **72**, 259–272.

Verhulst, J., Montaña, C., Mandujano, M.C. & Franco, M. (2008) Demographic mechanisms in the coexistence of two closely related perennials in a fluctuating environment. *Oecologia*, **156**, 95–105.

Vermeulen, P.J., Stuefer, J.F., Anten, N.P.R. & During, H.J. (2009) Carbon gain in the competition for light between genotypes of the clonal herb *Potentilla reptans*. *Journal of Ecology*, **97**, 508–517.

Verrecchia, E., Yair, A., Kidron, G.J. & Verrecchia, K. (1995) Physical properties of the psammophile cryptogamic crust and their consequences to the water regime of sandy soils, north-western Negev Desert, Israel. *Journal of Arid Environments*, **29**, 427–437.

Viketoft, M. (2008) Effects of six grassland plant species on soil nematodes: A glasshouse experiment. *Soil Biology & Biochemistry*, **40**, 906–915.

Viketoft, M., Palmborg, C., Sohlenius, B., Huss-Danell, K. & Bengtsson, J. (2005) Plant species effects on soil nematode communities in experimental grasslands. *Applied Soil Ecology*, **30**, 90–103.

Vilà, M., Maron, J.L. & Marco, L. (2005) Evidence for the enemy release hypothesis in *Hypericum perforatum*. *Oecologia*, **142**, 474–479.

Vile, D., Shipley, B. & Garnier, E. (2006) A structural equation model to integrate changes in functional strategies during old-field succession. *Ecology*, **87**, 504–517.

Vitousek, P.M. (2004) *Nutrient Cycling and Limitation: Hawai'i as a Model System*. Princeton University Press, Princeton, NJ.

Volkov, I., Banavar, J.R., Hubbell, S.P. & Maritan, A. (2009) Inferring species interactions in tropical forests. *Proceedings of the National Academy of Sciences of the U.S.A.*, **106**, 13854–13859.

Von Holle, B., Delcourt, H.R. & Simberloff, D. (2003) The importance of biological inertia in plant community resistance to invasion. *Journal of Vegetation Science*, **14**, 425–432.

Von Holle, B. & Simberloff, D. (2004) Testing Fox's assembly rule: does plant invasion depend on recipient community structure? *Oikos*, **105**, 551–563.

Vos, V.C.A., van Ruijven, J., Berg, M.P., Peeters, E. & Berendse, F. (2013) Leaf litter quality drives litter mixing effects through complementary resource use among detritivores. *Oecologia*, **173**, 269–280.

Vranckx, G. & Vandelook, F. (2012) A season- and gap-detection mechanism regulates seed germination of two temperate forest pioneers. *Plant Biology*, **14**, 481–490.

Wade, M.J. (1978) A critical review of the models of group selection. *Quarterly Review of Biology*, **53**, 101–114.

Walker, B.H. (1992) Biodiversity and ecological redundancy. *Conservation Biology*, **6**, 18–23.

Walker, B.H. (1993) Rangeland ecology: understanding and managing change. *Ambio*, **22**, 80–87.

Walker, B.H., Kinzig, A. & Langridge, J. (1999) Plant attribute diversity, resilience, and ecosystem function: the nature and significance of dominant and minor species. *Ecosystems*, **2**, 95–113.

Walker, D. (1970) Direction and rate in some British postglacial hydroseres. In *Studies in the Postglacial History of the British Isles* (eds D. Walker & R.G. West), pp. 117–140. Cambridge University Press, Cambridge, UK.

Walker, J., Thompson, C.H., Fergus, I.F. & Tunstall, B.R. (1981) plant succession and soil development in coastal dunes of Eastern Australia. In *Forest Succession: Concepts and Applications* (eds D.C. West, H.H. Shugart & D.B. Botkin), pp. 107–131. Springer, New York.

Walker, L.R. & Chapin, F.S.I. (1986) Physiological controls over seedling growth in primary succession on an Alaskan floodplain. *Ecology*, **67**, 1508–1523.

Walker, L.R., Clarkson, B.D., Silvester, W.B. & Clarkson, B.R. (2003a) Colonization dynamics and facilitative impacts of a nitrogen-fixing shrub in primary succession. *Journal of Vegetation Science*, **14**, 277–290.

Walker, L.R., Thompson, D.B. & Landau, F.H. (2001) Experimental manipulations of fertile islands and nurse plant effects in the Mojave Desert, USA. *Western North American Naturalist*, **61**, 25–35.

Walker, N.A., Henry, H.A.L., Wilson, D.J. & Jefferies, R.L. (2003b) The dynamics of nitrogen movement in an Arctic salt marsh in response to goose herbivory: a parameterized model with alternate stable states. *Journal of Ecology*, **91**, 637–650.

Walker, S. & Wilson, J.B. (2002) Tests for nonequilibrium, instability and stabilizing processes in semiarid plant communities. *Ecology*, **83**, 809–822.

Walker, S., Wilson, J.B., Steel, J.B., Rapson, G.L., Smith, B., King, W.M. & Cottam, Y.H. (2003c) Properties of ecotones: evidence from five coastal ecotones. *Journal of Vegetation Science*, **14**, 579–590.

Walley, K.A., Khan, M.S.I. & Bradshaw, A.D. (1974) The potential for evolution of heavy-metal tolerance in plants. I. Copper and zinc tolerance in Agrostis tenuis. *Heredity*, **32**, 309–319.

Wang, H.C., Wang, S.F., Lin, K.C., Shaner, P.J. & Lin, T.C. (2013) Litterfall and element fluxes in a natural hardwood forest and a Chinese-fir plantation experiencing frequent typhoon disturbance in central Taiwan. *Biotropica*, **45**, 541–548.

Wang, P., Stieglitz, T., Zhou, D.W. & Cahill, J.F. (2010) Are competitive effect and response two sides of the same coin, or fundamentally different? *Functional Ecology*, **24**, 196–207.

Wang, Y.-H., He, W.-M., Dong, M., Yu, F.-H., Zhang, L.-L., Cui, Q.-G. & Chu, Y. (2008) Effects of shaking on the growth and mechanical properties of Hedysarum laeve may be independent of water regimes. *International Journal of Plant Sciences*, **169**, 503–508.

Ward, C.M. (1988) Marine terraces of the Waitutu district and their relation to the late Cenozoic tectonics of the southern Fiordland region. *Journal of the Royal Society of New Zealand*, **18**, 1–28.

Wardle, D.A. (2001) Experimental demonstration that plant diversity reduces invasibility. Evidence of a biological mechanism or a consequence of sampling effect? *Oikos*, **95**, 161–170.

Wardle, D.A. (2002) *Communities and Ecosystems: Linking the Aboveground and Belowground Components*. Princeton University Press, Princeton, NJ.

Wardle, D.A., Walker, L.R. & Bardgett, R.D. (2004) Ecosystem properties and forest decline in contrasting long-term chronosequences. *Science*, **305**, 509–513.

Wardle, P. (1964) Facets of the distribution of forest vegetation in New Zealand. *New Zealand Journal of Botany*, **2**, 352–366.

Warming, E. (1909) *Oecology of Plants: an Introduction to the Study of Plant-Communities*. Oxford University Press, Oxford.

Waser, N.M. & Real, L.A. (1979) Effective mutualism between sequentially flowering plant-species. *Nature*, **281**, 670–672.

Watkins, A.J. & Wilson, J.B. (1992) Fine-scale community structure of lawns. *Journal of Ecology*, **80**, 15–24.

Watkins, A.J. & Wilson, J.B. (1994) Plant community structure, and its relation to the vertical complexity of communities: dominance/diversity and spatial rank consistency. *Oikos*, **70**, 91–98.

Watkins, A.J. & Wilson, J.B. (2003) Local texture convergence: a new approach to seeking assembly rules. *Oikos*, **102**, 525–532.

Watkinson, A.R. (1985) On the abundance of plants along an environmental gradient. *Journal of Ecology*, **73**, 569–578.

Watson, I.W., Burnside, D.G. & Holm, A.M. (1996) Event-driven or continuous; which is the better model for managers? *Rangeland Journal*, **18**, 351–369.

Watt, A.S. (1947) Pattern and process in the plant community. *Journal of Ecology*, **35**, 1–22.

Watt, A.S. (1955) Bracken versus heather, a study in plant sociology. *Journal of Ecology*, **43**, 490–506.

Watt, A.S. (1960) Population changes in acidophilous grass-heath in Breckland, 1936–1957. *Journal of Ecology*, **48**, 605–629.

Watt, A.S. (1981) Further observations on the effects of excluding rabbits from Grassland A in East Anglian Breckland: the pattern of change and factors affecting it (1936–1973). *Journal of Ecology*, **69**, 509–536.

Weaver, J.E. & Clements, F.E. (1929) *Plant Ecology*. McGraw-Hill, New York.

Wedin, D. & Tilman, D. (1993) Competition among grasses along a nitrogen gradient – initial conditions and mechanisms of competition. *Ecological Monographs*, **63**, 199–229.

Weiher, E., Clarke, G.D.P. & Keddy, P.A. (1998) Community assembly rules, morphological dispersion, and the coexistence of plant species. *Oikos*, **81**, 309–322.

Weiner, J., Wright, D.B. & Castro, S. (1997) Symmetry of below-ground competition between Kochia scoparia individuals. *Oikos*, **79**, 85–91.

Weisner, S.E.B. (1993) Long term competitive displacement of *Typha latifolia* by *T. angustifolia* in a eutrophic lake. *Oecologia*, **94**, 451–456.

Welbank, P.J. (1963) Toxin production during decay of *Agropyron repens* (couch grass) and other species. *Weed Research*, **3**, 205–214.

Went, F.W. (1955) The ecology of desert plants. *Scientific American*, **192**, 68–75.

Wheelwright, N.T. (1985) Competition for dispersers, and the timing of flowering and fruiting in a guild of tropical trees. *Oikos*, **44**, 465–477.

Whelan, R.J. (1995) *The Ecology of Fire*. Cambridge University Press, Cambridge, UK.

Whisenant, S.G. & Wagstaff, F.J. (1995) Successional trajectories of a grazed salt desert shrubland. *Vegetatio*, **94**, 133–140.

White, A.J., Wratten, S.D., Berry, N.A. & Weigmann, U. (1995) Habitat manipulation to enhance biological control of Brassica pests by hover flies (Diptera: Syrphidae). *Journal of Economic Entomology*, **88**, 1171–1176.

White, J.A. & Whitham, T.G. (2000) Associational susceptibility of cottonwood to a box elder herbivore. *Ecology*, **81**, 1795–1803.

White, P.S. (1979) Pattern, process, and natural disturbance in vegetation. *Botanical Review*, **45**, 229–299.

Whitford, W.G. (2002) *Ecology of Desert Systems*. Academic Press, San Diego.

Whitford, W.G., Anderson, J. & Rice, P.M. (1997) Stemflow contribution to the 'fertile island' effect in creosotebush, *Larrea tridentata*. *Journal of Arid Environments*, **35**, 451–457.

Whitham, T.G., Young, W.P., Martinsen, G.D., Gehring, C.A., Schweitzer, J.A., Shuster, S.M., Wimp, G.M., Fischer, D.G., Bailey, J.K., Lindroth, R.L., Woolbright, S. & Kuske, C.R. (2003) Community and ecosystem genetics: a consequence of the extended phenotype. *Ecology*, **84**, 559–573.

Whittaker, R.H. (1960) Vegetation of the Siskiyou Mountains, Oregon and California. *Ecological Monographs*, **30**, 279–338.

Whittaker, R.H. (1965) Dominance and diversity in land plant communities. *Science*, **147**, 250–260.

Whittaker, R.H. (1967) Gradient analysis of vegetation. *Biological Reviews*, **49**, 207–264.

Whittaker, R.H. (1975a) *Communities and Ecosystems*, 2nd edn. Macmillan, New York.

Whittaker, R.H. (1975b) The design and stability of some plant communities. In *Unifying Concepts in Ecology* (eds W.H. van Dobben & R.H. Lowe-McConnell), pp. 169–181. Junk, The Hague.

Whittaker, R.H. (1977) Evolution of species diversity in land communities. *Evolutionary Biology*, **10**, 1–67.

Whittaker, R.H. & Levin, S.A. (1975) *Niche: Theory and Application*. Dowden, Hutchinson and Ross, Stroudsburg, PA.

Whittaker, R.H. & Woodwell, G.M. (1972) Evolution of natural communities. In *Ecosystem Structure and Function* (ed. J.A. Wiens), pp. 137–155. Oregon State University Press, Corvallis.

Whittington, W.J. & O'Brien, T.A. (1968) A comparison of yields from plots sown with a single species or a mixture of grass species. *Journal of Applied Ecology*, **5**, 209–213.

Wiens, J.A. (1989) *The Ecology of Bird Communities, Vol. 1, Foundations and Patterns*. Cambridge University Press, Cambridge, UK.

Wilkinson, M.J. & Stace, C.A. (1991) A new taxonomic treatment of the *Festuca ovina* L. aggregate (Poaceae) in the British Isles. *Biological Journal of the Linnean Society*, **106**, 347–397.

Willby, N.J., Abernethy, V.J. & Demars, B.O.L. (2000) Attribute-based classification of European hydrophytes and its relationship to habitat utilization. *Freshwater Biology*, **43**, 43–74.

Williams, H.T.P. & Lenton, T.M. (2008) Environmental regulation in a network of simulated microbial ecosystems. *Proceedings of the National Academy of Sciences of the U.S.A.*, **105**, 10432–10437.

Williams, J.W., Shuman, B. & Webb, T. (2001) Dissimilarity analyses of late-Quaternary vegetation and climate in eastern North America. *Ecology*, **82**, 3346–3362.

Williamson, G.B. (1990) Allelopathy, Koch's postulates and the neck riddle. In *Perspectives on Plant Competition* (eds J.B. Grace & D. Tilman), pp. 143–162. Academic Press, San Diego.

Willig, M.R., Presley, S.J., Bloch, C.P., Castro-Arellano, I., Cisneros, L.M., Higgins, C.L. & Klingbeil, B.T. (2011) Tropical metacommunities along elevational gradients: effects of forest type and other environmental factors. *Oikos*, **120**, 1497–1508.

Wilsey, B.J. & Polley, H.W. (2002) Reductions in grassland species evenness increase dicot seedling invasion and spittle bug infestation. *Ecology Letters*, **5**, 676–684.

Wilson, D.S. (1992) Complex interactions in metacommunities, with implications for biodiversity and higher levels of selection. *Ecology*, **73**, 1984–2000.

Wilson, E.O. (1961) The nature of the taxon cycle in the Melanesian ant fauna. *American Naturalist*, **95**, 169–193.

Wilson, J.B. (1987) Group selection in plant populations. *Theoretical and Applied Genetics*, **74**, 493–502.

Wilson, J.B. (1988a) A review of evidence on the control of shoot:root ratio, in relation to models. *Annals of Botany*, **61**, 433–449.

Wilson, J.B. (1988b) The effect of initial advantage on the course of competition. *Oikos*, **51**, 19–24.

Wilson, J.B. (1988c) Shoot competition and root competition. *Journal of Applied Ecology*, **25**, 279–296.

Wilson, J.B. (1988d) Community structure in the flora of islands in Lake Manapouri. *Journal of Ecology*, **76**, 1030–1042.

Wilson, J.B. (1989a) Relations between native and exotic plant guilds in the Upper Clutha, New Zealand. *Journal of Ecology*, **77**, 223–235.

Wilson, J.B. (1989b) A null model of guild proportionality, applied to stratification of a New Zealand temperate rain forest. *Oecologia*, **80**, 263–267.

Wilson, J.B. (1990) Mechanisms of species coexistence: twelve explanations for Hutchinson's 'Paradox of the Plankton': evidence from New Zealand plant communities. *New Zealand Journal of Ecology*, **13**, 17–42.

Wilson, J.B. (1991) Methods for fitting dominance/diversity curves. *Journal of Vegetation Science*, **2**, 35–46.

Wilson, J.B. (1993) Macronutrient (NPK) toxicity and interactions in the grass Festuca ovina. *Journal of Plant Nutrition*, **16**, 1151–1159.

Wilson, J.B. (1994a) The 'Intermediate disturbance hypothesis' of species coexistence is based on patch dynamics. *New Zealand Journal of Ecology*, **18**, 176–181.

Wilson, J.B. (1994b) Who makes the assembly rules? *Journal of Vegetation Science*, **5**, 275–278.

Wilson, J.B. (1995) Fox and Brown's 'random data sets' are not random. *Oikos*, **74**, 543–544.

Wilson, J.B. (1997) An evolutionary perspective on the 'death hormone' hypothesis in plants. *Physiologia Plantarum*, **99**, 511–516.

Wilson, J.B. (1999a) Assembly rules in plant communities. In *Ecological Assembly Rules: Perspectives, Advances and Retreats* (eds E. Weiher & P.A. Keddy), pp. 130–164. Cambridge University Press, Cambridge, UK.

Wilson, J.B. (1999b) Guilds, functional types and ecological groups. *Oikos*, **86**, 507–522.

Wilson, J.B. (2002) The 'emergent property' of Anand and Li is a mathematical artefact. Emergent properties do not exist. *Community Ecology*, **3**, 47–48.

Wilson, J.B. (2003) The deductive method in community ecology. *Oikos*, **101**, 216–218.

Wilson, J.B. (2011) Cover plus: ways of measuring plant canopies and the terms used for them. *Journal of Vegetation Science*, **22**, 197–206.

Wilson, J.B. & Agnew, A.D.Q. (1992) Positive-feedback switches in plant communities. *Advances in Ecological Research*, **23**, 263–336.

Wilson, J.B., Agnew, A.D.Q. & Partridge, T.R. (1994) Carr texture in Britain and New Zealand: community convergence compared with a null model. *Journal of Vegetation Science*, **5**, 109–116.

Wilson, J.B., Agnew, A.D.Q. & Sykes, M.T. (2004) Ecology or mythology? Are Whittaker's "gradient analysis" curves reliable evidence of continuity in vegetation? *Preslia*, **76**, 245–253.

Wilson, J.B. & Allen, R.B. (1990) Deterministic versus Individualistic community structure: a test from invasion by Nothofagus menziesii in southern New Zealand. *Journal of Vegetation Science*, **1**, 467–474.

Wilson, J.B., Allen, R.B. & Lee, W.G. (1995c) An assembly rule in the ground and herbaceous strata of a New Zealand rainforest. *Functional Ecology*, **9**, 61–64.

Wilson, J.B. & Anderson, B.J. (2001) Species-pool relations: like a wooden light bulb? *Folia Geobotanica*, **36**, 35–44.

Wilson, J.B., Crawley, M.J., Dodd, M.E. & Silvertown, J. (1996c) Evidence for constraint on species coexistence in vegetation of the Park Grass experiment. *Vegetatio*, **124**, 183–190.

Wilson, J.B. & Gitay, H. (1995a) Limitations to species coexistence: evidence for competition from field observations, using a patch model. *Journal of Vegetation Science*, **6**, 369–376.

Wilson, J.B. & Gitay, H. (1995b) Community structure and assembly rules in a dune slack: variance in richness, guild proportionality, biomass constancy and dominance/diversity relations. *Vegetatio*, **116**, 93–106.

Wilson, J.B. & Gitay, H. (1999) Alternative classifications in the intrinsic guild structure of a New Zealand tussock grassland. *Oikos*, **86**, 566–572.

Wilson, J.B., Gitay, H. & Agnew, A.D.Q. (1987) Does niche limitation exist? *Functional Ecology*, **1**, 391–397.

Wilson, J.B., Gitay, H., Roxburgh, S.H., King, W.M. & Tangney, R.S. (1992c) Egler's concept of 'Initial Floristic Composition' in succession – ecologists citing it don't agree what it means. *Oikos*, **64**, 591–593.

Wilson, J.B., Gitay, H., Steel, J.B. & King, W.M. (1998) Relative abundance distributions in plant communities: the effects of species richness and of spatial scale. *Journal of Vegetation Science*, **9**, 213–220.

Wilson, J.B., Hubbard, J.C.E. & Rapson, G.L. (1988) A comparison of the realised niche relations of species in New Zealand and Britain. *Oecologia*, **76**, 106–110.

Wilson, J.B., James, R.E., Newman, J.E. & Myers, T.E. (1992a) Rock pool algae – species composition determined by chance. *Oecologia*, **91**, 150–152.

Wilson, J.B. & King, W.M. (1995) Human mediated switches as processes in landscape ecology. *Landscape Ecology*, **10**, 191–196.

Wilson, J.B. & Lee, W.G. (1989) Infiltration invasion. *Functional Ecology*, **3**, 379–380.

Wilson, J.B. & Lee, W.G. (1994) Niche overlap of congeners: a test using plant altitudinal distribution. *Oikos*, **69**, 469–475.

Wilson, J.B. & Lee, W.G. (2000) C-S-R Triangle theory: community-level predictions, tests, evaluation of criticisms, and relation to other theories. *Oikos*, **91**, 77–96.

Wilson, J.B. & Lee, W.G. (2012) Is New Zealand vegetation really problematic'? Dansereau's puzzles revisited. *Biological Reviews*, **87**, 367–389.

Wilson, J.B. & Newman, E.I. (1987) Competition between upland grasses: root and shoot competition between Deschampsia flexuosa and Festuca ovina. *Acta Oecologia Oecologia Generalis*, **8**, 501–511.

Wilson, J.B., Peet, R.K., Dengler, J. & Partel, M. (2012) Plant species richness: the world records. *Journal of Vegetation Science*, **23**, 796–802.

Wilson, J.B. & Roxburgh, S.H. (1994) A demonstration of guild-based assembly rules for a plant community, and determination of intrinsic guilds. *Oikos*, **69**, 267–276.

Wilson, J.B. & Roxburgh, S.H. (2001) Intrinsic guild structure: determination from competition experiments. *Oikos*, **92**, 189–192.

Wilson, J.B., Roxburgh, S.H. & Watkins, A.J. (1992b) Limitation to plant species coexistence at a point: a study in a New Zealand lawn. *Journal of Vegetation Science*, **3**, 711–714.

Wilson, J.B. & Smith, B. (2001) Methods for testing for texture convergence using abundance data: a randomisation test and a method for comparing the shape of distributions. *Community Ecology*, **2**, 57–66.

Wilson, J.B., Spijkerman, E. & Huisman, J. (2007b) Is there really insufficient support for Tilman's R* concept? A comment on Miller et al. *American Naturalist*, **169**, 700–706.

Wilson, J.B., Steel, J.B., Dodd, M.E., Anderson, B.J., Ullmann, I. & Bannister, P. (2000b) A test of community re-assembly using the exotic communities of New Zealand roadsides, in comparison to British roadsides. *Journal of Ecology*, **88**, 757–764.

Wilson, J.B., Steel, J.B., Newman, J.E. & King, W.M. (2000a) Quantitative aspects of community structure examined in a semi-arid grassland. *Journal of Ecology*, **88**, 749–756.

Wilson, J.B., Steel, J.B., Newman, J.E. & Tangney, R.S. (1995b) Are bryophyte communities different? *Journal of Bryology*, **18**, 689–705.

Wilson, J.B., Steel, J.B. & Steel, S.-L.K. (2007a) Do plants ever compete for space? *Folia Geobotanica*, **42**, 431–436.

Wilson, J.B. & Stubbs, W.J. (2012) Evidence for assembly rules: limiting similarity within a saltmarsh. *Journal of Ecology*, **100**, 210–221.

Wilson, J.B. & Sykes, M.T. (1988) Some tests for niche limitation by examination of species diversity in the Dunedin area, New Zealand. *New Zealand Journal of Botany*, **26**, 237–244.

Wilson, J.B., Sykes, M.T. & Peet, R.K. (1995a) Time and space in the community structure of a species-rich grassland. *Journal of Vegetation Science*, **6**, 729–740.

Wilson, J.B., Ullmann, I. & Bannister, P. (1996b) Do species assemblages ever recur? *Journal of Ecology*, **84**, 471–474.

Wilson, J.B. & Watkins, A.J. (1994) Guilds and assembly rules in lawn communities. *Journal of Vegetation Science*, **5**, 591–600.

Wilson, J.B., Wells, T.C.E., Trueman, I.C., Jones, G., Atkinson, M.D., Crawley, M.J., Dodd, M.E. & Silvertown, J. (1996a) Are there assembly rules for plant species abundance: an investigation in relation to soil resources and successional trends. *Journal of Ecology*, **84**, 527–538.

Wilson, J.B. & Whittaker, R.J. (1995) Assembly rules demonstrated in a saltmarsh community. *Journal of Ecology*, **83**, 801–807.

Wilson, S.D. & Keddy, P.A. (1985) Plant zonation on a shoreline gradient: physiological response curves of component species. *Journal of Ecology*, **73**, 851–860.

Wilson, S.D. & Tilman, D. (1991) Components of plant competition along an experimental gradient of nitrogen availability. *Ecology*, **72**, 1050–1065.

Wilson, S.D. & Tilman, D. (1993) Plant competition and resource availability in response to disturbance and fertilization. *Ecology*, **74**, 599–611.

Woodell, S.R.J., Mooney, H.A. & Hill, A.J. (1969) The behaviour of *Larrea divaricata* (Creosote Bush) in response to rainfall in North California. *Journal of Ecology*, **57**, 37–44.

Wootton, J.T. (2005) Field parameterization and experimental test of the neutral theory of biodiversity. *Nature*, **433**, 309–312.

Wright, I.J., Reich, P.B., Westoby, M., Ackerly, D.D., Baruch, Z., Bongers, F., Cavender-Bares, J., Chapin, T., Cornelissen, J.H.C., Diemer, M., Flexas, J., Garnier, E., Groom, P.K., Gulias, J., Hikosaka, K., Lamont, B.B., Lee, T., Lee, W., Lusk, C., Midgley, J.J., Navas, M.-L., Niinemets, U., Oleksyn, J., Osada, N., Poorter, H., Poot, P., Prior, L., Pyankov, V.I., Roumet, C., Thomas, S.C., Tjoelker, M.G., Veneklaas, E.J. & Villar, R. (2004) The worldwide leaf economics spectrum. *Nature*, **428**, 821–827.

Wright, S.J. (2002) Plant diversity in tropical forests: a review of mechanisms of species coexistence. *Oecologia*, **130**, 1–14.

Wright, S.J. & Calderon, O. (1995) Phylogenetic patterns among tropical flowering phenologies. *Journal of Ecology*, **83**, 937–948.

Wright, S.J., Muller-Landau, H.C., Condit, R. & Hubbell, S.P. (2003) Gap-dependent recruitment, realized vital rates, and size distributions of tropical trees. *Ecology*, **84**, 3174–3185.

Wright, S.J. & van Schaik, C.P. (1994) Light and the phenology of tropical trees. *American Naturalist*, **143**, 192–199.

Wurzburger, N. & Hendrick, R.L. (2009) Plant litter chemistry and mycorrhizal roots promote a nitrogen feedback in a temperate forest. *Journal of Ecology*, **97**, 528–536.

Wyse, S.V. & Burns, B.R. (2013) Effects of *Agathis australis* (New Zealand kauri) leaf litter on germination and seedling growth differs among plant species. *New Zealand Journal of Ecology*, **37**, 178–183.

Yamazaki, M., Iwamoto, S. & Seiwa, K. (2009) Distance- and density-dependent seedling mortality caused by several diseases in eight tree species co-occurring in a temperate forest. *Plant Ecology*, **201**, 181–196.

Yan, X., Wang, Z., Huang, L., Wang, C., Hou, R., Xu, Z. & Quao, X. (2009) Research progress on electrical signals in higher plants. *Progress in Natural Science*, **19**, 531–541.

Yeaton, R.I. (1978) A cyclical relationship between Larrea tridentata and Opuntia leptocaulis in the Northern Chihuahuan Desert. *Journal of Ecology*, **66**, 651–656.

Yeaton, R.I. & Cody, M.L. (1976) Competition and spacing in plant communities in the northern Mohave Desert. *Journal of Ecology*, **64**, 689–696.

Yi, H.-S., Heil, M., Adame-Álvarez, R.M., Ballhorn, D.J. & Ryu, C.-M. (2009) Airborne induction and priming of plant defenses against a bacterial pathogen. *Plant Physiology*, **151**, 2152–2161.

Yoder, C.K. & Nowak, R.S. (1999a) Hydraulic lift among native plant species in the Mojave Desert. *Plant and Soil*, **215**, 93–102.

Yoder, C.K. & Nowak, R.S. (1999b) Soil moisture extraction by evergreen and drought deciduous shrubs in the Mojave Desert during wet and dry years. *Journal of Arid Environments*, **42**, 81–96.

Yodzis, P. (1978) *Competition for Space and the Structure of Ecological Communities*. Springer, Berlin.

Yodzis, P. (1986) Competition, mortality, and community structure. In *Community Ecology* (eds J.M. Diamond & T.J. Case), pp. 480–491. Harper and Row, New York.

Young, H.S., McCauley, D.J., Pollock, A. & Dirzo, R. (2014) Differential plant damage due to litterfall in palm-dominated forest stands in a Central Pacific atoll. *Journal of Tropical Ecology*, **30**, 231–236.

Young, T.P., Stanton, M.L. & Christian, C.E. (2003) Effects of natural and simulated herbivory on spine lengths of Acacia drepanolobium in Kenya. *Oikos*, **101**, 171–179.

Zhang, J., Zhao, H., Zhang, T., Zhao, X. & Drake, S. (2005) Community succession along a chronosequence of vegetation restoration on sand dunes in Horqin Sandy Land. *Journal of Arid Environments*, **62**, 555–566.

Zhang, Q.G. & Zhang, D.Y. (2007) Colonization sequence influences selection and complementarity effects on biomass production in experimental algal microcosms. *Oikos*, **116**, 1748–1758.

Zhang, S.T. & Lamb, E.G. (2012) Plant competitive ability and the transitivity of competitive hierarchies change with plant age. *Plant Ecology*, **213**, 15–23.

Zimmerman, G.T. & Neuenschwander, L.F. (1984) Livestock grazing influences on community structure, fire intensity, and fire frequency within the Douglas-fir/Ninebark habitat type. *Journal of Range Management*, **37**, 104–110.

Zobel, K., Zobel, M. & Rosén, E. (1994) An experimental test of diversity maintenance mechanisms by a species removal experiment in a species-rich wooded meadow. *Folia Geobotanica et Phytotaxonomica*, **29**, 449–457.

Zohary, M. (1973) *Geobotanical Foundations of the Middle East*. Springer, Stuttgart.

Zonneveld, I.S. (1995) Vicinism and mass effect. *Journal of Vegetation Science*, **6**, 441–444.

Zuppinger-Dingley, D., Schmid, B., Petermann, J.S., Yadav, V., De Deyn, G.B. & Flynn, D.F.B. (2014) Selection for niche differentiation in plant communities increases biodiversity effects. *Nature*, **515**, 108–112.

Subject Index

Taxonomic Index